Lecture Notes in Computer Science 8252

Commenced Publication in 1973
Founding and Former Series Editors:
Gerhard Goos, Juris Hartmanis, and Jan van Leeuwen

Editorial Board

Sajal K. Das   Cristina Nita-Rotaru
Murat Kantarcioglu (Eds.)

Decision and Game Theory for Security

4th International Conference, GameSec 2013
Fort Worth, TX, USA, November 11-12, 2013
Proceedings

 Springer

Volume Editors

Sajal K. Das
Missouri University of Science and Technology, Department of Computer Science
500 West 15th Street, 325B Computer Science Building, Rolla, MO 65409, USA
E-mail: sdas@mst.edu

Cristina Nita-Rotaru
Purdue University, Department of Computer Science, LWSN 2142J
305 N. University Street, West Lafayette, IN 47907, USA
E-mail: cnitarot@purdue.edu

Murat Kantarcioglu
University of Texas at Dallas, Data Security and Privacy Lab
800 W. Campbell Road, MS EC31, Richardson, TX 75080, USA
E-mail: muratk@utdallas.edu

ISSN 0302-9743   e-ISSN 1611-3349
ISBN 978-3-319-02785-2   e-ISBN 978-3-319-02786-9
DOI 10.1007/978-3-319-02786-9
Springer Cham Heidelberg New York Dordrecht London

Library of Congress Control Number: 2013950386

CR Subject Classification (1998): C.2.0, J.1, D.4.6, K.4.4, K.6.5, F.1-2

LNCS Sublibrary: SL 4 – Security and Cryptology

Typesetting: Camera-ready by author, data conversion by Scientific Publishing Services, Chennai, India

Printed on acid-free paper

Springer is part of Springer Science+Business Media (www.springer.com)

Preface

Security is a multifaceted problem area that requires a careful appreciation of many complexities regarding the underlying technical infrastructure as well as of human, economic, and social factors. Securing resources involves decision making on multiple levels of abstraction while considering variable planning horizons. At the same time, the selection of security measures needs to account for limited resources available to both malicious attackers and administrators defending networked systems. Various degrees of uncertainty and incomplete information about the intentions and capabilities of miscreants further exacerbate the struggle to select appropriate mechanisms and policies.

The GameSec conference aims to bring together researchers working on the theoretical foundations and behavioral aspects of enhancing security capabilities in a principled manner. The successful editions of the conference series in the past three years took place in Berlin, Germany (2010), College Park, Maryland, USA (2011), and Budapest, Hungary (2012). Contributions to these meetings included analytic models based on game, information, communication, optimization, decision, and control theories that were applied to diverse security topics. In addition, researchers contributed papers that highlighted the connection between economic incentives and real-world security, reputation, trust, and privacy problems. We believe that such contributions will play an important role in defining and developing the science of cyber security.

The 4th International Conference on Decision and Game Theory for Security (GameSec 2013) took place in Fort Worth, Texas, USA, on November 11–12, 2013. In response to the general call for papers, many papers were received covering various economic aspects of security and privacy. The international Program Committee evaluated the submitted papers based on their significance, originality, technical quality, and exposition.

This edited volume of the conference proceedings contains five full papers, three short papers, and seven invited papers that constituted the conference program divided into several sessions held over two days. In addition, the conference program had two exciting keynote talks delivered by Prof. Andrew Odlyzko (University of Minnesota) and another from Dr. Cliff Wang (Army Research Office). We sincerely thank all the organizing members (listed here) for their hard work.

November 2013

Sajal K. Das
Cristina Nita-Rotaru
Murat Kantarcioglu

Organization

Program Committee

Tansu Alpcan	The University of Melbourne, Australia
Ross Anderson	Cambridge University, UK
John Baras	University of Maryland, USA
Tamer Basar	University of Illinois at Urbana-Champaign, USA
Alvaro Cardenas	Fujitsu Laboratories of America
Nicolas Christin	Carnegie Mellon University, USA
John Chuang	UC Berkeley, USA
Mark Felegyhazi	Budapest University of Technology and Economics (BME), Hungary
Jens Grossklags	Penn State University, USA
Celine Hoe	University of Texas at Dallas, USA
Benjamin Johnson	UC Berkeley, USA
Eduard Jorswieck	Technical University Dresden, Germany
Murat Kantarcioglu	University of Texas at Dallas, USA
Xiang-Yang Li	Illinois Institute of Technology, USA
Refik Molva	EURECOM, France
Tyler Moore	Southern Methodist University, USA
John Musacchio	University of California, Santa Cruz, USA
Cristina Nita-Rotaru	Purdue University, USA
Mehrdad Nojoumian	Southern Illinois University, USA
Andrew Odlyzko	University of Minnesota, USA
Alina Oprea	RSA Labs, USA
Stefan Schmid	TU Berlin & Telekom Innovation Laboratories, Germany
Radu State	University of Luxembourg
Nan Zhang	The George Washington University, USA

Organizing Committee

General Chair

Sajal K. Das	Missouri University of Science and Technology, USA

TPC Co-chairs

Murat Kantarcioglu	University of Texas at Dallas, USA
Cristina Nita-Rotaru	Purdue University, USA

Local Arrangements Chair

Matthew Wright   University of Texas at Arlington, USA

Finance and Registration Chair

Donggang Liu   University of Texas at Arlington, USA

Publicity Chair

Yingying (Jennifer) Chen   Stevens Institute of Technology, USA

Webmaster

Vaibhav Khadilkar   University of Texas at Dallas, USA

Table of Contents

On Communication over Gaussian Sensor Networks with Adversaries: Further Results

Emrah Akyol[1], Kenneth Rose[1], and Tamer Başar[2]

[1] University of California, Santa Barbara
{eakyol,rose}@ece.ucsb.edu
[2] University of Illinois, Urbana-Champaign
basar1@illinois.edu

Abstract. This paper presents new results on the game theoretical analysis of optimal communications strategies over a sensor network model. Our model involves one single Gaussian source observed by many sensors, subject to additive independent Gaussian observation noise. Sensors communicate with the receiver over an additive Gaussian multiple access channel. The aim of the receiver is to reconstruct the underlying source with minimum mean squared error. The scenario of interest here is one where some of the sensors act as adversary (jammer): they aim to maximize distortion. While our recent prior work solved the case where either all or none of the sensors coordinate (use randomized strategies), the focus of this work is the setting where only a subset of the transmitter and/or jammer sensors can coordinate. We show that the solution crucially depends on the ratio of the number of transmitter sensors that can coordinate to the ones that cannot. If this ratio is larger than a fixed threshold determined by the network settings (transmit and jamming power, channel noise and sensor observation noise), then the problem is a zero-sum game and admits a saddle point solution where transmitters with coordination capabilities use randomized linear encoding while the rest of the transmitter sensors is not used at all. Adversarial sensors that can coordinate generate identical Gaussian noise while other adversaries generate independent Gaussian noise. Otherwise (if that ratio is smaller than the threshold), the problem becomes a Stackelberg game where the leader (all transmitter sensors) uses fixed (non-randomized) linear encoding while the follower (all adversarial sensors) uses fixed linear encoding with the opposite sign.

Keywords: Game theory, sensor networks, source-channel coding, coordination.

1 Introduction

Communications over sensors networks is an active research area offering a rich set of problems of theoretical and practical significance, see e.g., [8] and the

S.K. Das, C. Nita-Rotaru, and M. Kantarcioglu (Eds.): GameSec 2013, LNCS 8252, pp. 1–9, 2013.

references therein. Game theoretic considerations, i.e., the presence of adversary and its impact on the design of optimal communication strategies have been studied for a long time [9,10]. In this paper, we extend our prior work on the game theoretic analysis of Gaussian sensor networks, on a particular model introduced in [7], by utilizing the results on the game theoretic analysis of the Gaussian test channel in [3–6].

In this paper, we consider the sensor network model illustrated in Figure 1 and explained in detail in Section 2. The first M sensors (i.e., the transmitters) and the receiver constitute Player 1 (minimizer) and the remaining K sensors (i.e., the adversaries) constitute Player 2 (maximizer). This zero-sum game does not admit a saddle-point in pure strategies (fixed encoding functions), but admits one in mixed strategies (randomized functions).

Our prior work considered two extremal settings [2], depending on the "coordination" capabilities of the sensors. Coordination here refers to the ability of using randomized encoders, i.e., all transmitter sensors and the receiver; and also the adversaries among themselves agree on some (pseudo)random sequence, denoted as $\{\gamma\}$ (for transmitters and the receiver) and $\{\theta\}$ (for adversaries) in the paper. The main message of our prior work is that "coordination" plays a pivotal role in the analysis and the implementation of optimal strategies for both the transmitter and adversarial sensors. Depending on the coordination capabilities of the the transmitters and the adversaries, we considered two extreme settings. In the first setting, we considered the more general case of mixed strategies and present the saddle-point solution in Theorem 1. In the second setting, encoding functions of transmitters are limited to the fixed mappings. This setting can be viewed as a Stackelberg game where Player 1 is the leader, restricted to pure strategies, and Player 2 is the follower, who observes Player 1's choice of pure strategies and plays accordingly.

In this paper, we consider a more practical setting where only a given subset of the transmitters and also the adversarial sensors can coordinate. Our main result is: if the number of transmitter sensors that can coordinate is large enough compared to ones that cannot, then the problem becomes a zero-sum game with a saddle point, where the coordination capable transmitters use randomized linear strategy and incapable transmitters are not used at all. Discarding these transmitter sensors is rather surprising but the gain from coordination compansates for this loss. Coordination is also important for the adversarial sensors. When transmitters coordinate, adversaries must also coordinate to generate identical realizations of Gaussian jamming noise. In contrast with transmitters, the adversarial sensors which cannot coordinate is of use: they generate independent copies of the identically distributed Gaussian jamming noise. Otherwise, i.e., the number of coordinating transmitters are not large enough, transmitters use deterministic (pure strategies) linear encoding, i.e., $g_T(X) = \alpha_T X$ and optimal adversarial strategy is also uncoded communications in the opposite direction of the transmitters, i.e., $g_A(X) = \alpha_A X$ for some $\alpha_T, \alpha_A \in \mathbb{R}^+$. For both settings, uncoded communication is optimal and separate source and channel coding is strictly suboptimal.

This paper is organized as follows. In Section 2, we present the problem definition. We review prior work, particularly [2] in Section 3. In Section 4, we present
our main result and finally we provide conclusions in Section 5.

2   Problem Definition

In general, lowercase letters (e.g., x) denote scalars, boldface lowercase (e.g.,
$\boldsymbol{x}$) vectors, uppercase (e.g., U, X) matrices and random variables, and boldface
uppercase (e.g., $\boldsymbol{X}$) random vectors. $\mathbb{E}(\cdot)$, $\mathbb{P}(\cdot)$ and $\mathbb{R}$ denote the expectation and
probability operators, and the set of real numbers respectively. $Bern(p)$ denotes
the Bernoulli random variable, taking values 1 with probability p and -1 with
$1-p$. Gaussian distribution with mean $\boldsymbol{\mu}$ and covariance matrix R is denoted as
$\mathcal{N}(\boldsymbol{\mu}, R)$.

The sensor network model is illustrated in Figure 1. The underlying source
$\{S(i)\}$ is a sequence of i.i.d. real valued Gaussian random variables with zero
mean and variance σ_S^2. Sensor $m \in [1:M+K]$ observes a sequence $\{U_m(i)\}$
defined as

$$U_m(i) = S(i) + W_m(i), \tag{1}$$

where $\{W_m(i)\}$ is a sequence of i.i.d. Gaussian random variables with zero mean
and variance $\sigma_{W_m}^2$, independent of $\{S(i)\}$. Sensor $m \in [1:M+K]$ can apply
arbitrary Borel measurable function $g_m^N : \mathbb{R}^N \to \mathbb{R}$ to the observation sequence
of length N, $\boldsymbol{U}_m$ so as to generate sequence of channel inputs $X_m(i) = g_m^N(\boldsymbol{U}_m)$
under power constraint:

$$\lim_{N \to \infty} \frac{1}{N} \sum_{i=1}^{N} \mathbb{E}\{X_m^2(i)\} \leq P_m \tag{2}$$

The channel output is then given as

$$Y(i) = Z(i) + \sum_{j=1}^{M+K} X_j(i) \tag{3}$$

where $\{Z(i)\}$ is a sequence of i.i.d. Gaussian random variables of zero mean
and variance σ_Z^2, independent of $\{S(i)\}$ and $\{W_m(i)\}$. The receiver applies a
Borel measurable function $h^N : \mathbb{R}^N \to \mathbb{R}$ to the received sequence $\{Y(i)\}$ to
minimize the cost, which is measured as mean squared error (MSE) between the
underlying source S and the estimate at the receiver $\hat{S}$ as

$$J(g_m^N(\cdot), h^N(\cdot)) = \lim_{N \to \infty} \frac{1}{N} \sum_{i=1}^{N} \mathbb{E}\{(S(i) - \hat{S}(i))^2\} \tag{4}$$

for $m = 1, 2, \ldots, M + K$.

The transmitters $g_m^N(\cdot)$ for $m \in [1:M]$ and the receiver $h^N(\cdot)$ seek to minimize
the cost (4) while the adversaries aim to maximize (4) by properly choosing $g_k^N(\cdot)$

for $k \in [M+1 : M+K]$. We focus on the symmetric sensor and symmetric source where $P_m = P_T$ and $\sigma^2_{W\,m} = \sigma^2_{W\,T}$, $\forall m \in [1 : M]$ and $\sigma^2_{W_k} = \sigma^2_{W_T}$ and $P_k = P_A$, $\forall k \in [M+1 : M+K]$.

A transmitter-receiver-adversarial policy $(g_m^{N*}, g_k^{N*}, h^{N*})$ constitutes a saddle-point solution if it satisfies the pair of inequalities

$$J(g_m^{N*}, g_k^N, h^N) \leq J(g_m^{N*}, g_k^{N*}, h^{N*}) \leq J(g_m^N, g_k^{N*}, h^N) \tag{5}$$

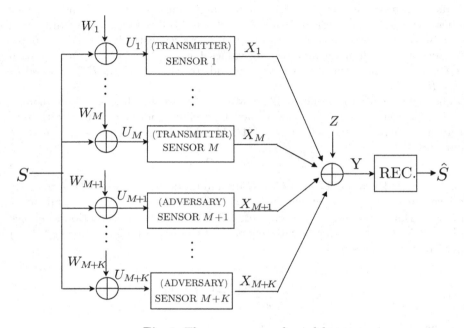

Fig. 1. The sensor network model

3   Review of Prior Work

3.1   Full Coordination

First scenario is concerned with the setting where "all" transmitter sensors have the ability to *coordinate*, i.e., all transmitters and the receiver can agree on an i.i.d. sequence of random variables $\{\gamma(i)\}$ generated, for example, by a side channel, the output of which is, however, not available to the adversarial sensors[1]. The ability of coordination allows transmitters and the receiver to agree on randomized encoding mappings. Surprisingly, in this setting, the adversarial sensors also need to coordinate, i.e., agree on an i.i.d. random sequence, denoted as $\{\theta(i)\}$, to generate the optimal jamming strategy. The saddle point solution of this problem is presented in the following theorem.

---

[1] An alternative practical method to coordinate is to generate the identical pseudo-random numbers at each sensor, based on pre-determined seed.

Theorem 1 ([2]). *The optimal encoding function for the transmitters is randomized uncoded transmission:*

$$X_m(i) = \gamma(i)\alpha_T U_m(i), \quad M \geq m \geq 1 \tag{6}$$

where $\gamma(i)$ is i.i.d. Bernoulli $(\frac{1}{2})$ over the alphabet $\{-1, 1\}$

$$\gamma(i) \sim Bern(\frac{1}{2}). \tag{7}$$

The optimal jamming function (for adversarial sensors) is to generate i.i.d. Gaussian output

$$X_k(i) = \theta(i), \quad M + K \geq k \geq M + 1 \tag{8}$$

where

$$\theta(i) \sim \mathcal{N}(0, P_A), \tag{9}$$

and is independent of the adversarial sensor input $U_k(i)$. The optimal receiver is the Bayesian estimator of S given Y, i.e.,

$$h(Y(i)) = \frac{M\alpha_T\sigma_S^2}{M\alpha_T^2\sigma_S^2 + M^2\alpha_T^2\sigma_{W_T}^2 + K^2 P_A + \sigma_Z^2} Y(i). \tag{10}$$

Cost at this saddle point as a function of the number of transmitter and adversarial sensors is:

$$J_C(M, K) = \sigma_S^2 \frac{M^2\alpha_T^2\sigma_{W_T}^2 + K^2 P_A + \sigma_Z^2}{M\alpha_T^2\sigma_S^2 + M^2\alpha_T^2\sigma_{W_T}^2 + K^2 P_A + \sigma_Z^2} \tag{11}$$

where $\alpha_T = \sqrt{\frac{P_T}{\sigma_S^2 + \sigma_{W_T}^2}}$.

The proof follows from verification of the fact that the mappings in this theorem satisfy the saddle point criteria given in (5).

Remark 1. Coordination is essential for adversarial sensors in the case of coordinating transmitters and receiver, in the sense that lack of adversarial coordination strictly decreases the overall cost.

3.2 No Coordination

In this section, we focus on the problem, where the transmitters do not have the ability to secretly agree on a random variable, i.e., "coordination" to generate their transmission function X_k. In this case, our analysis yields that the optimal transmitter strategy, which is almost surely unique, is uncoded transmission with linear mappings, while the adversarial optimal strategy for the (jamming) sensors is uncoded, linear mappings with the opposite sign of the transmitter functions. The following theorem presents our mail results associated with "no coordination" setting. A rather surprising observation is that the adversarial coordination is useless for this setting, i.e., even if the adversarial sensors can cooperate, the optimal mappings and hence the resulting cost at the saddle point does not change. Note however that, as we will show later, coordination capability of adversarial sensors is essential in the second extremal setting where transmitters are allowed to coordinate their choices.

Theorem 2 ([2]). *The optimal encoding function for the transmitters is uncoded transmission, i.e.,*

$$X_m(i) = \alpha_T U_m(i), \quad M \geq m \geq 1 \tag{12}$$

The optimal jamming function (for adversarial sensors) is uncoded transmission with the opposite sign of the transmitters, i.e.,

$$X_k(i) = \alpha_A U_k(i), \quad M + K \geq k \geq M + 1 \tag{13}$$

The optimal decoding function is the Bayesian estimator of S given Y, i.e.,

$$h(Y(i)) = \frac{\left[(M\alpha_T + K\alpha_A)\sigma_S^2\right] Y(i)}{(M\alpha_T + K\alpha_A)\sigma_S^2 + M^2\alpha_T^2\sigma_{W_T}^2 + K^2\alpha_A^2\sigma_{W_A}^2 + \sigma_Z^2}. \tag{14}$$

Cost as a function of M and K is

$$J_{NC}(M,K) = \sigma_S^2 \frac{M^2\alpha_T^2\sigma_{W_T}^2 + K^2\alpha_A^2\sigma_{W_A}^2 + \sigma_Z^2}{(M\alpha_T + K\alpha_A)\sigma_S^2 + M^2\alpha_T^2\sigma_{W_T}^2 + K^2\alpha_A^2\sigma_{W_A}^2 + \sigma_Z^2} \tag{15}$$

where $\alpha_T = \sqrt{\frac{P_T}{\sigma_S^2 + \sigma_{W_T}^2}}$ and $\alpha_A = -\sqrt{\frac{P_A}{\sigma_S^2 + \sigma_{W_A}^2}}$.

The proof of theorem, can be found in [2], involves detailed information theoretic analysis and is omitted here for brevity. This problem setting implies a Stackelberg game where transmitters and the receiver play first as the Player 1, as they select their encoding functions. Then, Player 2 (the adversarial sensors), knowing the choice of Player 1, chooses its strategy.

Remark 2. Note that in this setting, the coordination capability for the adversaries do not help, in sharp contrast to the previous setting where, both transmitters and adversaries coordinate.

4   Main Result

The focus of this paper is the setting between the two extreme scenarios of coordination, namely full or no coordination. We assume that $M\epsilon$ transmitter sensors can coordinate with the receiver while $M(1 - \epsilon)$ of them cannot coordinate. Similar to transmitters, only $K\eta$ of the adversarial sensors can coordinate while $K(1 - \eta)$ adversarial sensors cannot coordinate. Let us assume, without loss of generality, that first $M\epsilon$ transmitters and $K\eta$ adversaries can coordinate. Let us also define the quantity ϵ_0 as the solution to:

$$J_C(M\epsilon_0, \sqrt{K^2\eta^2 + K(1 - \eta)}) = J_{NC}(M,K) \tag{16}$$

The following theorem captures our main result.

Theorem 3. *If $\epsilon > \epsilon_0$, $M\epsilon$ capable transmitters use randomized linear encoding, while remaining $M(1 - \epsilon)$ transmitters are not used.*

$$X_m(i) = \gamma(i)\alpha_T U_m(i), \quad M\epsilon \geq m \geq 1 \tag{17}$$
$$X_m(i) = 0 \quad M \geq m \geq M\epsilon \tag{18}$$

where $\gamma(i)$ is i.i.d. Bernoulli ($\frac{1}{2}$) over the alphabet $\{-1, 1\}$

$$\gamma(i) \sim Bern(\frac{1}{2}). \tag{19}$$

The optimal jamming policy (for the capable adversarial sensors) is to generate the identical Gaussian noise

$$X_k(i) = \theta(i), \quad M + K\eta \geq k \geq M + 1 \tag{20}$$

while remaining adversaries will generate independent Gaussian noise

$$X_k(i) = \theta_k(i), \quad M + K \geq k \geq M + k\eta \tag{21}$$

where

$$\theta_k(i) \sim \theta(i) \sim \mathcal{N}(0, P_A), \forall k \tag{22}$$

are independent of the adversarial sensor input $U_k(i)$.

If $\epsilon < \epsilon_0$, then the optimal encoding function for all transmitters is deterministic linear encoding, i.e.,

$$X_m(i) = \alpha_T U_m(i), \quad M \geq m \geq 1 \tag{23}$$

The optimal jamming function (for adversarial sensors) is uncoded transmission with the opposite sign of the transmitters, i.e.,

$$X_k(i) = \alpha_A U_k(i), \quad M + K \geq k \geq M + 1 \tag{24}$$

where $\alpha_T = \sqrt{\frac{P_T}{\sigma_S^2 + \sigma_{W_T}^2}}$ and $\alpha_A = -\sqrt{\frac{P_A}{\sigma_S^2 + \sigma_{W_A}^2}}$.

Proof. The transmitters have two choices: i) All transmitters will choose not to use randomization. Then, the adversarial sensors do not need to use randomization since the optimal strategy is deterministic, linear coding with the opposite sign, as illustrated in Theorem 2. Hence, cost associate with this option is $J_{NC}(M, K)$. ii) Capable transmitters will use randomized encoding. This choice implies that remaining transmitters do not send information as they do not have access to randomization sequence $\{\gamma\}$, hence they are not used. The adversarial sensors which can coordinate generate identical realization of the Gaussian noise while, remaining adversaries generate independent realizations. The total effective noise adversarial power will be $((K\eta)^2 + (1 - \eta)K)P_A$, and the cost associated with this setting is $J_C(M\epsilon, \sqrt{K^2\eta^2 + K(1 - \eta)})$. Hence, the transmitter will choose between two options depending on their costs,

$J_C(M\epsilon, \sqrt{K^2\eta^2 + K(1-\eta)})$ and $J_{NC}(M, K)$. Since, J_C is a decreasing function in M and hence in ϵ, whenever $\epsilon > \epsilon_0$, transmitters use randomization (and hence so do the adversaries), otherwise problem setting becomes identical to "no coordination". The rest of the proof simply follows from the proofs of Theorem 1 and 2 and is omitted here for brevity.

Remark 3. Note that in the first regime ($\epsilon > \epsilon_0$), we have a zero-sum game with saddle point. In the second regime ($\epsilon < \epsilon_0$), we have a Stackelberg game where all transmitters and receiver constitute the leader and adversaries constitute the follower.

5   Conclusion

In this paper, we presented new results on the game theoretical analysis of optimal communication strategies over a sensor network. Our recent prior [2] work had solved two extreme coordination cases where either all or none of the sensors coordinate. In this work, we focused on the setting where only a subset of the transmitter and/or jammer sensors can coordinate. We showed that the solution crucially depends on the number of transmitters and adversaries that can coordinate. In one regime, then the problem is a zero-sum game and admits a saddle point solution where transmitters with coordination capabilities use randomized linear encoding while the remaining the transmitter sensors are not used at all. Adversarial sensors that can coordinate generate identical Gaussian noise while other adversaries generate independent Gaussian noise. In the other regime, the problem becomes a Stackelberg game where the leader (all transmitter sensors) uses fixed (non-randomized) linear encoding while the follower (all adversarial sensors) uses fixed linear encoding with the opposite sign.

Our analysis has uncovered an interesting result regarding the mixed setting considered in this paper. The optimal strategy for transmitters sensors can be to discard the ones that cannot coordinate. Note that the coordination aspect of the problem is entirely due to game-theoretic considerations, which are also highlighted in this surprising result.

Several questions are currently under investigation, including extensions of the analysis to vector sources and channels, the asymptotic (in the number of sensors M and K) analysis of the results presented here; and extension of our analysis to asymmetric and/or non-Gaussian settings. An initial attempt to extend the results associated with the Gaussian test channel to non-Gaussian setting can be fond in [1].

Acknowledgments. This work was supported in part by the NSF under grants CCF-1016861, CCF-1118075 and CCF-1111342.

References

1. Akyol, E., Rose, K., Başar, T.: On optimal jamming over an additive noise channel. draft, http://arxiv.org/abs/1303.3049

2. Akyol, E., Rose, K., Başar, T.: Gaussian sensor networks with adversarial nodes. In: IEEE International Symposium on Information Theory. IEEE (2013)
3. Başar, T.: The Gaussian test channel with an intelligent jammer. IEEE Trans. on Inf. Th. 29(1), 152–157 (1983)
4. Başar, T., Wu, Y.: A complete characterization of minimax and maximin encoder-decoder policies for communication channels with incomplete statistical description. IEEE Trans. on Inf. Th. 31(4), 482–489 (1985)
5. Başar, T., Wu, Y.: Solutions to a class of minimax decision problems arising in communication systems. Journal of Optimization Theory and Applications 51(3), 375–404 (1986)
6. Bansal, R., Başar, T.: Communication games with partially soft power constraints. Journal of Optimization Theory and Applications 61(3), 329–346 (1989)
7. Gastpar, M.: Uncoded transmission is exactly optimal for a simple Gaussian sensor network. IEEE Trans. on Inf. Th. 54(11), 5247–5251 (2008)
8. Gastpar, M., Vetterli, M.: Power, spatio-temporal bandwidth, and distortion in large sensor networks. IEEE Journal on Selected Areas in Communications 23(4), 745–754 (2005)
9. Kashyap, A., Başar, T., Srikant, R.: Correlated jamming on MIMO Gaussian fading channels. IEEE Trans. on Inf. Th. 50(9), 2119–2123 (2004)
10. Shafiee, S., Ulukuş, S.: Mutual information games in multiuser channels with correlated jamming. IEEE Trans. on Inf. Th. 55(10), 4598–4607 (2009)

A True Random Generator
Using Human Gameplay

Mohsen Alimomeni, Reihaneh Safavi-Naini, and Setareh Sharifian

University of Calgary, Department of Computer Science, Canada
{malimome,rei,ssharifi}@ucalgary.ca

Abstract. True Randomness Generators (TRG) use the output of an entropy source to generate a sequence of symbols that is sufficiently close to a uniformly random sequence, and so can be securely used in applications that require unpredictability of each symbol. A TRG algorithm generally consists of (i) an entropy source and (ii) an extractor algorithm that uses a random *seed* to extract the randomness of the entropy source. We propose a TRG that uses the user input in a game played between the user and the computer both as the output of an entropy source, and the random seed required for the extractor. An important property of this TRG is that the (randomness) quality of its output can be flexibly adjusted. We describe the theoretical foundation of the approach and design and implement a game that instantiates the approach. We give the results of our experiments with users playing the game, and analysis of the resulting output strings. Our results support effectiveness of the approach in generating high quality randomness. We discuss our results and propose directions for future work.

1   Introduction

Many security systems and in particular crypto-algorithms use random values as an input to the system. Cryptosystems need randomness for purposes such as key generation, data padding or challenge in challenge-response protocols. In nearly all cases unpredictability of random values is critical to the security of the whole systems. Generating true randomness however is not an easy task and needs a physical entropy source. Operating systems such as Windows and Linux use special sub-systems that combine randomness from different parts of the hardware and software system to collect entropy [Mic,GPR06]. Poor choices of randomness has lead to breakdown of numerous security systems. Important examples of such failures are, attack on Netscape implementation of the SSL protocol [GD] and weakness of entropy collection in Linux and Windows Pseudo-Random Generator [GPR06,DGP09]. A recent example of the need for careful treatment of randomness in cryptographic systems was highlighted in [HDWH12,LHA⁺12] where the same public and private keys were generated by key generation modules and this was partly attributed to randomness that was poorly generated in Linux kernel randomness generation subsystem.

An *entropy source* uses physical processes such as noise in electronic circuits, or software processes that are "unpredictable", to output a sequence over an

S.K. Das, C. Nita-Rotaru, and M. Kantarcioglu (Eds.): GameSec 2013, LNCS 8252, pp. 10–28, 2013.

alphabet that is highly "unpredictable", where unpredictability is measured by *min-entropy* (See Definition 3). Although the output of an entropy sequence can have high level of randomness, but the underlying distribution may be far from uniform. To make the output of an entropy source to follow a uniform distribution, a post processing step is usually used. *Randomness extractors* are deterministic or probabilistic functions that transform the output of an entropy source to uniform distribution using a mapping from n to m bits (usually $m \leq n$), extracting the entropy of the source.

To guarantee randomness of their output, randomness extractors need guarantee on the randomness property of (e.g. the min-entropy) their input entropy source. Extractors that can extract randomness from sources that satisfy a lower bound on their min-entropy, are probabilistic [Sha11]. A probabilistic extractor has two inputs: an entropy source together with a second input that is called *seed*. Good probabilistic extractors use a short seed (logarithmic in the input size) to extract all the randomness (close to the min-entropy) of the input entropy source. A *True Randomness Generator (TRG)* thus consists of two modules: (i) an entropy source that generates a sequence of symbols with a lower bound on its min-entropy, followed by, (ii) a randomness extractor. In practice one needs to estimate the min-entropy of the entropy source to be able to choose appropriate parameters for the extractor. The distribution of the entropy source and its min-entropy may fluctuate over time and so a TRG needs to use an extractor that provides sufficient tolerance for these fluctuations.

Human as Entropy Source: In [HN09], Halprin et al. proposed an innovative approach to construct an entropy source using human game play. Their work built on the results in experimental psychology. It is known that humans, if asked to choose numbers randomly, will do a poor job and their choices will be biased. Wagenaar [Wag72] used experiments in which participants were asked to produce random sequences and noted that in all experiments human choices deviated from uniform distribution. In [RB92], Rapport et al. through a series of experiments showed that if human plays a competitive zero-sum game with uniform choices as the best strategy, their choices will be close to uniform. In their experiment they used matching pennies game in which each player makes a choice between head or tail using an unbiased coin, and the first player wins if both players choose the same side and the second, if they choose different side. In this game the optimal strategy of users is random selection between head and tail. Their result showed that users almost followed uniform random strategy confirming that human can be a good source of entropy if they are engaged in a strategic game and entropy generation is an indirect result of their actions.

Human Game Play for Generating Randomness. Halprin et al. used these studies to propose an entropy source using human game play against a computer. In their work human plays a zero-sum game with uniform optimal strategy against the computer. The game is an extended matching pennies game (user has more than two choices) and is played many times. The sequence resulting from human choices is considered as the output of an entropy source, and is used as the input to a randomness extractor. The result is a TRG with an output that is a random sequence "close" to uniform. In addition to the human input sequence, the TRG uses a second source of perfect randomness to provide seed for the randomness extractor.

In this paper we propose an integrated approach where the game play between a human and the computer is used to implement the two phases of a TRG including randomness source and randomness extraction phase. That is the user's input provides the required randomness for the entropy source and the extractor both.

1.1   Our Contribution

We propose a TRG that uses human game play against a computer, as the *only* source of randomness. The game consists of a sequence of sub-games. Each sub-game is a simple two player zero-sum game between the user and the computer which can be seen as an extended matching pennies game. In each sub-game the human makes a choice among a number of alternatives, each with the same probability of winning, resulting in the user's best strategy to be random selection among their possible choices. The first game corresponds to the entropy generation step in TRG and subsequent sub-games correspond to steps of an extractor algorithm.

The TRG algorithm is based on a seeded extractor that is constructed using an *expander graph*. Expander graphs are highly connected d-regular graphs where each vertex is connected to d neighbours. This notion is captured by a measure called *spectral expansion*. It has been proved that random walks on these graphs can be used to extract randomness [AB09]. Assuming an initial probability distribution $\mathbf{p}$ on the set of vertices of the graph, it is proved [AB09] that by taking a random walk of ℓ steps from any vertex in the graph that is chosen according to $\mathbf{p}$, one ends up at a vertex that represents a distribution over the vertices that is ϵ-close to the uniform distribution. In other words starting from any distribution, each step of the random walk results in a new distribution over the vertices that is closer to uniform and so by taking sufficiently long walk, one can obtain a distribution that is ϵ-close to the uniform distribution.

We use human input to provide the required randomness in the framework above: that is for the initial distribution $\mathbf{p}$ as well as the randomness for each step of the walk. To obtain randomness from human, a sequence of games is presented to the user and the human input in the game is used in the TRG algorithm. In the first sub-game, the graph is presented to the user who will be asked to randomly choose a vertex. This choice represents a source symbol that is generated according to some unknown distribution $\mathbf{p}$; that is human choice is effectively a symbol of an entropy source. Human choices however, although have some entropy but cannot be assumed to be uniformly distributed. A subsequent random walk of length ℓ over the graph will be used to obtain an output symbol for TRG with close to uniform randomness guarantee.

To use human input for the generation of the random walk, on each vertex of the graph the user is presented with a simple game which effectively requires them to choose among the set of the neighbouring vertices. The game is zero-sum with uniform optimal strategy and so the human input would correspond to uniform selection, and consequently one *random step* on the graph. For a given ϵ and an estimate for the min-entropy of the initial vertex selection, one can determine the number of required random steps so that the output of the TRG has the required randomness.

In the above we effectively assume *human input in a uniform optimal strategy zero-sum game is close to uniform*. This assumption is reasonable when human is presented with a few choices, (based on the experiments in [RB92]). In practice however the human input will be close to uniform and so the proposed extraction process can be seen as approximating the random walk by a high min-entropy walk. Obtaining theoretical results on the quality of output in an expander graph extractor when the random walk is replaced with a walk with high min-entropy is an interesting theoretical question. We however demonstrate feasibility of this approach experimentally.

We designed and implemented a TRG that is based on a game on a 3-regular expander graph with 10 vertices. The game consists of a sequence of sub-games. A number of screen shots of the game are shown in Figure 1. In each sub-game the human and the computer make a choice from a set of vertices. If the choices coincide, computer wins and if they do not coincide, human wins. In our implementation the human first make a choice and then the computer's choice is shown. In the first sub-game user makes a choice among the 10 vertices of the graph, and in all subsequent sub-games, among the 3 neighbours of the current vertex. We perform a number of experiments to validate our assumptions.

Experiments. We implemented the above game and experimented with nine human users playing the games. We measured min-entropy of human input in the first sub-game, that is when human is an entropy source, and also subsequent sub-games when human input is used to emulate the random walk. For the former, that is to estimate the initial distribution $\mathbf{p}$, we designed a one round game which requires the user to choose a vertex of the graph and they win if their choice is not predictable by the computer. We used NIST [BK12] tests for estimating the min-entropy of both distributions. The details of experiments are given in Section 4. Our results once again shows that humans, once engaged in a two-party zero sum game with uniform optimal strategy, are good sources of randomness. The min-entropy of human choices in the first sub-game is 2.1 bits per symbol (10 symbol) in average and in the subsequent sub-games is 1.38 bits per symbol (3 symbols) in average. These compared to the maximum available entropy of the source on corresponding number of symbols, i.e. $\log_2 10 = 3.32$ and $\log_2 3 = 1.58$, indicate that indeed the human choices are close to uniformly random and the final output of TRG is expected to be close to random.

Applications. TRGs are an essential component of computing systems. User based TRG add an extra level of assurance about the randomness source: users know that their input has been used to generate randomness. An example application of this approach is generating good random keys using user's input. Asking a user to generate a 64 bit key that is random will certainly result in a biased string. Using the approach presented in this paper, the user can select an element of the space (say a password with 13 characters) randomly. The user choice will be used as the initial entropy source, and then subsequent games will ensure that the final 64 bits is close to uniform. Assuming a 3-regular expander graph with 10 vertices, one needs 3 steps in the expander graph to reach a 1/4-close to uniform. Section 4.1 further discusses how longer sequences can be generated.

1.2   Related Work

The idea of using a game to motivate human to generate unbiased randomness was proposed in [HN09]. Authors used the experimental results in psychology [Wag72,RB92] along with game theoretic approach to show humans playing matching pennies game generate a sequence which is close to uniform. Halprin et al. argued that this game when played between a computer and a human can be used as an entropy source. To increase the amount of randomness generated by human with each choice, Halprin et al. used an extension of this game that uses n choices to the player: the user is presented by an $n \times n$ matrix displayed on the computer screen and is asked to choose a matrix location. The user wins if their choice is the same as the square chosen by the computer. They noted that the visual representation of the game resulted in the user input to be biased as users avoided corner points and limiting squares. The sequence generated by human was used as the input to a seeded extractor (Definition5) to generate a sequence that is ϵ-close to uniform (Definition1). They provided visual representations of human choices that indicates a good spread of points. However statical and min-entropy evaluation of the system is restricted to using statistical tests on the output of the seeded extractor.

Extraction can use a general seeded extractor that will guarantee randomness of the output for *any distribution* with min-entropy (Definition 3) k, or an approach proposed in [BST03] in which the set of possible input sources is limited to a set of 2^t possible ones all with min-entropy k. The former approach requires a fresh random seed for each extraction but has the advantage that the input source can have any distribution with the required min-entropy. This latter approach however requires the input sequence to be one of the set of 2^t possible sources, but has the advantage that one can choose a function from a class of available extractors and hard code that in the systems. This means in practice no randomness is required. However no guarantee can be given about the output if the input sequence is not one of the 2^t that has been used for the design of the system and this property cannot be tested for an input sequence. Halprin et al. used the latter approach, using a t-universal hash function as the extractor. The randomness guarantee of the final result requires the assumption that the human input is one of the 2^t sources. In practice, t can not be arbitrarily large and must be small to guarantee a minimum output rate for randomness. This can pose a security risk that the actual distribution is not one of the 2^t distributions. Halprin et al. did not perform quantitative analysis of user sequences and used visual representation of the human choices to conclude the choices were random.

We note that simpler extractors such as the Von Neumann extractor [vN51] put strong requirements on their input sequence. For example Von Neumann extractor requires the input string to be a Bernoulli sequence with parameter p which is not satisfied by the sequence of human choices where successive choices may be dependent with unknown and changing distribution. The expander graph extractor works for all distributions whose min-entropy is lower bounded by a given value, and does not put extra requirements on the input sequence.

Using human input as entropy source has also been used in a different form. In computer systems, users' usage of input devices such as mouse and keyboard can be used for background entropy collection. This process is used for example in Linux based systems [GPR06]. [ZLwW+09] uses mouse movement and applies hash functions.

Paper Organization. Section 2 provides the background. Section 3 outlines our approach and Section 4 gives the results of our experiments. Section 5 provides concluding remarks.

2   Preliminaries

We will use the following notations. Random variables are denoted by capital letters, such as X. A random variable X is defined over a set $\mathcal{X}$ with a probability distribution $\Pr_X$, meaning that X takes the value $x \in \mathcal{X}$ with probability $\Pr_X(x) = \Pr[X = x]$. Uniform distribution over a set $\mathcal{X}$ is denoted by $U_{\mathcal{X}}$ or U_n if $\mathcal{X} = \{0,1\}^n$. The logarithms will be in base 2 throughout the paper.

Definition 1. *Consider two random variables X and Y taking values in $\mathcal{X}$. Statistical distance of the two variables is given by,*

$$\Delta(X;Y) = \frac{1}{2} \sum_{x \in \mathcal{X}} \left| \Pr_X(x) - \Pr_Y(x) \right|$$

We say that X and Y are ϵ-close if $\Delta(X;Y) \leq \epsilon$.

A *source* on $\{0,1\}^n$ is a random variable X that takes values in $\{0,1\}^n$ with a given distribution $\Pr_X$.

Definition 2. *Let C be a class of sources on $\{0,1\}^n$. A deterministic ϵ-extractor for C is a function $\mathsf{ext} : \{0,1\}^n \to \{0,1\}^m$ such that for every $X \in C$, $\mathsf{ext}(X)$ is "ϵ-close" to U_m.*

Deterministic extractors exist for a limited classes of sources [Sha11]. A small random seed is used in extractors to allow for more general classes of sources to be used as input to the extractor. The more general classes of sources are the sources with a guaranteed amount of randomness. To quantify the amount of randomness in a random variable min-entropy is used.

Definition 3. *The* min *entropy of a random variable X is : $H_\infty(X) = min_x \left\{ \log \frac{1}{Pr_X(x)} \right\}$.*

Definition 4. *A random variable X is a $k-$source if $H_\infty(X) \geq k$, i.e., if $Pr_X(x) \leq 2^{-k}$. The min-entropy rate is denoted by δ and is defined as $k = \delta n$, where n is the number of source output bits.*

Using probabilistic method, it is proved [Sha11] that there exists probabilistic extractors that can extract at least k bits of randomness from a k-source.

Definition 5. *A function* ext : $\{0,1\}^n \times \{0,1\}^d \rightarrow \{0,1\}^m$ *is a seeded (k,ϵ)-extractor if for every k-source X on $\{0,1\}^n$, $Ext(X,U_d)$ is ϵ-close to U_m, where U_m is the random variable associated with the uniform distribution on m bits.*

A relevant result on seeded extractors is the following.

Theorem 1. *[Sha11] For every $n \in \mathbb{N}$, $k \in [0,n]$ and $\epsilon > 0$, there exists a (k,ϵ)-extractor* ext : $\{0,1\}^n \times \{0,1\}^d \rightarrow \{0,1\}^m$ *with $m = k + d - 2\log(\frac{1}{\epsilon}) + O(1)$.*

2.1   Expander Graphs

Expander graphs are well connected graphs in the sense that to make the graph disconnected one needs to remove relatively large number of edges. Connectivity of a graph can be quantified using measures such as the minimum number of neighbouring vertices for all sub-graphs or minimum number of edges that leave all sub-graphs (minimums are taken over all subgraphs of certain size) [HLW06]. For a d-regular graph the second eigenvalues of the adjacency matrix captures the connectivity of the graph. This measure is referred to as *spectral expansion*.

   Normalized adjacency matrix of a d-regular graph with n vertices is an $n \times n$ binary matrix with $A_{i,j} = \frac{1}{d}$ if vertex i and j are connected by an edge, and zero otherwise.

Expander Graphs as Extractors. Given a graph and a starting vertex X, one can make a random walk of length ℓ, by randomly choosing one of the neighbours of X, say X_1, move to X_1, then randomly choose one of the neighbours of X_1, say X_2, and repeat this ℓ times.

   Let G denote an expander graph with normalized adjacency matrix A, and let $\mathbf{p}$ denote an initial distribution on the vertices of G. After one random step from each vertex, the distribution on the vertices is given by $A\mathbf{p}$ and becomes closer to uniform. That is, the statistical distance between the distribution on the graph vertices and the uniform distribution reduces. Continuing the random walk on the graph for ℓ steps, the distribution on the vertices becomes $A^\ell \mathbf{p}$ and gets closer to the uniform distribution. The rate of convergence to uniform distribution for d-regular expander graphs is determined by the second eigenvalue of the normalized adjacency matrix of the graph which is denoted by λ from now on.

Lemma 1. *[AB09, lemma 21.3] Let G be a regular graph over n vertices and $\mathbf{p}$ be an initial distribution over G's vertices. Then we have:*

$$\left\| A^l \mathbf{p} - U_n \right\|_2 \leq \lambda^l.$$

where $\|.\|_2$ is the l2 norm defined as $\|X\|_2 = \sqrt{\sum_{i=1}^{n} x_i^2}$, considering X to be the vector $(x_1, x_2, \ldots, x_n)$. Note that U_n and $A^l \boldsymbol{p}$ are considered as vectors in above.

The random walk on an expander graph explained above gives the following extractor construction.

Lemma 2. *[AB09, lemma 21.27] Let $\epsilon > 0$. For every n and $k \leq n$, there exists an explicit (k, ϵ)-extractor* ext $: \{0, 1\}^n \times \{0, 1\}^t \rightarrow \{0, 1\}^n$ *where $t = O(n - k + \log 1/\epsilon)$.*

The above lemma assumes an expander graph with $\lambda = 1/2$, but in general for an arbitrary λ and min-entropy k, we can derive the following theorem from the above lemmas:

Theorem 2. *Let U_n be the uniform distribution and X be a k-source with probability distribution $\boldsymbol{p}$ over $\{0, 1\}^n$. Let G be a d-regular expander graph over 2^n vertices with normalized adjacency matrix A. For a random-walk of length l over the graph starting from a vertex selected according to distribution $\boldsymbol{p}$, we have*

$$\Delta(A^l \boldsymbol{p}; U_n) \leq \frac{1}{2} \lambda^l \sqrt{n}(2^{-k/2} + 2^{-n/2})$$

The proof of the above theorem follows from the proof of Lemmas 1 and 2.

$$\Delta(A^l \boldsymbol{p}; U_n) = \frac{1}{2} \sum_{a \in \{0,1\}^n} \left| \Pr[A^l \boldsymbol{p} = a] - \Pr[U_n = a] \right| \tag{1}$$

$$\leq \frac{1}{2} \sqrt{n} \left\| A^l \boldsymbol{p} - U_n \right\|_2 \tag{2}$$

$$\leq \frac{1}{2} \sqrt{n} \lambda^l \left\| \boldsymbol{p} - U_n \right\|_2 \tag{3}$$

$$= \frac{1}{2} \sqrt{n} \lambda^l (2^{-k/2} + 2^{-n/2}) \tag{4}$$

where equation (1) follows from the definition of statistical distance, equation (2) is followed from the relation between l2 and l1 norms, i.e. $|V| \leq \sqrt{n} \|V\|_2$, equation (3) comes from the proof of lemmas 1 and 2, and equation (4) follows from linear algebra facts ($\|\boldsymbol{p} - U_n\|_2 \leq \sqrt{\|\boldsymbol{p}\|_2^2 + \|U_n\|_2^2}$) and that min-entropy of $\boldsymbol{p}$ is k, which gives $\|\boldsymbol{p}\|_2^2 \leq 2^{-k}$.

$\square$

For an expander graph G, given ϵ and k as the min-entropy of the initial distribution on vertices, we can compute the number of required steps of the random-walk on the expander graph so that the distribution on the graph vertices becomes ϵ-close to uniform distribution. Note that min-entropy of the initial vertex distribution results in closeness to uniform distribution, but the random walk will amplify this closeness.

Let $\lambda = 2^{-\alpha}$ and $\epsilon = 2^{-\beta}$. To be ϵ-close to uniform, we must have $\frac{1}{2}\sqrt{n}\lambda^l(2^{-k/2} + 2^{-n/2}) \leq \epsilon$. This gives us the following lower bound on l:

$$l \geq \frac{1}{\alpha}[\beta + \log(\sqrt{n}) + \log(2^{-k/2} + 2^{-n/2})] \tag{5}$$

The above bound requires the value λ for the graph. Equation 5 shows that for a given ϵ and min-entropy, smaller λ correspond to shorter random walk. So one needs to find graphs with smaller λ.

The following theorem shows that regular graphs have small λ.

Theorem 3. *[AB09, section 21.2.1] For a constant $d \in \mathbb{N}$, any d-regular, N-vertex graph G satisfies $\lambda \geq 2\sqrt{d-1}/d(1 - o(1))$ where the $o(1)$ term vanishes as $N \to \infty$.*

Ramanujan graphs are d-regular graphs that achieve $\lambda \geq 2\sqrt{d-1}/d$ and so are excellent as spectral expanders. For a fixed d and large N, the d-regular N-vertex Ramanujan graph minimizes the λ. There are several explicit constructions of Ramanujan graphs. Here we explain one of the simpler constructions.

2.2   A Simple Explicit Construction for Expander Graphs

There are explicit constructions of expander graphs that can be efficiently generated. That is vertices are indexed by $i \in I$ and there is an algorithm that for any i, generates the index of its neighbours. For example the *p-cycle with inverse chords* construction gives us a 3-regular expander graph with p vertices, p is prime, in which a vertex X is labelled by $x \in \{0, p-1\}$ and the neighbour vertices have indexes $x - 1$, $x + 1$ and x^{-1}. Here all arithmetic are mod p and 0^{-1} is defined to be 0. The spectral expansion of this graph is very close to the above-mentioned bound. The construction is due to Lubotzky-Phillips-Sarnak [LPS86] and the proof that the construction is a Ramanujan graph, uses deep mathematical results. The λ for this graph is upper bounded by 0.94. Other explicit constructions of expander graphs use graph product techniques such as Zig-Zag product and replacement product [RVW00].

2.3   Game Theoretic Definitions

A game consists of a set of *players*, each with a set of available *actions* and a specification of payoffs for each pair of actions. An *action profile* is a tuple of all players' actions. An assumption in game theory is that players play rationally; that is they aim to maximize their payoff, given some belief about the other players' actions. Thus, each player has *preferences* about the action profile. For describing players' preferences we use *utility function*. An strategic game is defined as follow.

Definition 6. *A strategic game with ordinal preferences consists of,*

- *A finite set of players,* $N = \{1, ..., N\}$*;*
- *A set of actions for player i denoted by* A_i*;*
  Action profile: $a = (a_1, ..., a_n) \in A = A_1 \times ... \times A_n$;
- *Preferences over the set of action profiles based on the* Utility Function *that captures payoffs.*
  Utility function *for player i:* $u_i : A \to R$, *which R is the set of real numbers. We say that player i prefers a to b iff* $u_i(a) > u_i(b)$

A *pure strategy* specifies the actions of a player in all possible situations that he will be in.

Mixed strategy of a player in a strategic game means more than one action is played with a positive probability by the player. The set of actions with non-zero probability form the *support of mixed strategy*. Mixed strategy is specified by a set of probability distributions. The following theorem shows the importance of mixed strategies [Osb04].

Theorem 4. *Every strategic game in which each player has finitely many actions has a mixed strategy Nash equilibrium.*

A finite two-player strategic game can be represented by a table. In such a representation "Row" player is player 1 and "column" player is player 2. That is, rows correspond to actions $a_1 \in A_1$ and columns correspond to actions $a_2 \in A_2$. Each cell in the table includes a pair of payoffs, starting with the row player followed by the column player. For a two-party game, each player action is a pure strategy and a mixed strategy is a probability distribution on the set of actions available to them.

Halprin et al. used an extended form of *"matching pennies"* using a two dimensional array defining the choices of the players. In basic matching pennies game with Head and Tail as possible actions for each player the payoff table is as follow:

-	Head	Tail
Head	(1,-1)	(-1,1)
Tail	(-1,1)	(1,-1)

Matching pennies game is a zero-sum game and for any action profile a we have $u_1(a) + u_2(a) = 0$. From the table it can be seen that there is no pure strategy Nash equilibrium for matching pennies. It is easy however to show that *the best strategy for both players in the game is uniform distribution on the set of possible actions.* This is easy to see intuitively and also show formally using a distribution p and $1 - p$ to denote the probability that the first player chooses Head or Tail, respectively, and then requiring p to be chosen such that the player 2 remains completely indifference about the choice of player 1. That is, $u_2(Head) = u_2(Tail)$ which gives $-p + 1(1 - p) = p - (1 - p)$ and $p = 1/2$. A similar argument shows that the best strategy of player 2 is uniform also.

In our work we use *extended matching pennies* game where each player has a set of m possible actions. Using a similar approach one can show that the best strategy for players is uniform random selection among the n alternatives. Extended matching pennies game was also used in [HN09] to increase the amount of randomness from each user input. Traditional matching pennies game provides at most 1 bit randomness for each user's input. Using extended matching pennies, this can be increased to $\log_2 m$ bits per action. This however is an upper bound and depending on the set up and the actual value of m, the min-entropy of user input can be different. Our experimental results for $m = 10$ and $m = 3$ corresponding to the size and degree of the expander graph used in our work, showed 2.1 and 1.32 bits per input, respectively.

3   TRG Using Human Input in Games

We propose an integrated approach to construct TRGs using human input in games. Important properties of this approach are, (i) the only source of randomness for the TRG is the human input, and (ii) the final results have guaranteed and adjustable level of randomness. This latter property means that the user can improve the quality of the output randomness at any time and to any level of closeness to uniform randomness, simply by adjusting the length of the random walk. In comparison, in the construction of Halprin et al. (i) the entropy source is based on human input and a second external source of perfect randomness is required to provide seed for the extractor, and (ii) the size and quality of the final output depends on the extractor that is used after obtaining the output of the entropy source. Here changing the quality of the final output, requires replacing the extractor and performing the calculations from scratch.

The outline of the approach is given in Section 1.1 and includes, (i) choosing an expander graph with appropriate parameters, and (ii) designing an appropriate game that is "attractive" to play and has the required property (uniform optimal strategy for each sub-game).

Choosing the Expander Graph. The starting point is choosing a d-regular expander graph with an appropriate number of vertices 2^n, each vertex labelled with a binary string of length n. The two parameters n and d will be directly related to the computational efficiency of the system in generating random bits: larger n means longer output string and more random bits, and larger d means faster convergence to the uniform distribution and so shorter walk to reach the same level of closeness to the uniform distribution (See Theorem 3). In practice however, because the graph is the basis of the game visual presentation to the user, one needs to consider usability of the system and so the choice of n and d will take this factor into account. Another important requirement is that steps of the random walk must correspond to an independent and uniformly distributed random variable. Experiments in human psychology [RB92] shows that bias will increase in human choices for larger sets of possible choices. and thus the random walk generated by human input will be farther from uniformly random. In Section 4.1 we discuss issues that arise in choosing the graph and extending the approach when longer sequences of random bits are required.

The Game. The TRG algorithm uses a game between human and computer. The game consists of a sequence of sub-games. In all sub-games the human makes a choice and wins if their choice is not correctly guessed by the computer. Each sub-game is an extended matching pennies game which is known to have uniform optimal strategy for both players. The main difference between the initial and subsequent sub-games is the number of choices available to the user, and the way these choices are used. The computer choices are made by a pseudorandom number generator.

- *Initial Sub-game.* In the initial sub-game the user chooses a vertex from the set of 2^n possible vertices in the graph, and wins if the computer cannot "guess" their choice.
- *Walk Sub-games.* Each subsequent sub-game correspond to a single step of the random walk. At vertex V, the player can choose any of the vertices that are adjacent to V. Using a d-regular graph ensures that the number of choices at every vertex is the same.

The game proceeds in ℓ steps where ℓ is determined by the required quality of the final output randomness and the min-entropy of the input. In practice one needs an estimate of user min-entropy of the initial game, when choosing among 2^n vertices, to be used with d the degree of the graph, in the equation 5 to obtain an estimate for the number of steps on the graph for a chosen value of ϵ.

To estimate min-entropy of user in the initial game, we developed a second game that simply asks user to choose a vertex and win if their choice is not correctly guessed by the computer. In our experiments the suggested min-entropy for a graph of 10 vertices is 2.32 bits which is although lower than the min-entropy of uniform distribution ($\log_2 10 \approx 3.58$) , but shows a high level of min-entropy.

Although human play differently but assuming they behave "similarly", one can obtain an average value for the min-entropy of the initial choice for a graph of size n, over a large population of users and use that for estimating the number of steps.

4   Experiments

We designed a game between human and computer and asked each player to play the game for at least 1000 rounds. This is sufficient to run the required min-entropy and statistical tests.

Our objectives in the experiments are the following:

1. Estimate the min-entropy of human choices in sub-game 1, which is required for the extraction component of the TRNG. As noted earlier, expander graphs can be used for extracting randomness if the min-entropy of the input is bounded below.
2. Examine the statistical properties (e.g. min-entropy) of the walk to verify if the walk by human is a good approximation of a random walk.
3. Examine the statistical properties of the final output,

We run the above three tests for all participants and will present the results.

4.1 The Game

Our basic game is a combination of hide and seek and the Roshambo (Rock-paper-scissors) games. The hide and seek game corresponds to the initial sub-game where human should choose from a number of vertices to start the game. The Roshambo game corresponds to the second sub-game (i.e. random walk). For the choice of expander graph, we used Peterson graph which is a 3-regular graph with 10 vertices.

In the first sub-game, the human player starts the game by selecting a vertex from the graph G. This will place a sheep on the vertex (Figure 1-b). The computer responds by placing a wolf on a random vertex. This is similar to the hide and seek game. The player loses if the wolf and sheep are on the same vertex. If the player wins, then the game will highlight the vertices that are adjacent to the selected vertex by the player (marked by a star in Figures). The same sequence of steps (user's choice (Figure 1-c), followed by the computer choice (Figure 1-d)) is now played using the highlighted vertices instead of all vertices in the graph, and the human will win (Figure 1-c) if the choices are different and lose otherwise. The winner is the player with higher final score.

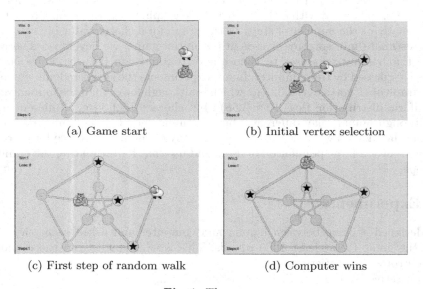

(a) Game start   (b) Initial vertex selection

(c) First step of random walk   (d) Computer wins

Fig. 1. The game

The Game Design. We implemented the game using HTML 5 technology so that it can be run and played on any system having an Internet browser and even on touch screen devices such as tablets or smart phones. The game is simple to learn and intuitive and can be accessed online [Ali13].

Game Design for Generating Random Keys. The graph has 10 vertices and so can generate at most $\log_2 3$ bits of entropy. In a real world design, a large 3-regular expander graph (e.g. with roughly 2^{64} vertices) can be easily constructed using the construction in section 2.2. To construct such a graph, the largest prime number smaller than 2^{64} is chosen for the number of vertices. For such a graph, λ is estimated to be $2\sqrt{3-1}/3 = 0.94$. The choice of 3-regular graphs is based on the psychological experiments [RB92] showing that too many choices would increase the bias of human selection. To prevent this, we only recommend using 3-regular graphs for the construction of the expander graph.

Now let $\epsilon = 2^{-15}$, and considering the min-entropy of the initial selection be 32 (i.e. 0.5 per bit), then we need 12 steps of random walk on the graph to get 2^{-15}-close to uniform distribution:

$$l \geq \frac{1}{\alpha}\beta + \log(\sqrt{n}) + \log(2^{-k/2} + 2^{-n/2}) \approx 11.23.$$

where $\alpha \approx 0.089, \beta = 15, k = 32, n = 64$.

Note that although we cannot find the exact value of the min-entropy of the initial selection, it is possible to find a good estimate experimentally. This is by considering many samples from different human players and using statistical models to estimate average min-entropy. Equation 5 can be used to find the minimum number of steps that is required for the final distribution to be ϵ-close to uniform distribution.

The graph will appear as a continuous circle. For the first sub-game, the human chooses a random vertex from the graph, or equivalently a random point on the circle. Using a circular representation of the graph has the advantage that all vertices appear with the same value (importance) and no end points will be left out because of its location and this will avoid human tendency to avoid corners, as observed in [Wag72,HN09].

After the initial vertex selection, the graph will be displayed locally (vertices adjacent to the current selected vertex of the user) and then the user is asked to play the second sub-game.

4.2 Measuring Min-Entropy

To measure the min-entropy (or Shannon entropy) of a source ones needs to assume certain structure in the source distribution. For a list of n samples $\{s_i\}_{i=1}^n$ from a source S over the finite set $\mathbb{S}$, if we assume that the source S is i.i.d., that is samples are independently and identically distributed, then having enough samples from the source allows us to estimate the probability distribution of the source with high confidence and find the entropy as in [BK12] (Section 9.2).

NIST draft [BK12] gives the requirements of entropy sources and also proposes a number of tests to estimate the min-entropy of the source in i.i.d. and non-i.i.d. cases, both. The testing method first checks whether the source can be considered i.i.d. NIST suggests the following set of tests for i.i.d. sources (Section 9.1 of [BK12]): 6 shuffling tests, Chi-square test, test for independence and goodness of fit. If all of the test are passed, the source is assumed to be i.i.d. and then *a conservative method* is used to estimate the entropy of i.i.d. source. If any of the tests are not passed however, the source is considered to be non-i.i.d., and a number of tests are used to estimate the min-entropy. These second group of tests are collision test, partial collection test, Markov test, Compression test and the Frequency test, each outputting a value as the estimation of the min-entropy. The final min-entropy will be the minimum over all these estimated values. While i.i.d. and non-i.i.d. tests provide an estimation of the entropy, they may fail to detect anomalies in the source. Therefore, NIST defines a set of sanity checks that will make sure this does not happen. The sanity checks contain two tests: Compression test and collision test. If the sanity checks fail, no estimation will be given.

For our experiments, we obtained an unofficial version of the code (the code is not released yet) and used it to estimate the min-entropy of our source. We ran tests to verify whether our estimations are meaningful (our sanity checks), and also check consistency in the min-entropy estimation for a data set from /dev/random in Linux. Our analysis showed that the estimation from NIST set of tests are sound, but are very conservative (admitted in Section 9.3.1 of [BK12]). For example, we expect to have roughly the same approximation of min-entropy for the data in /dev/random. But the approximation from the NIST tests is very dependent on the number of samples given to the tests (which is quite intuitive and acceptable). This caused very low estimates for a subset of our users with smaller sample size and in general, *min-entropy estimation in our experiments were conservative.*

4.3   Measuring Statistical Property of a Source

To examine statistical properties of a sequence, we use the statistical tests in a battery of tests called Rabbit [LS07b]. Rabbit set of tests includes tests from NIST [ea10] and DIEHARD [Mar98] with modified parameters tuned for smaller sequences and hardware generators. We used an implementation of these test in [LS07a].

4.4   Measures of Randomness for Our Game

In our first experiment, we measured the min-entropy of the player's initial selection of a vertex on the graph. To measure min-entropy, we used the method in Section 4.2, and the results is as follows: The table summarizes the data we collected from 9 users. For each user, the first row is the number of total plays, the second row is the number of times the human choice collided with the computer's choice (the wolf was placed on the sheep), the third row is the probability that a collision occurred in game plays (the number of collisions divided by the total number of plays) and the last row is the min-entropy of the player's choices per bit. We then counted the number of times each vertex is selected by human. The expected behaviour is that the number of times each vertex is selected must be roughly the same.

Table 1. Min-entropy of users in first sub-game

User	1	2	3	4	5	6	7	8	9
Total Shots	1770	2141	3980	3439	2021	652	1685	905	983
Collisions	150	322	365	348	149	70	167	45	112
Probability	0.08	0.14	0.09	0.1	0.07	0.1	0.1	0.05	0.11
Min-entropy	0.45	0.49	0.61	0.65	0.52	0.49	0.47	0.55	0.51

In conclusion for the first sub-game,the lowest min-entropy is 0.45 per bit, and on average we expect the min-entropy of 0.52 per bit.

Random walk game:

In the second experiment, we collected data from game play of participants over a long sequence of game play. We expect the choice of neighbours to be uniformly random selection over the set of adjacent vertices. We map the set of adjacent vertices to the set $\{1, 2, 3\}$ based on an ordering on the vertices (e.g. an ordering on their labels which are numbers) To examine this in practice, we applied the statistical tests in section 4.3 to measure the statistical properties of this sequence. The results are summarized in table 2. The table gives a summary of a subset of tests and their corresponding p-value for the output. The data to generate this table is from a sample player. We also examined the data from other players with all statistical tests being passed with p-values far from 0 or 1.

We also calculated the min-entropy of the random walk (human choices for each step) to further confirm the random properties of this sequence. The table 3 summarizes the results:

The results show that the min-entropy of players is more when the choices are less (here 3). The min-entropy of the initial selection of vertices can be measured - assume it is k. Using the results in lemma 2, the final output of the game would be ϵ-close to uniform if the walk is uniformly random. Using the above results, we note that the walk is *close* to a random walk that is uniformly distributed.

Table 2. Result of statistical test

Statistical Test	result
LinearComplexity	0.87
LempelZiv	0.59
SpectralFourier	0.63
Kolmogorov-Smirnov	0.47
PeriodsInString	0.51
HammingWeight	0.85
HammingCorrelation	0.23
HammingIndependence	0.78
RunTest	0.13
RandomWalk	0.29

Table 3. Min-entropy of users in second sub-game

User	1	2	3	4	5	6	7	8	9
Total Shots	370	369	1009	1786	560	1071	4821	1190	1065
Collisions	118	118	237	595	140	373	1624	379	334
Probability	0.31	0.29	0.23	0.33	0.25	0.34	0.33	0.31	0.31
Min-entropy	0.49	0.51	0.64	0.75	0.78	0.72	0.83	0.61	0.79

Table 4. Result of statistical test

Statistical Test	result
LinearComplexity	0.51
LempelZiv	0.73
SpectralFourier	0.34
Kolmogorov-Smirnov	0.23
PeriodsInString	0.13
HammingWeight	0.55
HammingCorrelation	0.69
HammingIndependence	0.29
RunTest	0.60
RandomWalk	0.67

To examine the effect of this discrepancy, we ran the set of statistical tests on the final output and the results is summarized in the following table:

The numbers in table 4 are the p-values of each test. The test is passed if these values are far from 0 or 1. A margin of 0.001 is usually accepted for the test to be passed.

Overall the experiments shows viability of the approach in practice.

5   Concluding Remarks

TRGs are an essential component of security systems. Using human as an entropy source is particularly attractive in shared environments such as cloud services where traditional sources of entropy (computing hardware and software) are shared among users and extra caution must be used to ensure randomness extracted from the entropy sources do not result in correlated randomness for users that are sharing the services. There are a number of extensions and directions for future work. On theoretical side, analysis of randomness extractors that are based on expander graphs when the random walk is replaced by a walk with a guaranteed min-entropy is an interesting open question. On the implementation and experimental side, we noted that for generating larger strings, for example a 64 or 80 bit strings, the full graph cannot be presented to the user. Here one needs to find ways of enabling the user to make the initial selection of the string with "good" initial min-entropy and then use portions of the graph corresponding to the neighbours of the vertex, to receive the user's input for the random step at that vertex. The choice of the graph and creating an "interesting" game and interface to encourage random selection will improve effectiveness of the approach and usability of the system. We analysed strings generated by nine users. Wider user studies are required to measure min-entropy and statistical properties of strings at different stages (entropy source, random walk and final output) as well as usability of the system. Using human input also improves trustworthiness of the generated randomness. Hardware faults or malicious tampering with entropy sources may result in biases in the randomness that are not detectable. Our approach is protected against such faults or malicious tampering.

Our work is the first construction of a full TRG that uses only human game play. Using human users to construct TRGs with higher rate is an interesting direction for future work.

References

AB09.   Arora, S., Barak, B.: Computational complexity: a modern approach, vol. 1. Cambridge University Press, Cambridge (2009)

Ali13.   Alimomeni, M.: Sheep-wolf game to generate true random numbers by human (2013), http://pages.cpsc.ucalgary.ca/~malimome/expander/

BK12.   Barker, E., Kelsey, J.: Recommendation for the entropy sources used for random bit generation (August 2012), http://csrc.nist.gov/publications/drafts/800-90/draft-sp800-90b.pdf

BST03.   Barak, B., Shaltiel, R., Tromer, E.: True random number generators secure in a changing environment. In: Walter, C.D., Koç, Ç.K., Paar, C. (eds.) CHES 2003. LNCS, vol. 2779, pp. 166–180. Springer, Heidelberg (2003)

DGP09.   Dorrendorf, L., Gutterman, Z., Pinkas, B.: Cryptanalysis of the random number generator of the windows operating system. ACM Trans. Inf. Syst. Secur. 13(1), 10:1–10:32 (2009)

ea10.   Rukhin, et al.: A statistical test suite for the validation of random number generators and pseudo random number generators for cryptographic applications (2010), http://csrc.nist.gov/groups/ST/toolkit/rng/documents/SP800-22rev1a.pdf

GD.   Goldberg, I., David, W.: Randomness and the netscape browser, http://www.cs.berkeley.edu/~daw/papers/ddj-netscape.html

GPR06.   Gutterman, Z., Pinkas, B., Reinman, T.: Analysis of the linux random number generator. In: 2006 IEEE Symposium on Security and Privacy, p. 15. IEEE (2006)

HDWH12. Heninger, N., Durumeric, Z., Wustrow, E., Alex Halderman, J.: Mining your ps and qs: detection of widespread weak keys in network devices. In: Proceedings of the 21st USENIX Conference on Security Symposium, Security 2012, p. 35. USENIX Association, Berkeley (2012)

HLW06.   Hoory, S., Linial, N., Wigderson, A.: Expander graphs and their application. Bulletin of the AMS 43(4), 439–561 (2006)

HN09.   Halprin, R., Naor, M.: Games for extracting randomness. In: Proceedings
            of the 5th Symposium on Usable Privacy and Security, p. 12. ACM (2009)

LHA⁺12.   Lenstra, A.K., Hughes, J.P., Augier, M., Bos, J.W., Kleinjung, T.,
            Wachter, C.: Public keys. In: Safavi-Naini, R., Canetti, R. (eds.) CRYPTO
            2012. LNCS, vol. 7417, pp. 626–642. Springer, Heidelberg (2012)

LPS86.   Lubotzky, A., Phillips, R., Sarnak, P.: Explicit expanders and the ramanu-
            jan conjectures. In: Proceedings of the Eighteenth Annual ACM Sympo-
            sium on Theory of Computing, pp. 240–246. ACM (1986)

LS07a.   L'Ecuyerand, P., Simard, R.: Testu01 (August 2007),
            http://www.iro.umontreal.ca/~simardr/testu01/tu01.html

LS07b.   L'Ecuyer, P., Simard, R.: Testu01: A c library for empirical testing of ran-
            dom number generators. ACM Trans. Math. Softw. 33(4) (August 2007)

Mar98.   Marsaglia, G.: Diehard (1998), http://www.stat.fsu.edu/pub/diehard/

Mic.   Microsoft. Windows random number generator,
            http://msdn.microsoft.com/en-us/library/aa379942.aspx

Osb04.   Osborne, M.J.: An introduction to game theory, vol. 3. Oxford University
            Press, Oxford (2004)

RB92.   Rapoport, A., Budescu, D.V.: Generation of random series in two-person
            strictly competitive games. Journal of Experimental Psychology: Gen-
            eral 121(3), 352 (1992)

RVW00.   Reingold, O., Vadhan, S., Wigderson, A.: Entropy waves, the zig-zag graph
            product, and new constant-degree expanders and extractors. In: Proceed-
            ings of 41st Annual Symposium on Foundations of Computer Science,
            pp. 3–13 (2000)

Sha11.   Shaltiel, R.: An introduction to randomness extractors. Automata, Lan-
            guages and Programming, 21–41 (2011)

vN51.   von Neumann, J.: Various techniques used in connection with random
            digits. J. Research Nat. Bur. Stand., Appl. Math. Series 12, 36–38 (1951)

Wag72.   Wagenaar, W.A.: Generation of random sequences by human subjects: A
            critical survey of literature. Psychological Bulletin 77(1), 65 (1972)

ZLwW⁺09.   Zhou, Q., Liao, X., wo Wong, K., Hu, Y., Xiao, D.: True random number
            generator based on mouse movement and chaotic hash function. Informa-
            tion Sciences 179(19), 3442–3450 (2009)

A Game Theoretic Analysis of Collaboration in Wikipedia

S. Anand[1], Ofer Arazy[2], Narayan B. Mandayam[1], and Oded Nov[3]

[1] Wireless Information Networks Laboratory (WINLAB), Rutgers University
{anands72,narayan}@winlab.rutgers.edu
[2] Alberta School of Business, Edmonton
ofer.arazy@ualberta.ca
[3] Department of Technology Management and Innovation,
Polytechnic Institute of New York University
on272@nyu.edu

Abstract. Peer production projects such as Wikipedia or open-source software development allow volunteers to collectively create knowledge-based products. The inclusive nature of such projects poses difficult challenges for ensuring trustworthiness and combating vandalism. Prior studies in the area deal with descriptive aspects of peer production, failing to capture the idea that while contributors collaborate, they also compete for status in the community and for imposing their views on the product. In this paper, we investigate collaborative authoring in Wikipedia, where contributors append and overwrite previous contributions to a page. We assume that a contributor's goal is to maximize ownership of content sections, such that content owned (i.e. originated) by her survived the most recent revision of the page. We model contributors' interactions to increase their content ownership as a non-cooperative game, where a player's utility is associated with content owned and cost is a function of effort expended. Our results capture several real-life aspects of contributors interactions within peer-production projects. Namely, we show that at the Nash equilibrium there is an inverse relationship between the effort required to make a contribution and the survival of a contributor's content. In other words, majority of the content that survives is necessarily contributed by experts who expend relatively less effort than non-experts. An empirical analysis of Wikipedia articles provides support for our model's predictions. Implications for research and practice are discussed in the context of trustworthy collaboration as well as vandalism.

Keywords: Peer production, Wikipedia, collaboration, non-cooperative game, trustworthy collaboration, vandalism.

1   Introduction

Recent years have seen the emergence of a web-based peer-production model for collaborative work, whereby large numbers of individuals co-create knowledge-based goods, such as Wikipedia, and open source software [30], [12], [23], [14],

S.K. Das, C. Nita-Rotaru, and M. Kantarcioglu (Eds.): GameSec 2013, LNCS 8252, pp. 29–44, 2013.
© Springer International Publishing Switzerland 2013

[18], [24], [19]. Increasingly, individuals, companies, government agencies and other organizations rely on peer-produced products, stressing on the need to ensure trust worthiness of collaboration (e.g., deter vandalism) as well as the quality of end products.

Our focus in this study is Wikipedia, probably the most prominent example of peer-production. Wikipedia has become one of the most popular information sources on the web, and the quality of Wikipedia articles has been the topic of recent public debates. Wikipedia is based on wiki technology. Wiki is a web-based collaborative authoring tool that allows contributors to add new content, append existing content, or delete and overwrite prior contributions. Wikis track the history of revisions similarly to version control systems used in software development allowing users to revert a wiki page to an earlier revision [25], [36], [15].

Peer production projects face a key tension between inclusiveness and quality assurance. While such projects need to draw in a large group of contributors in order to leverage "the wisdom of the crowd," there is also requirement for accountability, security, and quality control [17], [20], [35]. Quality assurance measures are necessary not only in cases of vandalism; conflicts between contributors could also result from competition. For example, contributors to Wikipedia may wrestle to impose their own viewpoints on an article especially for controversial topics or attempt to dominate subsets of the peer-produced product. Another example is when contributors seeking status within the community compete to make the largest contribution, and in the process overwrite others' previous contributions. The result of such competitions is often "edit wars" where articles are changed back-and-forth between the same contributors.

Prior studies investigating an individual's motivation for contributing content to Wikipedia have identified a large number of motives driving participation [12], [22], including motives that are competitive in nature, such as ego, reputation enhancement, and the expression of one's opinions [22], [26]. However, studies investigating individuals did not consider the competitive dynamics emerging from motives such as reputation. Research into group interactions at Wikipedia, have tended to emphasize the collaborative (rather than competitive nature of interactions) [12]. Other studies investigated threats to security and trustworthiness resulting from malicious attacks (i.e. vandalism) [22] and the organizational mechanisms used by Wikipedia to combat such attacks [26]; yet these studies do not consider threats resulting from benevolent contributors. A relevant strand of the literature has looked at conflicts of opinions between contributors [13], [12]. However, the focus is on the result of these conflicts on content quality rather than the competitive mechanisms driving them. In summary, while peer-production projects, and in particular Wikipedia, have attracted significant attention within the research community, to the best of our knowledge, the competitive dynamics have not been investigated.

In order to better understand collaboration in Wikipedia and capture the competitive nature of interactions, we turn to game theory. Our underlying assumption is that a contributor's goal is to maximize ownership of content sections, such that content "owned" (i.e. originated) by that user survived the most recent revision of the page. We model contributors' interactions to increase their content ownership, as a non-cooperative game. A contributor's motivation for trying to maximize her ownership within a certain topical page could be based on the need to express one's views or to increase her status in the community; and competition could be the result of battles between opposing viewpoints (e.g. vandals and those seeking to ensure trustworthiness of content) or consequences of power struggles. The utility of each contributor in the non-cooperative game is the ownership in the page, defined as the fraction of content owned by the contributor in the page. Each contributor suffers a cost of contribution which is the effort expended towards making the contribution. The objective is then to determine the optimal strategies, i. e., the optimal number of contributions made by each contributor, so that her *net utility* is maximized. Here, the net utility is the difference between the utility (a measure of the ownership) and the cost (a measure of the effort expended). The optimal strategies are determined by determining the Nash equilibrium of the non-cooperative game that models the interactions between the contributors. We determine the conditions under which the Nash equilibrium of the game can be achieved and find its implications on the contributors' expertise levels on the topic. We report of an empirical analysis of Wikipedia that validates the model's predictions. The key results brought forth by our analysis include

- The ownership of contributors increases with the decreasing levels of effort expended by the contributor on the topic.
- Contributors expending equal amount of effort end up with equal ownership.

The rest of the paper is organized as follows. The non-cooperative game that models the interactions between contributors is described in Section 2. We then use Wikipedia data to validate the modeling in Section 3. We then discuss in Section 4 the relevance of our analysis and modeling to trust worthy collaboration and vandalism. Conclusions are drawn in Section 5 along with pointers to future directions.

2   User Contribution as a Non-cooperative Game

We model the interactions of the N content contributors to a page (i.e., users) as a non-cooperative game. The strategy set for each contributor is the amount and type of contribution she makes. Table 1 describes the notations used in our analysis. and their descriptions.

Let x_i represent the content owned by the ith user in the current version of the page. We define the utility, u_i, as the fraction of content owned by the ith contributor, and is given by

$$u_i = \frac{x_i}{\sum_{j=1}^{N} x_j}.$$

(1)

The objective of contributor i is to determine the optimal x_i so that u_i is maximum.

Table 1. Variables used in the analysis in this paper

Notation/Variable	Description
N	Number of users or content contributors
x_i	The amount of content owned by the ith user
β_i	Effort expended by user i to make unit contribution
u_i	The fractional ownership held by the ith user
n_i	Net utility of contributor i
1	The all-one vector
I	The identity matrix

It is observed from (1) that the optimal x_i that maximizes u_i is $x_i = \infty$. This is because the utility function is an increasing function of x_i. Intuitively, this result occurs because every time the ith user makes a contribution, his/her ownership increases. However this results in reduction in the ownership of other contributors, to counter which, they attempt to make additional contributions (by increasing their respective x_k's). This, in turn, reduces the ownership of contributor i, thereby causing him/her to further increase x_i to increase ownership. This process continues ad infinitum resulting in $x_i \to \infty$, $\forall\ i$. This degenerate scenario can be mitigated as follows.

Suppose the ith contributor expends an effort, β_i, to make a unit contribution. For instance, β_i can be the cost incurred by the ith user, in terms of time and effort spent in learning the topic and in posting content on a Wiki page. Therefore, the ith contributor expends a total effort $\beta_i x_i$, to achieve x_i amount of content ownership in the page. The net utility experienced by the ith contributor, n_i, can be written as the difference between utility of contributor i, given by (1) and the total effort expended by contributor i, i. e.,

$$n_i = u_i - \beta_i x_i = \frac{x_i}{\sum_{j=1}^{N} x_j} - \beta_i x_i. \tag{2}$$

It is observed that the net utility obtained by the ith contributor not only depends on the strategy of the ith contributor (i.e., x_i), but also on the strategies of all the other contributors (i.e., x_j, $j \neq i$). This results in the non-cooperative game of complete information [29] between the contributors. The optimal x_i, $\forall\ i$ (termed as x_i^*), which is determined by maximizing n_i in (2), is then the Nash equilibrium of the non-cooperative game where no contributor can make a unilateral change.

Applying the first order necessary condition to (2), x_i^* is obtained as the solution to

$$\frac{\partial n_i}{\partial x_i}\bigg|_{x_i = x_i^*} = \frac{\sum_{\substack{k=1 \\ k \neq i}}^{N} x_k^*}{\left(\sum_{j=1}^{N} x_j^*\right)^2} - \beta_i = 0, \forall i \tag{3}$$

subject to the constraints $x_i^* \geq 0$, $\forall\ i$. From (3), we obtain $\frac{\partial^2 n_i}{\partial x_i^2} = -\frac{2 \sum_{\substack{k=1 \\ k \neq i}}^{N} x_k}{\left(\sum_{j=1}^{N} x_j\right)^3} < 0$, $\forall\ i$, when $x_i \geq 0$. Thus, n_i is a concave function of x_i and x_i^*, which solves

(3) subject to $x_i^* \geq 0$, $\forall\, i$, is a local as well as a global maximum point. In other words, according to [28], *the non-cooperative game has a unique Nash equilibrium*, $\mathbf{x}^* = \begin{bmatrix} x_1^* & x_2^* & \cdots & x_N^* \end{bmatrix}^T$, obtained by numerically solving the system of N non-linear equations specified by (3). However, to study the effect of the effort levels (β_i's) on the strategies of the contributors, it is desirable to obtain an expression that relates the vectors, $\mathbf{x}^*$, $\mathbf{x} = [x_i]_{1 \leq i \leq N}$ and $\boldsymbol{\beta} = [\beta_i]_{1 \leq i \leq N}$.

Re-writing (3),

$$\left(\sum_{j=1}^{N} x_j^* \right)^2 - \alpha_i \sum_{\substack{j=1 \\ j \neq i}}^{N} x_j^* = 0, \forall N, \tag{4}$$

where $\alpha_i \triangleq 1/\beta_i$. Eqn. (4) can be written as

$$(\mathbf{x}^*)^T \mathbf{1}\mathbf{1}^T \mathbf{x}^* \mathbf{1} - \mathbf{D}_\alpha \left(\mathbf{1}\mathbf{1}^T - \mathbf{I} \right) \mathbf{x}^* = \mathbf{0}, \tag{5}$$

where $(.)^T$ represents the transpose of a vector or a matrix, $\mathbf{D}_\alpha$ is the diagonal matrix $\mathbf{diag}\,(\alpha_1, \alpha_2, \cdots, \alpha_N)$, $\mathbf{1}$ is the column vector in which all entries are one, $\mathbf{0}$ is the column vector in which all entries are zero and $\mathbf{I}$ is the identity matrix.

It can be easily verified the vectors, $\mathbf{y}_1 = \begin{bmatrix} \frac{1}{\sqrt{N}} & \frac{1}{\sqrt{N}} & \frac{1}{\sqrt{N}} & \frac{1}{\sqrt{N}} & \cdots & \frac{1}{\sqrt{N}} \end{bmatrix}^T$ and for $j = 2, 3, \cdots, N$, $\mathbf{y}_j = \begin{bmatrix} y_{1j} & y_{2j} & y_{3j} & \cdots & y_{(N-1)j} & y_{Nj} \end{bmatrix}^T$, where

$$y_{kj} = \begin{cases} \dfrac{1}{\sqrt{j(j-1)}} & k < j \\[2mm] -\dfrac{j-1}{\sqrt{j(j-1)}} & k = j \\[2mm] 0 & k > j, \end{cases} \tag{6}$$

form a set of orthonormal eigen vectors to the matrix, $\mathbf{1}\mathbf{1}^T$. The eigen value corresponding to $\mathbf{y}_1$ is N and those corresponding to $\mathbf{y}_2, \cdots, \mathbf{y}_N$ are 0s. Let $\mathbf{P} = [\mathbf{y}_1|\mathbf{y}_2|\cdots|\mathbf{y}_N]$. Then, $\mathbf{P}$ is an orthogonal matrix and by orthogonality transformation,

$$\mathbf{P}^T \mathbf{1}\mathbf{1}^T \mathbf{P} = \mathbf{D} = \mathrm{diag}\,(N, 0, 0, \cdots, 0). \tag{7}$$

Let $\mathbf{z} = \begin{bmatrix} z_1 & z_2 & z_3 & \cdots & z_{N-1} & z_N \end{bmatrix}^T$. Since the eigen vectors of a matrix form a basis for the $N-$dimensional sub-space [27], the vector, $\mathbf{x}^*$, can be written as $\mathbf{x}^* = \mathbf{P}\mathbf{z}$. A similar expression has been solved in [10] in the context of pricing in wireless networks and we outline here the key steps to determine the optimal $\mathbf{x}^*$.

– Using $\mathbf{x}^* = \mathbf{P}\mathbf{z}$ in (5) and (7), we obtain

$$\mathbf{z}^T \mathbf{D}\mathbf{z}\mathbf{1} - \mathbf{D}_\alpha \left(\mathbf{1}\mathbf{1}^T - \mathbf{I} \right) \mathbf{P}\mathbf{z} = \mathbf{0}. \tag{8}$$

– The above is a set of non-linear equations in $\mathbf{z}$, in which the kth equation depends on z_1 and z_j, $k \leq j \leq N$. Solving the non-linear equations by backward substitution [27], z_k, $2 \leq k \leq N$ can be written in terms of z_1 as

$$\frac{z_k}{\sqrt{k(k-1)}} = \frac{N z_1^2}{k(k-1)} \left[\frac{k}{\alpha_k} + \sum_{j=k+1}^{N} \frac{1}{\alpha_j} \right] - \frac{z_1}{\sqrt{N}} \frac{N(N-1)}{k(k-1)}. \tag{9}$$

- Using (9) to replace all z_k's in terms of z_1 in the set of non-linear equations in (8), z_1 can be obtained as

$$z_1 = \frac{N-1}{\sqrt{N}} \frac{1}{G},$$ (10)

where

$$G \triangleq \sum_{j=1}^{N} \frac{1}{\alpha_j}.$$ (11)

- Combining (9) and (10),

$$\frac{z_k}{\sqrt{k(k-1)}} = \frac{(N-1)^2}{k(k-1)} G^{-1} \left[G^{-1} \left(\frac{k}{\alpha_k} + \sum_{j=k+1}^{N} \frac{1}{\alpha_j} \right) - 1 \right] \quad 2 \leq k \leq N.$$ (12)

- Using the facts $\mathbf{x}^* = \mathbf{Pz}$, and $\alpha_i = \frac{1}{\beta_i}$ in (10) and (12), the unique Nash equilibrium of the non-cooperative game can be obtained as

$$x_i^* = \frac{\sum_{j=1}^{N} \beta_j - (N-1)\beta_i}{\left(\sum_{j=1}^{N} \beta_j \right)^2}.$$ (13)

Note that the unique Nash equilibrium $\mathbf{x}^*$, is feasible, *i.e.*, $x_i^* > 0$, $\forall\, i$ if and only if

$$(N-1)\beta_i < \sum_{j=1}^{N} \beta_j.$$ (14)

The utility (ownership) of contributor i at the Nash equilibrium, u_i^*, can then be obtained from (1) and (13) as,

$$u_i^* = \left[1 - \left(\frac{(N-1)\beta_i}{\sum_{j=1}^{N} \beta_j} \right) \right]^+,$$ (15)

where $x^+ = \max(x,0)$. It is observed that the ownership u_i^* is non-zero if and only if (14) is satisfied, i.e., if the Nash equilibrium is feasible. The condition in (14) and the expression in (15) have the following interesting implications.

- From (15), the ownership of contributors depend on the β_j of *all the contributors*. This is intuitively correct in a peer production project like Wikipedia because contributions are made by multiple users and the ownership held by a user will depend on the effort of all the users that worked together in making contributions to the page.
- The expression in (15) indicates that contributors who expend smaller effort have larger ownership and those who expend larger effort have low ownership, i.e., the fractional content ownership is a decreasing function of the effort expended.

– Asymptotically, i.e., as the number of contributors, N, becomes large, the ownership, u_i^* in (15), can be written as

$$u_i* = \left(1 - \frac{\beta_i}{E[\beta]}\right)^+,\qquad(16)$$

where $E[\beta] \overset{\triangle}{=} \frac{1}{N}\sum_{j=1}^N \beta_j$, is the *average effort* of all the users that make contributions to the page. From (16), only those contributors for whom $\beta_i < E[\beta]$, i.e., only those contributors whose effort is below the average effort expended in posting content to a page, end up with non-zero ownership. In other words, given the effort involved in making a contribution, and the ease in which others can overwrite one's contributions, only those who expend less effort in making their contributions than the average effort required, end up with non-zero ownership.

3   Empirical Validation with Data

While the non-cooperative game theoretic models developed in Section 2 are based on intuitive notions of ownership and effort, it is necessary to validate these with real data from contributions to Wikipedia articles. We require a set of Wikipedia articles with data on: (a) the content "owned" by contributors at each revision (which can be analogous to the utility, u_i^* in (15) and (b) the cumulative effort exerted by each contributor (including all of his/her contributions) up to each revision, which can represent the effort, β_i, used in the expressions in (12) and (15). We use the data set from Arazy *et al* [13], who explored automated techniques for estimated Wikipedia contributors' relative contributions. The data set in [13] includes nine articles randomly selected from English Wikipedia. Each article was created over an average period of 3.5 years. Section 3.1 presents the details of the data set in [13] and Section 3.2 provides a validation of the same against the models developed in Section 2.

3.1   Extracting Data from Wikipedia Articles

The content "owned" by contributors at the end date of each article period was calculated using the method described in [13]. A sentence was employed as the unit of analysis, and each full sentence was initially owned by the contributor who added it. As content on a wiki page evolves, a contributor may lose a sentence when more than 50% of that sentence was deleted or revised. A contributor making a major revision to a sentence can take ownership of that sentence. The algorithm tracks the evolution of content, recording the number of sentences owned by each contributor at each revision, until the study's end date. The effort exerted by each contributor was, too, based on the method and data set described in [13]. Two research assistants worked independently to analyze every "edit" made to the 9 articles in the sample set and record: contributor's ID; the type of each "edit" to the wiki page (the categories used included: add

Table 2. The list of articles for the data set in [13] and their attributes

Article title	Start Date (MM/DD/YYYY)	End Date (MM/DD/YYYY)	Duration (years)	Edits	Unique Editors
Aikodo [1]	11/29/2001	06/13/2004	2.5	72	62
Angel [2]	11/30/2001	12/09/2005	4.0	341	277
Baryon [3]	02/25/2002	08/25/2005	3.5	73	62
Board Game [4]	11/04/2001	12/30/2004	3.2	220	155
Buckminster Fuller [5]	12/13/2001	07/14/2004	2.6	65	55
Center for Disease Control and Prevention [6]	10/16/2001	03/05/2006	4.4	65	58
Classical Mechanics [7]	06/06/2002	08/13/2006	4.2	202	165
Dartmouth College [8]	10/01/2001	08/26/2004	2.9	70	55
Erin Borockowich [9]	09/24/2001	02/02/2006	4.4	59	54
Total			31.7	1167	943
Average			3.5	129.7	104.8

new content, improve navigation, delete content, proofread, and add hyperlink); the scope of each edit (on a 5-point scale, from minor to major). For example, a particular edit might be categorized as major addition of new content. The two assessors reviewed the "History" section of articles (where Wikipedia keeps a log of all changes to a page), comparing subsequent versions. Once the assessors completed their independent work, and inter-rater agreement levels were calculated (yielding very high levels of agreement), the average of the two assessors was used in the analysis. Finally, the above data set was used to obtain the following parameters on each Wikipedia article listed in Table 2:

- The number of exclusive contributors/users (N)
- The total effort expended by the ith user $(1 \leq i \leq N)$, s_i
- The number of edits made by the ith user $(1 \leq i \leq N)$, e_i
- The number of sentences owned by the ith user $(1 \leq i \leq N)$, p_i.

The following subsection provides a detailed explanation on how we use these parameters to verify the game theoretic analysis described in Section 2.

3.2   Numerical Verification of the Analysis

Using the set of parameters obtained from the pages in Table 2, listed in Section 3.1, we compute the effort expended by user i for unit contribution, β_i, as

$$\beta_i = \frac{s_i}{e_i}. \tag{17}$$

Using the β_i's thus obtained, we use the expression in (15) to determine the estimated fractional ownership on the Wikipedia page, that will be held by each contributor. We compare this with the fraction $\frac{p_i}{\sum_{j=1}^{N} p_j}$ Figs. 1, 2 and 3 show the comparison between the ownership obtained according to the game theoretic analysis described in Section 2 and that given by the data set in [13] for

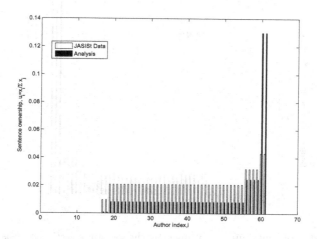

Fig. 1. Ownership of contributors obtained by the game theoretic analysis presented in Section 2 (described by the legend, "Analysis") and the data obtained from Wikipedia pages according to the algorithm in [13] (described by the legend, "JASIST data"), for the page, "Aikido". Contributors/Authors are indexed according to the decreasing order of effort, β_i's.

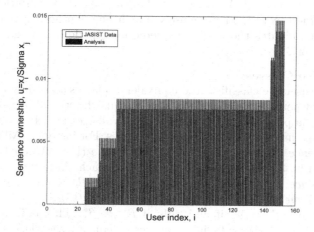

Fig. 2. Ownership of contributors obtained by the game theoretic analysis presented in Section 2 (described by the legend, "Analysis") and the data obtained from Wikipedia pages according to the algorithm in [13] (described by the legend, "JASIST data"), for the page, "Board Game". Contributors/Authors are indexed according to the decreasing order of effort, β_i's.

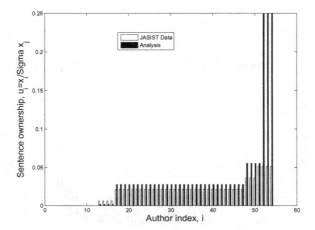

Fig. 3. Ownership of contributors obtained by the game theoretic analysis presented in Section 2 (described by the legend, "Analysis") and the data obtained from Wikipedia pages according to the algorithm in [13] (described by the legend, "JASIST data"), for the page, "Erin Borockowich". Contributors are indexed according to the decreasing order of effort, β_i's.

the Wikipedia pages, "Aikido", "Board Game" and "Erin Borockowich", respectively. For this first analysis, we anonymized the data set, indexing the users in the decreasing order of β_is. We find that the patterns in the empirical data and that of the game-theoretic model closely match one another. In particular, the empirical data validates the following predictions made by the game theoretic model in Section 2[1].

1. **Equivalence Classes:**
   (a) Let the users be classified into equivalence classes according to their fractional ownership, i.e., all users having equal fractional ownership in the Wikipedia page belongs to the same equivalence class. It is observed that each page has five to six equivalence classes. For instance, Aikido, has five equivalence classes (Fig. 1) and Board game (Fig. 2) and Erin Borockowich (Fig. 3), have six equivalence classes each. *Note that the number of equivalence classes obtained from the data is the same as that predicted by the game theoretic analysis described in Section 2.*
   (b) From (15), $u_i^* = u_j^*$ if and only if $\beta_i^* = \beta_j^*$. This indicates that the distribution of the data into number of equivalence classes applies not only to fractional ownership, but also to the effort expended by users. In other words each Wiki page is expected to have five to six categories of contributors/users. A more detailed analysis of the distribution suggests

---

[1] These trends were observed not only for the three articles shown in Figs. 1-3 but also for all the nine articles listed in Table 2. We show results for three articles here due to lack of space.

that the majority of users fall into the equivalence middle classes, while the classes on the extreme representing very low and very high levels of effort (and content ownership) comprise of relatively few users. *While the above can be inferred from the data alone, the game theoretic analysis provides a mathematical framework that validates this observation.*

2. **Non-zero Ownership:** It is observed from (15) that $u_i^* = 0$ if and only if the condition in (14) is violated. The number of users in our sample data with zero ownership matches the number predicted by the game-theoretic model thus providing validation for the condition (14) (at least for the Wikipedia pages included in our analysis). *Again, it is observed that the relation between the number of users with zero ownership and their corresponding β_i's could have been inferred from the data alone, the game theoretic analysis presented in Section 2 provides a mathematical framework to model this phenomenon.*

After establishing that the general trend (i.e. anonymized data) for the empirical data and the model's predictions match one another, we perform a more detailed analysis where we pay attention to users' identities. That is, we organize both data sets, namely the fractional ownership data taken directly from [13] and the ownership values our model in Section 2 predicted, for each user. We then calculate the correlation between the two data sets, using the Pearson's correlation coefficient [23]. The result of the analysis for the nine articles in our data set is presented in Fig. 4. As could be seen from the figure, correlation coefficients range between 0.47 and 0.88, representing moderate-high correlation. When combining the entire data from the nine articles into a single data set, the Pearson correlation was 0.65 (with a $p-$value, $p \approx 0.04$). Therefore, we now proceed to verify if the discrepancies in the values of the ownership obtained by the game theoretic analysis and that obtained from the data can be offset by establishing a linear fit that maps the set of values obtained by analysis to the ones obtained from the data.

Let $\mathbf{a} \triangleq \left[a_1\ a_2\ a_3\ \cdots\ a_N \right]$ represent the ownership of the contributors obtained by the game theoretic analysis and let $\mathbf{d} \triangleq \left[d_1\ d_2\ d_3\ \cdots\ d_N \right]$ represent the ownership of the contributors obtained from the data as described in [13]. For each page, we fit a function

$$\hat{d}_i = \rho a_i + \delta\ 1 \le i \le N, \tag{18}$$

where the parameters ρ and δ are obtained by the method of least squares [27]. We use the values for the pages "Aikido", "Angel", "Baryon" and "Board Game" as the training data to obtain ρ and δ. We then use the values of ρ and δ thus obtained to determine $\hat{d}_i$ for the other five pages. We then compare $\hat{d}_i$ and d_i and compute the estimation error for each page. Fig. 5 shows the estimation error for the remaining five pages. It is observed that the error is between 7-9%. The error for the training set of data was found to be around 5%. *This indicates that the game theoretic analysis presented in Section 2 models the contributor interactions in peer production projects like Wikipedia accurately upto a linear scaling factor.*

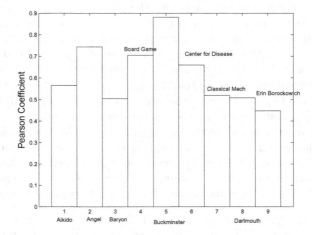

Fig. 4. Pearson correlation co-efficient between the values of the fractional ownership, u_i^*, obtained from the data in [13] and that obtained by the analysis in Section 2

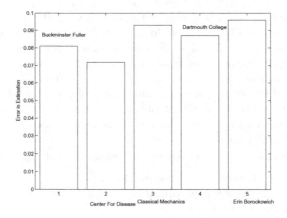

Fig. 5. Estimation error for the linear fit by the method of least squares. The values of the fractional ownership, u_i^*, obtained for the pages, "Aikido", "Angel", "Baryon" and "Board Game" are used as training data for the linear fit.

4   Trustworthy Collaboration and Vandalism

An important insight provided by our non-cooperative game model (and validated by our empirical analysis) is that only contributors with below-average effort levels are able to maintain fractional ownership on wiki pages. That is, by and large only the edits made by contributors who exert little effort survive the collaborative authoring process. In Section 1, we referred to two key concerns that are associated with trustworthy collaboration in peer-production projects:

(a) a risk that non-experts will contribute content of low quality, and (b) a threat that malicious participants would vandalize Wikipedia pages. In spite of these serious concerns, the content on Wikipedia articles is generally of high quality and Wikipedia maintains the status as one of the most reliable sources of information on the web [20]. How then, does Wikipedia maintain high-quality content in the face of threats of low-quality or malicious contributions? Our results can have important implications for investigation of trustworthy collaboration on Wikipedia (and more broadly, in peer-production projects). In the sections that follow, we provide two interpretations of our results that help explain how the threats highlighted above are mitigated.

1. **Trustworthiness/Quality of Wikipedia pages:** The first interpretation of the model and its empirical validation involves the concern of non-expert, low quality contributions eroding the trustworthiness of peer-produced product. This interpretation suggests that low effort is associated with greater likelihood of content survival due to a skill advantage: contributors who are experts in their field of contribution expend less effort, and their contributions are of higher quality [11]. Thus, the effort associated with contribution is inversely related with its quality and consequently with its likelihood of survival of subsequent editing.

2. **Vandalism:** The second interpretation concerns the danger of vandalism activities reducing the trustworthiness of the peer-produced products. Since the underlying Wiki mechanisms allow any editor to easily revert the edits of other contributors, the effort involved in vandalistic edits is higher than the effort of reverting such edits. Thus, high effort is associated with vandalism and relatively lower effort is linked to correction of vandalism. The game theoretic analysis presented in Section 2 predicts that the contributions made by users expending large effort do not survive the edit process and end up with zero ownership. Therefore, most vandalistic edits would not survive over time, as also observed in [21], [33], [32].

In summary, following on the intuition observed in [21], [33], [32], we modeled competition between players as a non-cooperative game, where a player's utility is associated with surviving fractional content owned, and cost is a function of effort exerted. Broader design implications emerging from this interpretation include the need to make version control mechanisms not only highly usable, but also highly open and egalitarian, and accessible to participants in a peer-production process. In addition, these insights suggest the importance of concurrent use of other quality control mechanisms, including user-designated alerts (where users are notified when changes are made to an article, or other part of the collaboratively-created product); watch lists (where users can track certain articles); and IP or user blocking in cases where repeated attacks from the same source are deemed to be acts of vandalism. The combination of these mechanisms make three important contributions to the trustworthiness of peer-production projects: first, their existence deter potential vandals; second, they reduce the costs of identifying and responding quickly to attacks; and third, they enable users to easily revert the consequences of vandalism .

5   Conclusion

To better understand the success of peer production, we developed a non-cooperative game theoretic model of the creation of Wikipedia articles. The utility of a contributor was her relative ownership of the peer-produced product that survived a large number of iterations of collaborative editing. The work presented here contributes to better understanding of the trustworthiness of peer-production by

- Solving the game and demonstrating the conditions under which a Nash equilibrium exists, showing that asymptotically only users with below average effort would maintain ownership
- Empirically validating the model, demonstrating that only users with below average effort would maintain ownership, as well as showing editors' equivalence classes
- Offering interpretations and implications for research on trustworthy peer-production (in terms of expertise and vandalism).

To the best of our knowledge, this is the first modeling of user interactions on Wikipedia as a non-cooperative game. Our analysis points to the benefits of deploying multiple mechanisms which afford the combination of large-scale and low-effort quality control as way to ensure the trustworthiness of products created through web-based peer-production. Further research is needed to analyze the effectiveness of each of these mechanisms, and to address other aspects of peer production through game theoretic analysis.

Acknowledgment. This work was supported in part by a National Academies Keck Futures Initiative (NAKFI) grant.

References

1. http://en.wikipedia.org/wiki/Aikido
2. http://en.wikipedia.org/wiki/Angel
3. http://en.wikipedia.org/wiki/Baryon
4. http://en.wikipedia.org/wiki/Board_Game
5. http://en.wikipedia.org/wiki/Buckminster_Fuller
6. http://en.wikipedia.org/wiki/Centers_for_Disease_Control_and_Prevention
7. http://en.wikipedia.org/wiki/Classical_Mechanics
8. http://en.wikipedia.org/wiki/Dartmouth_College
9. http://en.wikipedia.org/wiki/Erin_Borockowich
10. Anand, S., Sengupta, S., Chandramouli, R.: Price bandwidth dynamics for WSPs in heterogeneous wireless networks. Technical Report, Stevens Institute of Technology (2011)
11. Anthony, D., et al.: The quality of open source production:Zealots and good samaritans in the case of Wikipedia. Rationality and Society (2007)
12. Arazy, O., Nov, O., Patterson, R., Yeo, L.: Information quality in Wikipedia: The effects of group composition and task conflict. Journal of Management Information Systems 27, 71–98 (2011)

13. Arazy, O., Stroulia, E., Ruecker, S., Arias, C., Fieurontino, C., Ganev, V., Yao, T.: Recongnizing contributions in Wikis: Authorship categories, algorithms and visualizations. Journal of American Society for Information Science and Technology 61(6), 1166–1179 (2010)

14. Benkler, Y.: The Wealth of Networks: How Social Production Transforms Markets and Freedom. Yale University Press (2006)

15. Wagner, C., Majchrzak, A.: Enabling customer-centricity using Wikis and the Wiki way. Journal of Management Information Systems 23, 17–43 (2007)

16. Eisenhardt, K.M., Martin, J.A.: Dynamic capabilities: What are they? Strategic Management Journal 21, 1105–1121 (2000)

17. Forte, A., Park, T.: How people assess cooperatively authored information resources. In: Proceedings of the Eighth Annual International Symposium on Wikis and Open Collaboration (2012)

18. Grewal, R., Lilien, G.L., Mallapragada, G.: Location, location, location: How network embeddedness affects project success in open source systems. Management Science 52, 1043–1056 (2006)

19. Hippel, E.V., Krogh, G.V.: Open source software and the "private collective" innovation model: Issues for organization science. Organization Science 14, 209–223 (2003)

20. Kittur, A., et al.: Can you ever trust a Wiki?: Impacting perceived trustworthiness in Wikipedia. In: Proceedings of the ACM Conference on Computer Supported Cooperative Work (2008)

21. Kittur, A., Suh, B., Pendleton, B.A., Chi, E.H.: He says, she says: Conflict and coordination in Wikipedia. In: Proceedings of the ACM SIGCHI Conference on Human Factors in Computing Systems, pp. 453–462 (April 2007)

22. Krogh, G.V., Hippel, E.V.: The promise of research on open source software. Management Science 52, 975–983 (2006)

23. Lakhani, K., Wolf, R.: Why hackers do what they do: Understanding motivation effort? In: Feller, B.F.J., Hissam, S., Lakhani, K. (eds.) Perspectives in Free and Open-Source Software. MIT Press (2005)

24. Lee, G.K., Cole, R.E.: From a firm-based to a community-based model of knowledge creation: The case of the Linux Kernel development. Organization Science 14, 633–649 (2003)

25. Leuf, B., Cunningham, W.: Quick Collaboration on the Web: Reading. Addison-Wesley, Massachusetts (2001)

26. Majchrzak, A.: Where is the theory in Wikis? MIS Quarterly 33, 18–20 (2009)

27. Meyer, C.D.: Matrix Theory and Applied Linear Algebra. SIAM (1972)

28. Nash, J.: Equilibrium points in N-person games. National Academy of Sciences 36, 48–49 (1950)

29. Neumann, J.V., Morgenstern, O.: Theory of Games and Economic Behavior. Princeton University Press (1944)

30. Nov, O., Kuk, G.: Open source content contributors' response to free-riding: The effect of personality and context. Computers in Human Behavior 24(6), 2848–2861 (2008)

31. Peng, D.X., Schroeder, R.G., Shah, R.: Linking routines to operations capabilities: A new perspective. Journal of Operations Management 26, 730–748 (2008)

32. Stvilia, B., Twidale, M.B., Smith, L., Gasser, L.: Information quality work organization in Wikipedia. Journal of the American Society for Information Science and Technology 59(6), 983–1001 (2008)

33. Suh, B., Chi, E.H., Kittur, B.A.P.A.: Us vs them: Understanding social dynamics in Wikipedia with revert graph visualizations. In: IEEE Symposium on Visual Analytics Science and Technology (VAST 2007), pp. 163–170 (October 2007)
34. Teece, D.J., Pisano, G., Shuen, A.: Dynamic capabilities and strategic management. Strategic Management Journal 18, 509–533 (1997)
35. Towne, W.B., et al.: Your process is showing: Controversy management and perceived quality in Wikipedia. In: Proceedings of the ACM Conference on Computer Supported Cooperative Work (2013)
36. Wagner, C.: Wiki: A technology for conversational knowledge management and group collaboration. Communications of the Association for Information Systems 13, 265–289 (2004)
37. Winter, S.G.: Understanding dynamic capabilities. Strategic Management Journal 24, 991–995 (2003)

Controllability of Dynamical Systems: Threat Models and Reactive Security

Carlos Barreto[1], Alvaro A. Cárdenas[1], and Nicanor Quijano[2]

[1] Department of Computer Science,
University of Texas at Dallas
[2] Department of Electrical Engineering,
Universidad de Los Andes, Colombia

Abstract. We study controllability and stability properties of dynamical systems when actuator or sensor signals are under attack. We formulate a detailed adversary model that considers different levels of privilege for the attacker such as read and write access to information flows. We then study the impact of these attacks and propose reactive countermeasures based on game theory. In one case-study we use a basic differential game, and in the other case study we introduce a heuristic game for stability.

1 Introduction

The security of cyber-physical control systems has received significant attention in the last couple of years [1,2,3,4,5,6]. In this document we focus on controllability and stability properties of dynamical systems and discuss the theoretical background to analyze how these properties behave under attacks.

The first part of the paper focuses on defining a threat model and risk-assessment analysis based in the theory of linear state space systems and is general enough to be applicable to a wide-range of cyber-physical systems.

The second part of the paper covers reactive security—when the control signal of the defender changes in response to attacks—as a game between a controller and an attacker.

2 System Model

Probably the most general and widely used framework in control systems is the theory of Linear Time Invariant (LTI) state space systems. In this setting we consider a forced (i.e., non-homogeneous) system of linear differential equations:

$$\dot{x}(t) = Ax(t) + Bu(t)$$
$$y(t) = Cx(t) + Du(t) \tag{1}$$

where $x(t) \in \mathbb{R}^n$ is a vector of physical quantities representing the state of the system at time t, $u(t) \in \mathbb{R}^p$ is the control input at time t, $y(t) \in \mathbb{R}^q$ is a vector

S.K. Das, C. Nita-Rotaru, and M. Kantarcioglu (Eds.): GameSec 2013, LNCS 8252, pp. 45–64, 2013.
© Springer International Publishing Switzerland 2013

of sensor measurements at time t, and A, B, C and D are matrices representing the dynamics of the system.

Even when the system under control exhibits non-linear dynamics, non-linear systems can usually be approximated by linear systems to study their properties near their region of operation (we will show an example of this linearization in one of the case studies in the paper).

3   Control/Security Properties

Similar to security properties such as confidentiality, integrity, and availability, there are several control properties that a system designer or plant operator would like to maintain, even under attack.

In the theory of linear state space systems, two fundamental (and dual) properties are controllability and observability. In this paper we focus on controllability properties.

Controllability means that the state of the system can be driven to any arbitrary place by using the manipulated variables (i.e., the control input). An LTI system is controllable iff

$$\text{rank}([B \quad AB \quad \cdots \quad A^{n-1}B]) = n \tag{2}$$

Another important property of a control system is **stability**. There are several notions of stability (asymptotic stability, Lyapunov stability, BIBO stability, etc.); however, they all intuitively describe the notion that the state $x(t)$ of the system will converge (or remain relatively close) to a desired set point x^* after disturbances.

Stabilizability is a weaker notion of controllability, and it is satisfied if the uncontrollable modes of the system are stable.

4   Attack Model

In this section we define an attack model for control systems containing three part: goals of the attacker, offline information, and online information.

4.1   Goals of an Attacker

While in a general setting an attacker can have many different objectives, in this paper we focus on attackers that try to manipulate the controllability or stability of the system.

Obtain Control: One goal can be to obtain controllabilty of the system with the minimal attack effort: find the smallest set of controller compromises u_a such that the system becomes fully controllable by the attacker.

Disrupt Control: A weaker objective can be to simply make the system uncontrollable to the defender (even if the system is also uncontrollable by the attacker).

Make the System Unstable: If the control strategy of the defender is fixed, a different objective available to the attacker is to make the system unstable.

4.2   Offline Information Available to the Attacker

Well-informed attackers can create more precise attacks and can determine with confidence the effect of their actions. In this paper we assume the attacker has access to the following information:

System Parameters: Matrices A, B, C, D. Without knowledge of these matrices, attacks will have random effects and the consequences will be unpredictable by the attacker.

Control Algorithm: The attacker knows the output $u(t)$ the controller will give to any sensor values $y(t)$. One simple example is when $y(t) = x(t)$ and $u(t) = Kx(t)$. In this example, if the attacker knows the control algorithm, it means the attacker has knowledge of the matrix K.

4.3   Online Information (and Access) Available to the Attacker

Table 1. Online Capabilities of the Attacker

	Impact	Explanation
Read-Only $y(t)$	The attacker can get information on state of the system. It can estimate the state if the system is observable or partially estimate some modes.	The attacker can eavesdrop on $y(t)$ but cannot send fake $y_a(t)$ values (i.e., it cannot impersonate itself as the controller to the actuator).
Write-Only $y(t)$	The attacker can launch deception (also known as false data injection) attacks to the controller, but without having knowledge of the state of the system.	The attacker can impersonate itself to the controller, but cannot eavesdrop on legitimate sensor readings $y(t)$.
Read-Write $y(t)$	The attacker can try to estimate the state of the system and use that information to launch deception attacks.	The attacker can eavesdrop on $u(t)$ and send false sensor readings $y_a(t)$ to the controller.
Read-Only $u(t)$	If the attacker knows the initial state x_0 and has access to $u(t)$ since time t_0, it can estimate the state of the system.	Attacker can eavesdrop on $u(t)$ but cannot send fake $u_a(t)$ values (i.e., it cannot impersonate itself as the controller to the actuator).
Write-Only $u(t)$	The attacker can manipulate the output of the actuator.	The attacker can impersonate itself to the actuator, but cannot eavesdrop on legitimate $u(t)$ commands.
Read-Write $u(t)$	The attacker can manipulate the output of the actuator and has information of the original intended control signal.	The attacker can eavesdrop on $u(t)$ and send false $u_a(t)$ commands to the actuator.

In this section we propose two tables describing a new attacker model that considers the information attackers have online, and the privilege access they have over the **regulatory control loop**–in this paper we leave out of the scope supervisory, hierarchical, and human-machine interface attacker models.

Table 2. Examples of Attacks. Empty blocks can be considered as combinations of attacks described in the first row and the first column. In practice, an attacker that compromises a PLC can potentially change (depending on the architecture of the PLC) the sensor readings and send them to other PLCs or Human Machine Interfaces. We do not consider this case here as our focus is regulatory control.

	No Acces to $u(t)$	Read-Only $u(t)$	Write-Only $u(t)$	Read-Write $u(t)$
No Access to $y(t)$	No Attacks.	The attacker has physical access to the actuator and can read the analog signals it receives.	The attacker installs its own actuators.	Man-in-the-Middle between PLC and actuator.
Read-Only $y(t)$	The attacker installs its own sensors.			The attacker compromises a PLC.
Write-Only $y(t)$	Attacker uses physical attack (e.g., move a temperature sensor to a refrigerator).		Attacker changes configuration parameters of PLC.	
Read-Write $y(t)$	The attacker obtains the secret keys of the sensors.			

We assume attackers can control a subset of sensors or actuators, but they will have different level of access depending on the model assumed. For example an attacker might be able to get read access to the sensors but not write access; or it can get write-only access to the actuators but not read-access. We think this level of granularity is very important to model precisely what the attackers can do to the system and we argue that this level of granularity has been missing in a lot of the previous work for the security of control systems. The proposed information and privilege-level of attackers during run-time can be seen in Table 1. Examples of when do these assumptions make sense are given in Table 2

5   Attacking Controllability

Using the attacker model defined in the last section, we now turn to the problem of how controllability and stability can be attacked. This analysis can be used for risk assessment by identifying the resiliency of the system to attacks or to identify the actuators and sensors that are most valuable to the system.

5.1   Attacking Controllability with $u(t)$

When an attacker has **Write-Only** or **Read-Write** access to a subset of control signals u_a (it does not matter which), because the ordering of the vector x is arbitrary, we can always partition the system in the following form:

$$\dot{x} = Ax + \begin{bmatrix} B_{11} & B_{12} \\ B_{21} & B_{22} \end{bmatrix} \begin{bmatrix} u_s \\ u_a \end{bmatrix}$$

$$\dot{x} = Ax + \begin{bmatrix} B_{11} \\ B_{21} \end{bmatrix} u_s + \begin{bmatrix} B_{12} \\ B_{22} \end{bmatrix} u_a$$

$$\dot{x} = Ax + B_s u_s + B_a u_a \tag{3}$$

where $B_s = [B_{11} \ B_{21}]^T$, $B_a = [B_{12} \ B_{22}]$, u_s represents the first s rows of the vector u and u_a the remaining rows (which we assume are under the control of the adversary).

Table 3. Assuming the original system is controllable, we can study the impact to the system when the attacker obtains write-access to the subset of actuators $u_a(t)$. For the third and last rows, the attacker needs to change the parameter of the controller K_a or needs to get access to sensor readings to estimate $x(t)$.

Defender $u_s(t)$	Attacker $u_a(t)$	Attacker Goal	Result
$Kx(t)$	arbitrary	Obtains controllability iff:	$\operatorname{rank}([B_a \quad A_s B_a \quad \cdots \quad A_s^{n-1}B_a]) = n$
arbitrary	0 (DoS)	Disrupts controllability iff:	$\operatorname{rank}([B_s \quad AB_s \quad \cdots \quad A^{n-1}B_s]) < n$
arbitrary	$K_a x(t)$	Disrupts controllability iff:	$\operatorname{rank}([B_s \quad A_a B_s \quad \cdots \quad A_a^{n-1}B_s]) < n$
arbitrary	arbitrary	Maximize cost for defender at minimum effort	Differential game. Section 6
$K_s x(t)$	$K_a x(t)$	Destabilize the system	Heuristic stability game. Section 8

With this model we can now ask questions regarding the vulnerability of the system under attack. We study three cases as described in Table 3. These cases are just a few examples of the type of questions we can ask about the system under attack and are not exhaustive. In the next paragraphs we explain how we obtained the results of Table 3.

Attacker Goal: Obtain Control with Integrity Attacks. A common control strategy is to use state-feedback, where the original control signal (the control signal without attack) looks like $u = Kx$.

If the system is under attack (as described in Eq. (3)), only a portion of these control signals will maintain their integrity, thus:

$$u = \begin{bmatrix} u_s \\ u_a \end{bmatrix} = \begin{bmatrix} K_s x \\ u_a \end{bmatrix} \tag{4}$$

where K_s are the first s rows of the original matrix K.

With state feedback, the dynamical system under attack becomes:

$$\begin{aligned}
\dot{x} &= Ax + B_s K_s x + B_a u_a \\
\dot{x} &= (A + B_s K_s)x + B_a u_a \\
\dot{x} &= A_s x + B_a u_a
\end{aligned} \tag{5}$$

where $A_s = (A + B_s K_s)$. Therefore an attacker can obtain complete state controllability of the system iff

$$\text{rank}([B_a \quad A_s B_a \quad \cdots \quad A_s^{n-1} B_a]) = n \tag{6}$$

Attacker's Goal: Disrupt Control with Denial-of-Service Attacks. Because this is a set of forced differential equations, a denial-of-service means the forcing function would not be available. Therefore, if the attacker performs a denial-of-service attack on the system defined by Eq. (3), we get $u_a(t) = 0$ and thus the resulting dynamical system is:

$$\dot{x} = Ax + B_s u_s \tag{7}$$

In this setting, an attacker can disrupt complete state controllability of the system via DoS attacks iff

$$\text{rank}([B_s \quad AB_s \quad \cdots \quad A^{n-1}B_s]) < n \tag{8}$$

Attacker's Goal: Disrupt Control with a State Feedback Integrity Attack. In some cases the attacker can disrupt controllabilty of the system via simple state feedback attacks with an appropriate attack gain matrix K_a. Replacing this state feedback attack in Eq. (3), we get:

$$\begin{aligned}
\dot{x} &= Ax + B_s K_a x + B_s u_s \\
\dot{x} &= (A + B_a K_a)x + B_s u_s \\
\dot{x} &= A_a x + B_s u_s.
\end{aligned} \tag{9}$$

In this setting, an attacker can disrupt complete state controllability of the system via state-feedback integrity attacks iff

$$\text{rank}([B_s \quad A_a B_s \quad \cdots \quad A_a^{n-1}B_s]) < n \tag{10}$$

5.2   Attacking Controllability with $y(t)$

In this section we show that when the defender uses state feedback control, the adversary can use sensor measurements to reduce the problem to that of Eq. (3), and therefore, we can reproduce all the results we summarized in Table 3.

In a general LTI setting, the sensor measurements $y(t)$ are a function of the state and the inputs, and to deal with attackers that compromise the integrity of sensors we would need to consider *observability* and the design of *state estimators*. In this paper we make the assumption that $y(t) = x(t)$ (a valid assumption in many practical cases) and we leave the more general problem of attacks to the observability of the system for future work.

As in the previous section, we assume that the defender is using a state-feedback control law, i.e., $u(t) = Kx(t)$. Therefore, if the sensors are not compromised, the evolution of the system will follow the equation:

$$\dot{x} = (A + BK)x \tag{11}$$

However, if some of the sensors are compromised with **Write-Only** or **Read-Write** access to a subset of sensor signals $y_a(t) = x_a(t)$, then the evolution of the dynamical system will change to the following equation:

$$
\begin{aligned}
\dot{x} &= & Ax + \begin{bmatrix} B_{11}K_{11} & B_{12}K_{12} \\ B_{21}K_{21} & B_{22}K_{22} \end{bmatrix} \begin{bmatrix} x_s \\ x_a \end{bmatrix} \\
\dot{x} &= Ax + \begin{bmatrix} B_{11}K_{11} & 0 \\ B_{21}K_{21} & 0 \end{bmatrix} x + \begin{bmatrix} B_{12}K_{12} \\ B_{22}K_{22} \end{bmatrix} x_a \\
\dot{x} &= & Ax + (BK)_s x + (BK)_a x_a \\
\dot{x} &= & (A + (BK)_s)x + (BK)_a x_a \\
\dot{x} &= & A_s x + B_a u_a \tag{12}
\end{aligned}
$$

Note that Eq. (12) has the same form as Eq. (3), however, in this case

$$A_s = A + \begin{bmatrix} B_{11}K_{11} & 0 \\ B_{21}K_{21} & 0 \end{bmatrix}, \quad B_a = \begin{bmatrix} B_{12}K_{12} \\ B_{22}K_{22} \end{bmatrix}, \quad \text{and} \quad u_a = x_a \tag{13}$$

where the **fake** sensor measurement $x_a(t)$ becomes in practice, the control signal of the attacker $u_a(t)$.

Thus, any controllability question we can make with Eq. (3)—in particular the ones summarized in Table 3—can be reproduced in this new setting by analyzing Eq. (12) with the appropriate matrices.

6   Reactive Security: Differential Games

The primary line of defense for any system are its proactive security mechanisms. Therefore, in practice we must use the threat model to identify the most valuable targets for an adversary and invest in protecting them. In this paper, however, we focus on reactive security mechanisms; that is, we focus on algorithms for responding to attacks.

If an attack is detected, the defender can respond with different actions. Some of the possible responses include reconfiguration of the system, attack isolation, or even a system shutdown (for safety reasons). In this work we are interested in defenses that respond to attacks by changes in their control actions; thus creating a game-theory problem where the actions of the players are the control signals each of them has access to. In particular, we assume that if the system is not under attack, the system will operate with a *vanilla* control signal $u(t)$; however, when the system detects an attack, it changes to a *reactive* control signal $u_s(t)$ to maintain the system under the best possible conditions. This creates a differential game between the defender and the attacker.

The theory of noncooperative differential games considers a general dynamical system

$$\dot{x}(t) = f(t, x, u_d, u_a), \quad x(0) = x_0 \tag{14}$$

with two (or more in some cases) control signals $u_d(t)$ and $u_a(t)$, each of them controlled by a player in the game, and where each player has a utility function it wants to minimize.

Solutions for a Nash equilibrium in differential games usually consider two types of solutions, (1) open-loop solutions, and (2) closed-loop solutions.

In open-loop solutions the control signals $u_i(t)$ ($i = d, a$) are independent of the current state of the system $x(t)$. Open-loop solutions can be computed by using Pontryagin maximum principle, which results in a system of ordinary differential equations with two-point boundary value problems.

In closed-loop solutions the control signals $u_i(t, x)$ depend on time, and also on the state of the system $x(t)$. Closed-loop solutions are derived by using the principle of dynamic programming, which results in a system of nonlinear Hamilton-Jacobi partial differential equations. These equations can be ill-posed in general and thus closed-loop solutions are usually considered under Linear-Quadratic (LQ) differential games.

An LQ differential game has linear-dynamics and quadratic utility functions. The dynamical system considered in 2-player LQ games matches Eq. (3):

$$\dot{x}(t) = Ax(t) + B_s u_s(t) + B_a u_a(t)$$

$$\tag{15}$$

while the utility function (for the finite-time case) has the form:

$$J_i(u_d, u_a) = \int_0^T \left[x^T(t) Q_i(t) x(t) + u_i^T(t) R_{ii} u_i(t) + u_j^T(t) R_{ij} u_j(t) \right] dt + x^T(T) Q_{T,i} X(T)$$

where $i = d, a$ and $j \neq i$. This utility function is a natural extension to the traditional optimal control problem.

6.1   Threat Model and Differential Games Solutions

In this section we use our threat model to analyze solutions to differential games. In particular, we note that open-loop strategies make sense only if an attacker has **Write-Only** $u(t)$ and **Write-Only** $y(t)$. The write-only access to the sensors (and actuators) will prevent an attacker from estimating the state of the system, while allowing the attacker to use its control signal to affect the state of the system. From the defender point of view, since the attacker has write access to

$y(t)$, the defender cannot trust the sensor readings and will turn to open-loop control policies as well.

Now we turn our attention to closed-loop strategies. Recall that for closed-loop control strategies we assume that both players have access to the state of the system and use it for deciding their next control actions. Therefore these strategies make sense for an attacker that has **Read-only** $y(t)$, **Write-only** $u(t)$: If the system is observable, or in particular, if $y(t) = x(t)$, then **read-only** $y(t)$ allows the attacker to get access to $x(t)$ but does not allow the attacker to modify the sensor readings. This ensures the defender that the sensor readings are trustworthy and can be used to obtain $x(t)$ accurately.

7   Differential Game Example

We use a recent model for data integrity attacks in demand-response programs for the smart grid [7]. The model considers actuator attacks as an aggregate effect for multi-agent systems that all receive the same input control signal.

The system can be modeled as a scalar differential equation where p denotes the percentage of agents receiving the real pricing signal, and $1 - p$ denotes the percentage of compromised agents that receive a fake u_a the attack signal.

$$\dot{x} = ax + pbu_d + (1-p)bu_a, \quad x(0) = x_0 \tag{16}$$

As discussed before, we consider a game between a defender that wants to minimize a utility penalizing the deviation of x from the steady state 0 and the amount of control (the additional price of electricity u_d).

$$J_d(x, u_d) = \frac{1}{2} \int_0^T [\alpha x^2 + \beta(u_d)^2] dt \tag{17}$$

And an attacker that wants to maximize the state trajectory deviation from the target subject with the minimum amount of effort:

$$J_a(x, u_a) = \frac{1}{2} \int_0^T [-\alpha x^2 + \beta(u_a)^2] dt \tag{18}$$

In general, parameters α and β for the objectives of the attacker and the defender can be different, but we assume they are the same to simplify notation.

We consider open-loop solutions. To find the necessary conditions for optimality of the game (a Nash equilibrium between the two players) we need to use Pontryagin's minimum principle.

First we start by defining the Hamiltonian for the defender:

$$H_d(x, u_d, u_a, \lambda_d) = \frac{1}{2}(\alpha x^2 + \beta(u_d)^2) + \lambda_d(ax + pbu_d + (1-p)bu_a) \tag{19}$$

and the Hamiltonian for the attacker

$$H_a(x, u_d, u_a, \lambda_a) = \frac{1}{2}(-\alpha x^2 + \beta(u_a)^2) + \lambda_a(ax + pbu_d + (1-p)bu_a) \tag{20}$$

The necessary conditions for an optimal solution need to satisfy several constraints. First we find the partial derivative of the Hamiltonian with respect to the control inputs:

$$\beta u_d^* + \lambda_d^* pb = 0$$
$$\beta u_a^* + \lambda_a^*(1-p)b = 0$$

Therefore, the optimal control action by the defender is:

$$u_d^*(t) = -\frac{\lambda_d^*(t)pb}{\beta} \tag{21}$$

$$\tag{22}$$

and the optimal control action by the attacker is

$$u_a^*(t) = -\frac{\lambda_a^*(t)(1-p)b}{\beta} \tag{23}$$

To find the evolution of $\lambda_a^*(t)$ and $\lambda_d^*(t)$ we find the costate equations:

$$-\dot{\lambda}_d^* = \alpha x^* + a\lambda_d^*, \quad \lambda_d^*(T) = 0$$
$$\implies -\frac{\dot{\lambda}_d^*}{\alpha} = x^* + a\frac{\lambda_d^*}{\alpha}, \quad \lambda_d^*(T) = 0 \tag{24}$$

and

$$-\dot{\lambda}_a^* = -\alpha x^* + a\lambda_a^*, \quad \lambda_a^*(T) = 0$$
$$\implies -\left(-\frac{\dot{\lambda}_a^*}{\alpha}\right) = x^* + a\left(-\frac{\lambda_a^*}{\alpha}\right), \quad \lambda_a^*(T) = 0 \tag{25}$$

We can simplify the analysis by noting that Eq. (24) and Eq. (25) can be modeled by the following differential equation:

$$-\dot{q} = ax^* + aq, \quad q(T) = 0. \tag{26}$$

Once we solve for $q(t)$ w eknow that $\lambda_d^*(t) = \alpha q(t)$ and $\lambda_a^*(t) = -\alpha q(t)$.

As shown by Bensoussan [8], Eq. (16) and Eq. (26) can be solved explicitly by a decoupling argument, resulting in:

$$x^*(t) = x_0 \frac{s(e^{s(T-t)} + e^{-s(T-t)}) - a(e^{s(T-t)} - e^{-s(T-t)})}{s(e^{sT} + e^{-sT}) - a(e^{sT} - e^{-sT})} \tag{27}$$

where

$$s = \sqrt{a^2 + \frac{\alpha(pb)^2 - \alpha((1-p)b)^2}{\beta}} \tag{28}$$

(as long as s is not a complex number)

$$q(t) = H(t)x^*(t) \tag{29}$$

and

$$\frac{1}{H(t)} = -a + s\frac{e^{2s(T-t)} + 1}{e^{2s(T-t)} - 1} \tag{30}$$

7.1   Simulation Results

In this section we illustrate the behavior of the open loop differential game described by Eq. (16). Let us consider the system parameters $a = -4$, $b = 1$, $\alpha = 10$, $beta = 1$, $x(0) = 2$. Note that the solution $x^*(t)$ of the system is real for any time $t \in [0, T]$ if $x(0) \in \Re$ and the parameter s (see Eq. (28)) is real for any $p \in [0, 1]$.

The behavior of the system state x, as well as the control actions of each player for different values of p are depicted in Fig. 1. Although the system converges to zero for any value of p, the time of convergence is affected by the value of p. Specifically, when the attacker has control over the majority of the system inputs, i.e., when $1 - p > 0.5$, it is able to delay the convergence of the system to the equilibrium. However, the attacker tends to make more effort in its control signal as $1 - p$ is increased, i.e., $\int_0^T |u_a(t)|dt$ increases as $1 - p$ increases.

The defender experiences a similar behavior. Particularly, if the participation of the defender p is increased, then the system approaches the stable state faster, but with a higher cost in resources for the defender, represented by $\int_0^T |u_d(t)|dt$.

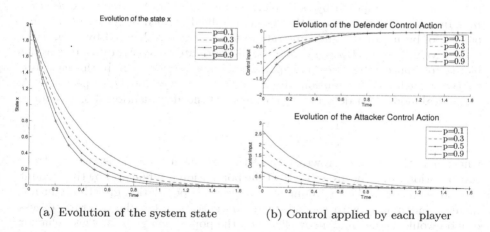

(a) Evolution of the system state   (b) Control applied by each player

Fig. 1. Evolution of the differential game in Eq. (16) considering different values of p

While it is clear that in most practical cases, the defender control action has a cost associated with it, in practice, the attacker action should not have these cost constraints. When an attacker compromises a control signal, the player paying for this control action is the defender, not the attacker. However, if we remove the $\beta(u_a)^2$ from the utility function of the attacker we get an ill-posed problem, where u_a does not have any constraints and cannot be find with the maximum principle.

8   Heuristic Stability Game

One of the problems with differential games is that the utility functions of the players usually need to satisfy very specific properties in order to have well-defined solutions. These properties usually limit the general applicability of these games, in particular by placing artificial limitations on what the attacker signal $u_a(t)$ can do.

In this section we discuss a stability game where the goal of the defender is to make the system stable while the goal of the attacker is to make the system unstable. This binary utility function does not allow this system to be formally analyzed for equilibrium strategies; however, our goal is to start exploring the design space to allow more realistic settings that do not impose artificial limitations on what the attack control signal $u_a(t)$ needs to satisfy.

A bioreactor is a system designed to provide some environmental conditions required to carry out a biochemical process. For example, a bioreactor might be used for processing some pharmaceuticals or food that involve the use of micro-organisms or substances derived from them. Particularly, processes focused on the growth of organisms (also called biomass) should provide a batch of organisms with food in order to promote the population growth. In such cases, the process is regulated by means of the *substrate feed* (or food income) and the output mass flow (composed by both biomass and substrate), namely the *dilution rate*. In this sense, the state of the system can be described by means of the proportion of both biomass and substrate in the bioreactor, denoted by x_1 and x_2, respectively. The substrate feed and the dilution rate are control variables of the bioreactor that can be used to regulate the biomass production. In this case, the substrate feed and the dilution rate are denoted by x_{2f} and D, respectively. A dynamical model that describes the behavior of the aforementioned bioreactor is

$$\dot{x}_1(t) = (\mu - D)x_1(t),$$
$$\dot{x}_2(t) = D(x_{2f} - x_2(t)) - \tfrac{\mu}{Y}x_1(t),$$
$$(31)$$

where μ is the average growth rate of the organisms, and Y is the substrate consumption rate. The growth rate of the population is influenced by the amount of food available in the environment; however, the population growth might not increase indefinitely with the substrate concentration. On the contrary, excess of food would induce inhibitory effects on the population growth. This behavior is modeled by means of the growth rate

$$\mu = \mu_{max}\left(\frac{x_2}{k_m + x_2 + k_1 x_2^2}\right),$$
$$(32)$$

where k_m and k_1 are constants. On the other hand, the substrate consumption rate is defined as ratio of the change in the population mass and the change in the substrate mass inside the bioreactor, that is $Y := -\frac{\dot{x}_1}{\dot{x}_2}$.

Table 4. Rest points of the system in Eq. (31)

x_1^*	x_2^*	
0	4	Stable
0.95103	1.512243	Unstable
1.530163	0.174593	Stable

Now, let us consider a process with the following parameters [9]: $\mu_{max} = 0.53hr^{-1}$, $k_m = 0.12$ g/liter, $k_1 = 0.4545$ liter/g, $Y = 0.4$. If we use constant inputs $D = 0.3$ and $x_{2f} = 4$, then the system in Eq. (31) is characterized by three rest points shown in Table 4. In this case, we assume that the system designer wants to stabilize the system at the unstable equilibrium point $x^* = [0.95103, 1.512243]^\top$. In particular, the control law is designed by means of state feedback. Accordingly, the design procedure involves 1) the linearization of the system around the desired equilibrium point, and 2) the design of the control law that stabilizes the system. Later, the vulnerabilities of the system are going to be analyzed.

8.1 Linearization and Control Design

The linear model of the system in Eq. (31) at the unstable equilibrium point is

$$\dot{z} = Az + Bu, \tag{33}$$

where $z = x - x^*$, $u = [u_1, u_2]^\top = [D, x_{2f}]^\top - [0.3, 4]^\top$, and

$$A = \begin{bmatrix} 0 & -0.068 \\ -0.75 & -0.13 \end{bmatrix}, \quad B = \begin{bmatrix} -0.994 & 0 \\ 2.488 & 0.3 \end{bmatrix}.$$

Now, using a state feedback control of the form $u = -Fz$, where $F = [F_1, F_2]^\top$ is a matrix in $\Re^{2 \times 2}$ and F_1 and F_2 are vectors in $\Re^{2 \times 1}$, we have

$$\dot{z} = (A - BF)z. \tag{34}$$

The previous expression can be rewritten in terms of each control input as

$$\dot{z} = (A - B_1 F_1^\top - B_2 F_2^\top)z, \tag{35}$$

where B_1 and B_2 are the columns of B that determine the influence that each input has in the system and $F_i^\top$ is the feedback control law at the ith input, for $i \in \{1, 2\}$. In this case, the pair (A, B) is controllable, as well as $(A, B1)$ and $(A, B2)$.

In particular, we consider a situation in which the designer fixes the jth input: u_j to a constant value, while it controls the ith input, denoted by u_i: i.e., the designer only controls $u_i = -F_i^\top z$, resulting

$$\dot{z} = (A - B_i F_i^\top)z + B_j u_j. \tag{36}$$

The selection of F_i should guarantee that the system is stable. According to the theorem of pole shifting, when the system is controllable, it is possible to find a matrix F_i such that the system in Eq. (36) is globally asymptotically stable [10]. This is achieved if the eigenvalues λ_h of $A - B_iF_i$ have negative real part.

Particularly, the design of the feedback control law can be designed by applying a coordinate transformation of the form $\tilde{z} = P_i z$. This linear transformation let us express the system in Eq. (36) in terms of the matrices in canonical form

$$A^\dagger = \begin{bmatrix} 0 & -\alpha_0 \\ 1 & -\alpha_1 \end{bmatrix}, \quad B_i^\dagger = \begin{bmatrix} 1 \\ 0 \end{bmatrix}, F_i^\dagger = [f_1, f_0],$$

where $A^\dagger = P_i^{-1}AP_i$, $B_i^\dagger = P_i^{-1}B_i$, $F_i^\dagger = F_iP_i$. On the other hand, α_0 and α_1 are the coefficients of the characteristic polynomial of A, i.e., $det(sI - A) = s^2 + \alpha_1 s + \alpha_0$ [10]. Particularly, in linear systems with one input, P_i is the controllability matrix of (A, B_i). Without loss of generality, we assume that the feedback F_i is designed to obtain poles λ_1 and λ_2 with negative real part, i.e., $\mathbf{Re}(\lambda_h) < 0$, for $h \in \{1, 2\}$. Therefore, a valid feedback law F_i must satisfy

$$det\Big(sI - (A - B_iF_i)\Big) = s^2 + (\alpha_1 + f_1)s + (\alpha_0 + f_0)$$
$$= (s - \lambda_1)(s - \lambda_2). \quad (37)$$

Note that the coefficients α_0 and α_1 are positive, since λ_1 and λ_2 have negative real part.

Under these conditions, the designer is able to stabilize the system through the feedback law F_i. The design of the feedback vector F_i can be made minimizing a cost criteria, by means of the Linear Quadratic Regulator (LQR) problem (which has a cost-function similar to the one considered in LQ differential games). Since the feedback control is only applied in one of the inputs, we have that $F = [F_i, \mathbf{0}]^\top$, where $\mathbf{0}$ is a vector of zeros.

8.2   Attacker Perspective

Now, let us consider a situation in which an attacker obtains control over the jth input u_j. In this case, we assume that the attacker is able to implement a control law $u_j = -F_j z$ that seeks to destabilize the system. Note that the attacker might succeed on its purpose, since (A, B_j) is controllable, for $j \in \{1, 2\}$.

It is important to note that if the attacker is able to observe the state of the system, then it would design the feedback vector F_j, such that the system in Eq. (35) is unstable. Since the defender trusts $x(t)$, we assume the attacker has **read-only** access to the sensors $y(t) = x(t)$.

In this sense, applying the similarity matrix P_j, we obain that the matrices in canonical form $A^\ddagger = P_j^{-1}(A - B_iF_i)P_j$, $B^\ddagger = P_j^{-1}B_j$, and $F = F^\ddagger P_j^{-1}$ that satisfy

$$A^\ddagger = \begin{bmatrix} 0 & -\alpha_0 - f_0 \\ 1 & -\alpha_1 - f1 \end{bmatrix}, B_j^\ddagger = \begin{bmatrix} 1 \\ 0 \end{bmatrix},$$

where P_j is the controllability matrix of $(A - B_i F_i, B_j)$. In this case, the characteristic polynomial of $(A - B_i F_i)$ is given by Eq. (37). Now, the attacker can break the system by modifying the coefficients of the characteristic polynomial through F_j. In particular, if we consider $F_j = F^\ddagger P_j^{-1}$, where $F^\ddagger = [g_1, g_0]$, then the characteristic polynomial is transformed into

$$det\Big(sI - (A - B_i F_i - B_j F_j)\Big) = s^2 + (\alpha_1 + f_1 + g_1)s$$

$$+ (\alpha_0 + f_0 + g_0). \quad (38)$$

For example, let us fix $g_1 = 0$ and set $g_0 = -(\alpha_0 + f_1 + \delta)$, with $\delta > 0$. Consequently, the characteristic polynomial of Eq. (38) becomes

$$det\Big(sI - (A - B_i F_i - B_j F_j)\Big) = s^2 + (\alpha_1 + f_1)s + -\delta.$$

Since there is a change of sign in the coefficients of the characteristic polynomial, the system in unstable under the attacker feedback law

$$F_j = [0, -(\alpha_0 + \delta)]P_j^{-1}. \quad (39)$$

In particular, the system has one unstable equilibrium point. Note that if the attacker knows the eigenvalues of the system, then it can calculate the value of $\alpha_0 + f_0$ by meas of

$$\alpha_0 + f_0 = \prod_{i=1}^{n} \lambda_h. \quad (40)$$

This relation can be extracted from the expanded characteristic polynomial in terms of the eigenvalues λ_h, for $h \in \{1, \dots, n\}$.

8.3   System Defense

We are interested in analyzing the actions that the system designer can take in order to stabilize the system when an attacker influences u_j. Intuitively, the designer would want to implement a feedback $\hat{F}_i$ that cancels the feedback law of the attacker F_j. Therefore, $\hat{F}_i$ must satisfy

$$(A - B[\hat{F}_i, F_j]^\top)x = (A - B[F_i, \mathbf{0}]^\top)x.$$

This can be rewritten as

$$B([\hat{F}_i, \mathbf{0}]^\top + [\mathbf{0}, F_j]^\top) = B[F_i, \mathbf{0}]^\top.$$

However, if $F_j \neq \mathbf{0}$, then it is not possible to find a feedback law $\hat{F}_i$ that satisfies the previous equality. This happens because the columns of $[\hat{F}_i, \mathbf{0}]^\top$ are linear independent from the columns of $[\mathbf{0}, F_j]^\top$.

Therefore, the designer must take actions that compensate, rather cancel, the attacks. In this sense, the designer would implement the feedback law $\hat{F}_1$ that shifts the poles of the attacked system to the negative imaginary semi plane. This can be done by repeating the feedback design procedure exposed above, having into account a system of the form

$$\dot{z} = (\hat{A} - B_i \hat{F}_i)z,$$

where $\hat{A} = A - B_j F_j$ is the observed system by the defender. $\hat{A}$ can be considered by the designer as a given of the control design problem. Note that the response of the designer is subject to 1) the knowledge that the system is attacked and 2) the knowledge of the effects of the attack on the system.

8.4   Simulations

In this section we analyze the defense and attack actions of two agents that seek to stabilize and destabilize a system.

We consider the case in which both defender and attacker update their feedback gains according to the actions the opponent. This can be seen as a repeated game, in which the players are the defender and the attacker. We assume that each player requires some constant time to update its action as a response to the move by the other player. Therefore, the game is repeated after a period T. Furthermore, the speed of response of the defender to attacks is measured in terms a time fraction of T, namely $DC \in [0,1]$ (for duty-cycle). In this sense, DC indicates that in each period T, the defender makes the system stable during the interval $[0, DCT]$ and the attacker disrupt the system during (DCT, T).

Simulations are made assuming that agents have knowledge about the actions of each other. In particular, the defender designs the feedback law F_i using a Linear-quadratic (LQ) state-feedback regulator that minimizes the cost function

$$J = \int_0^\infty (x^\top Q x + u^\top R u)\,dt,$$

where

$$Q = \begin{bmatrix} 1 & 0 \\ 0 & 0.1 \end{bmatrix} \quad R = \begin{bmatrix} 1 & 0 \\ 0 & 1 \end{bmatrix}.$$

On the other hand, the attacker actions are calculated according to Eq. (39) and (40).

Control of the Dilution Rate. First, consider the situation in which the defender manipulates the dilution rate D and the attacker controls the substrate feed x_{2f}. In this case, the natural behavior of the system under the influence of the defender's control signal is to reach the origin and remain there. However, the system becomes unstable if the attacker manipulates the substrate feed input and there is no response by the system designer. It can be seen that the system become unstable after the attack, that takes place at $t = 10s$.

Fig. 2 and 3 show the evolution of the game for $DC = 0.8$ and $DC = 0.1$, respectively. Note that although the attacker actions tend to affect the system, the defender is able to stabilize the system, even with a slow reaction to the attacker actions, i.e., the case with $DC = 0.1$.

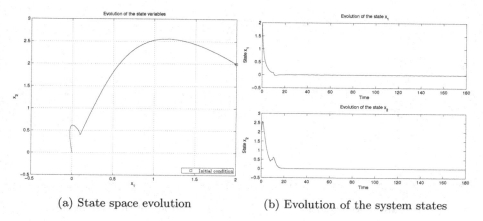

(a) State space evolution   (b) Evolution of the system states

Fig. 2. Evolution of the system when both attacker and defender compete by turns. $DC = 0.8$.

Control of the Substrate Feed. Here we consider the evolution of the system states when both defender and attacker play a game of control by manipulating the substrate feed and the dilution rate, respectively. It can be seen that an attack on the dilution rate is able to make unstable the system, with a notable effect on the biomass level x_1, with respect to the previous scenario.

Fig. 4 and 5 show the evolution of the game for $DC = 0.8$ and $DC = 0.5$, respectively. It can be seen that the defender is able to stabilize the system for $DC = 0.8$. However, the defender requires a lower reaction time to control the system, in contrast to the case when it controls the dilution rate. Specifically, when the defender controls the dilution rate (scenario 1), it is able to stabilize the system with a DC of 0.1 (see Fig. 3). Now, when the defender controls the substrate feed, it is not able to stabilize the system with a DC of 0.5.

Stability Experiments. Now we analyze the system behavior as a function of the parameter DC for a particular period T. We present numerical experiments to observe properties of the system.

Since the control law has jumps each time a player updates its strategy, obtaining explicit stability solutions is difficult (although in future work we plan to use the theory of hybrid systems to better characterize the stability of these systems). We are interested in the stability of the output variable defined as $y = x_1$, i.e., the stability of the biomass concentration. Specifically, our analysis is made by approximating an exponential function of the form $h(t) = e^{\sigma t}$ to the output $y(t)$. If the parameter σ is positive, then we conclude the system is unstable for a particular DC. On the other hand, if the system is stable, the variable $y(t)$ would be approximated by means of a decreasing exponential function.

In particular, scenario 1, in which the defender and the attacker control the dilution rate and the substrate feed, with $T = 100$ is stable for any DC in the interval $[0.0120, 1]$. On the other hand, in scenario 2 the attacker can destabilize

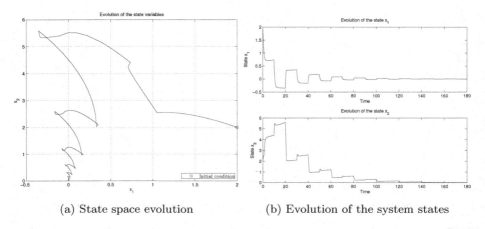

(a) State space evolution   (b) Evolution of the system states

Fig. 3. Evolution of the system when both attacker and defender compete by turns. $DC = 0.1$.

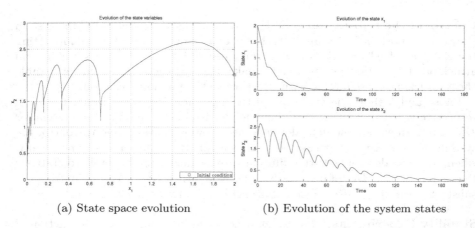

(a) State space evolution   (b) Evolution of the system states

Fig. 4. Evolution of the system when both attacker and defender compete by turns. $DC = 0.8$.

the system with a DC in the interval $[0.1, 0.32]$ and $T = 10$. The dependence of the stability of $y(t)$ with respect to DC for the scenario 2 is shown in the Fig. 6.

Note that in scenario 1, the defender is able to stabilize the system for very low reaction times, with respect to the scenario 2. This implies that in scenario 1 the attacks have to be effective for a larger period of time to destabilize the system. Hence, the control using the dilution rate is more robust to attacks than the control using the substrate feed. In this case, the measurement of the reaction time is relative to the period T.

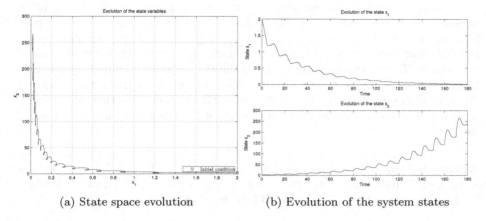

(a) State space evolution   (b) Evolution of the system states

Fig. 5. Evolution of the system when both attacker and defender compete by turns. $DC = 0.5$.

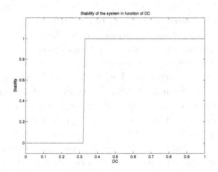

Fig. 6. Stability of the repeated game as a function of DC for scenario 2

9   Future Work

In this paper we have formulated new threat models for controllability and stability of dynamical systems and discussed some ideas on how to model reactive security games between a defender and an attacker. There are many open problems and several directions for future research.

As mentioned in the last section, one particular improvement that can be done to our analysis of the heuristic game is to leverage the theory of hybrid systems to analyze the stability of the game. Hybrid systems can also be used for the study of reachability, which is analogous to the problem of controllability in LTI systems. Computational reachability analysis of systems might be a good tool for analyzing more realistic control problems with bounded states or control actions.

Similarly the concept of proactive security needs further study. An intuitive idea for selecting the most valuable actuators is to consider the following problem: let $B = [\mathbf{b}_1\ \mathbf{b}_2\ \cdots\ \mathbf{b}_m]$, then for every input u_i there is a column vector $\mathbf{b}_i \in \mathbb{R}^n$ that uniquely defines how the actuator will affect the physical state of the system. To find the actuator that has the ability to control more states (and thus the most valuable target for the attacker) we can perform the following analysis: $\arg\max_i rank\ [\mathbf{b}_i\ A\mathbf{b}_i\ ...\ A^{n-1}\mathbf{b}_i]$. The defender can then invest more in protecting the information flow of the actuators depending on the ranking of each actuator.

Acknowledgments. This research was supported by The MITRE Corporation as part of an independent research project. The authors wish to thank Roshan K. Thomas and Adam L. Hahn of MITRE for their insights.

References

1. Amin, S., Schwartz, G.A., Shankar Sastry, S.: Security of interdependent and identical networked control systems. Automatica (2012)
2. Mo, Y., Sinopoli, B.: False data injection attacks in control systems. In: Preprints of the 1st Workshop on Secure Control Systems (2010)
3. Pasqualetti, F.: Secure control systems: A control-theoretic approach to cyber-physical security. Ph.D. dissertation, University of California (2012)
4. LeBlanc, H.J., Zhang, H., Sundaram, S., Koutsoukos, X.: Consensus of multi-agent networks in the presence of adversaries using only local information. In: Proceedings of the 1st International Conference on High Confidence Networked Systems, pp. 1–10. ACM (2012)
5. Shoukry, Y., Araujo, J., Tabuada, P., Srivastava, M., Johansson, K.H.: Minimax control for cyber-physical systems under network packet scheduling attacks. In: Proceedings of the 2nd ACM International Conference on High Confidence Networked Systems, pp. 93–100. ACM (2013)
6. Teixeira, A., Shames, I., Sandberg, H., Johansson, K.H.: Revealing stealthy attacks in control systems. In: 2012 50th Annual Allerton Conference on Communication, Control, and Computing (Allerton), pp. 1806–1813. IEEE (2012)
7. Tan, R., Krishna, V.B., Yau, D.K., Kalbarczyk, Z.: Impact of integrity attacks on real-time pricing in smart grids. In: ACM Computer and Communications Security (CCS) (2013)
8. Bensoussan, A.: Explicit solutions of linear quadratic differential games. In: Stochastic Processes, Optimization, and Control Theory: Applications in Financial Engineering, Queueing Networks, and Manufacturing Systems, pp. 19–34. Springer (2006)
9. Bequette, B.W.: Process Control: Modeling, Design and Simulation, vol. 6. Prentice Hall, Upper Saddle River (2003)
10. Sontag, E.: Mathematical Control Theory. In: Deterministic Finite-Dimensional Systems, 2nd edn. Texts in Applied Mathematics, vol. 6, Springer, New York (1998)

Adaptive Regret Minimization
in Bounded-Memory Games*

Jeremiah Blocki, Nicolas Christin, Anupam Datta, and Arunesh Sinha

Carnegie Mellon University, Pittsburgh, PA
{jblocki,nicolasc,danupam,aruneshs}@cmu.edu

Abstract Organizations that collect and use large volumes of personal information often use security audits to protect data subjects from inappropriate uses
of this information by authorized insiders. In face of unknown incentives of employees, a reasonable audit strategy for the organization is one that minimizes its
regret. While regret minimization has been extensively studied in repeated games,
the standard notion of regret for repeated games cannot capture the complexity
of the interaction between the organization (defender) and an adversary, which
arises from dependence of rewards and actions on history. To account for this
generality, we introduce a richer class of games called *bounded-memory games*,
which can provide a more accurate model of the audit process. We introduce
the notion of *k-adaptive regret*, which compares the reward obtained by playing
actions prescribed by the algorithm against a hypothetical *k-adaptive adversary*
with the reward obtained by the best expert in hindsight against the same adversary. Roughly, a hypothetical *k*-adaptive adversary adapts her strategy to the
defender's actions exactly as the real adversary would within each window of
k rounds. A *k*-adaptive adversary is a natural model for temporary adversaries
(e.g., company employees) who stay for a certain number of audit cycles and are
then replaced by a different person. Our definition is parameterized by a set of
experts, which can include both fixed and adaptive defender strategies. We investigate the inherent complexity of and design algorithms for adaptive regret
minimization in bounded-memory games of perfect and imperfect information.
We prove a hardness result showing that, with imperfect information, any *k*-
adaptive regret minimizing algorithm (with fixed strategies as experts) must be
inefficient unless NP = RP even when playing against an oblivious adversary.
In contrast, for bounded-memory games of perfect and imperfect information we
present approximate 0-adaptive regret minimization algorithms against an oblivious adversary running in time $n^{O(1)}$.

---

* This work was partially supported by the U.S. Army Research Office contract "Perpetu-
ally Available and Secure Information Systems" (DAAD19-02-1-0389) to Carnegie Mellon
CyLab, the NSF Science and Technology Center TRUST, the NSF CyberTrust grant "Privacy, Compliance and Information Risk in Complex Organizational Processes," the AFOSR
MURI "Collaborative Policies and Assured Information Sharing," and HHS Grant no. HHS
90TR0003/01. Jeremiah Blocki was also partially supported by a NSF Graduate Fellowship.
Arunesh Sinha was also partially supported by the CMU CIT Bertucci Fellowship. The views
and conclusions contained in this document are those of the authors and should not be interpreted as representing the official policies, either expressed or implied, of any sponsoring
institution, the U.S. government or any other entity.

S.K. Das, C. Nita-Rotaru, and M. Kantarcioglu (Eds.): GameSec 2013, LNCS 8252, pp. 65–84, 2013.

1 Introduction

Online learning algorithms that minimize regret provide strong guarantees in situations that involve repeatedly making decisions in an uncertain environment. There is a well developed theory for regret minimization in repeated games [1]. The goal of this paper is to study regret minimization for a richer class of settings. As a motivating example consider a hospital (defender) where a *series* of temporary employees or business affiliates (adversary) access patient records for legitimate purposes (e.g., treatment or payment) or inappropriately (e.g., out of curiosity about a family member or for financial gain). The hospital conducts audits to catch the violators, which involves expending resources in the form of time spent in human investigation. On the other hand, violations that are missed internally and caught externally (by government audits, patient complaints, etc.) also result in various losses such as reputation loss, loss due to litigation, etc. The hospital wants to minimize its overall loss by balancing the cost of audits with the risk of externally detected violations. In these settings with unknown adversary incentives, a reasonable strategy for the defender is one that minimizes her regret.

Modeling this interaction as a repeated game of imperfect information is challenging because this game has two additional characteristics that are not captured by a repeated game model: (1) *History-dependent rewards*: The payoff function depends not only on the current outcome but also on previous outcomes. For example, when a violation occurs the hospital might experience a greater loss if other violations have occurred in recent history. (2) *History-dependent actions*: Both players may *adapt* their strategies based on history. For example, if many violations have been detected and punished in recent history then a rational employee might choose to lay low rather than committing another violation.

Instead, we capture this form of history dependence by introducing *bounded-memory games*, a subclass of stochastic games.[1]. In each round of a two-player bounded-memory-m game, both players simultaneously play an action, observe an outcome and receive a reward. In contrast to a repeated game, the payoffs may depend on the state of the game. In contrast to a *general* stochastic game, the rewards may *only* depend on the outcomes from the last m rounds (e.g., violations that were caught in the last m rounds) as well as the actions of the players in the current round.

In a bounded-memory game, the standard notion of regret for a repeated game is not suitable because the adversary may adapt her actions based on the history of play. To account for this generality, we introduce (in Section 4) the notion of k-*adaptive regret*, which compares the reward obtained by playing actions prescribed by the algorithm against a hypothetical k-*adaptive adversary* with the reward obtained by the best expert in hindsight against the same adversary. Roughly, a hypothetical k-adaptive adversary is constructed by taking snapshots of the real adversary's strategy every k rounds. This means that the hypothetical k-adaptive adversary adapts her strategy to the defender's actions exactly as the real adversary would during each window of k rounds in the real game.

---

[1] Stochastic games [2] are expressive enough to model history dependence. However, one can prove that there is no regret minimization algorithm for the *general class of stochastic games* While we do not view this result as surprising or novel, we include it in the full version [3] of this paper for completeness.

When $k = 0$, this definition coincides with the standard definition of an *oblivious adversary* considered in defining regret for repeated games. When $k = \infty$ we get a *fully adaptive adversary*. A k-adaptive adversary is a natural model for temporary employees (e.g., residents, contractors) who stay for a certain number of audit cycles and are then replaced by a different person. Our definition is parameterized by a set of experts, which can include both fixed and adaptive defender strategies. In Section 5 we use the examples of a police chief enforcing the speed limit at a popular tourist destination, and of a hospital auditing accesses to patient records made by hospital employees, to illustrate the power of k-adaptive regret minimization when the defender plays against a series of temporary adversaries.

Next, we investigate the inherent complexity of and design algorithms for adaptive regret minimization in bounded-memory games of perfect and imperfect information. Our results are summarized in Table 1. We prove a hardness result (Section 6; Theorem 1) showing that, with imperfect information, any k-adaptive regret minimizing algorithm (with fixed strategies as experts) must be inefficient unless $\mathsf{NP} = \mathsf{RP}$ even when the real adversary is oblivious, and even if we use the notion of 0-adaptive regret. In fact, the result is even stronger and applies to any γ-approximate k-adaptive regret minimizing algorithm (ensuring that the regret bound converges to γ rather than 0 as the number of rounds $T \to \infty$) for $\gamma < \frac{1}{8n^\beta}$ where n is the number of states in the game and $\beta > 0$. Our hardness reduction from MAX3SAT uses the state of the bounded-memory game and the history-dependence of rewards in a critical way.

We present an inefficient k-adaptive regret minimizing algorithm by reducing the bounded-memory game to a repeated game. The algorithm is inefficient for bounded-memory games when the number of experts is exponential in the number of states of the game (e.g., if all fixed strategies are experts). In contrast, for bounded-memory games of perfect information, we present an efficient $n^{O(1/\gamma)}$ time γ-approximate 0-adaptive regret minimization algorithm against an oblivious adversary for any constant $\gamma > 0$ (Section 7; Theorem 4). We also show how this algorithm can be adapted to get an efficient γ-approximate 0-adaptive regret minimization algorithm for bounded-memory games of imperfect information (Section 7; Theorem 5). The main novelty in these algorithms is an implicit weight representation for an exponentially large set of adaptive experts, which includes all fixed strategies.

Table 1. Regret Minimization in Bounded Memory Games

	Imperfect Information	Perfect Information
Oblivious Regret ($k = 0$)	Hard (Theorem 1) APX (Theorem 5)	APX (Theorem 4)
k-Adaptive Regret ($k \geq 1$)	Hard (Theorem 1)	Hard (Full Version [3])
Fully Adaptive Regret ($k = \infty$)	X (Full Version [3])	X (Full Version [3])

X - no regret minimization algorithm exists
Hard - unless $\mathsf{NP} = \mathsf{RP}$ no regret minimization algorithm is efficiently computable
APX - efficient approximate regret minimization algorithms exist.

2   Related Work

A closely related work is the Regret Minimizing Audit (RMA) mechanism of Blocki et al. [4], which uses a repeated game model for the audit process. RMA deals with history-dependent rewards under certain assumptions about the defender's payoff function, but it does not consider history-dependent actions. While RMA provides strong performance guarantees for the defender against a byzantine adversary, the performance of RMA may be far from optimal when the adversary is rational (or nearly rational). In subsequent work, the same authors [5] introduced a model of a nearly rational adversary who behaves in a rational manner most of the time. A nearly rational adversary can usually be deterred from committing policy violations by high inspection and punishment levels. They suggested that the defender commit to his strategy before each audit round (e.g., by publicly releasing its inspection and punishment levels) as in a Stackelberg game [6]. However, the paper gives no efficient algorithm for computing the Stackelberg equilibrium.

More recent work by Blocki et al. introduced the notion of *audit games* [7] – a simplified game theoretic model of the audit process in which the adversary is purely rational (unlike the nearly rational adversary of [5]). Audit games generalize the model of *security games* [8] by including punishment level as part of the defenders action space. The punishment parameter introduces quadratic constraints into the optimization problem that must be solved to compute the Stackelberg equilibria, making it difficult to find the Stackelberg equilibria. The primary technical contribution of [5] is an efficient algorithm for computing the Stackelberg equilibrium of audit games. There are two potential advantages of the k-adaptive regret framework compared with the Stackelberg equilibria approach: (1) The k-adaptive regret minimization algorithm can be used even if the adversary's incentives are unknown, and (2) a k-adaptive adversary is a better model for a short term adversary (e.g., contractors, tourists) who may not informed about the defender's policy; and therefore may not even know what the "rational" best response is in a Stackelberg game. See section 5 for additional discussion.

Stochastic games were defined by Shapley [2]. Much of the work on stochastic games has focused on finding and computing equilibria for these games [2, 9]. There has been lot of work in regret minimization for repeated games [1]. Regret minimization in stochastic games has not been the subject of much research. Papadimitriou and Yannakakis showed that many natural optimization problems relating to stochastic games are hard [10]. These results do not apply to bounded-memory games. Golovin and Krause recently showed that a simple greedy algorithm can be used when a stochastic optimization problem satisfies a property called adaptive submodularity [11]. In general, bounded-memory games do not satisfy this property. Even-Dar et al., show that regret minimization is possible for a class of stochastic games (Markov Decision Processes) in which the adversary chooses the reward function at each state but does not influence the transitions [12]. They also prove that if the adversary controls the reward function and the transitions, then it is NP-Hard to even approximate the best fixed strategy. Mannor and Shimkin [13] show that if the adversary completely controls the transition model (a Controlled Markov Process) then it is possible to separate the stochastic game into a series of matrix games and efficiently minimize regret in each matrix game. Bounded-memory games are a different subset of stochastic games where the transitions and rewards are influenced by both players. While our hardness proof shares techniques with Even-Dar et al., [12], there are significant differences that arise from the bounded-memory nature of the game. We provide a detailed comparison in Section 6.

In a more /recent paper, Even-Dar et al., [14] handle a few specific global cost functions related to load balancing. These cost functions depend on history. In their setting, the adversary obliviously plays actions from a joint distribution. In contrast, we consider arbitrary cost functions with bounded dependence on history and adaptive adversaries.

Takimoto and Warmuth [15] developed an efficient online shortest path algorithm. In their setting the experts consists of all fixed paths from the source to the destination. Because there may be exponentially many paths their algorithm must use an implicit weight representation. Awerbuch and Kleinberg later provided a general framework for online linear optimization [16]. In our settings, an additional challenge arises because *experts adapt to adversary actions*. See Section 7 for a more detailed comparison.

Farias et al., [17] introduce a special class of adversaries that they call "flexible" adversaries. A defender playing against a flexible adversary can minimize regret by learning the average expected reward of every expert. Our work differs from theirs in two ways. First, we work with a stochastic game as opposed to a repeated game. Second, our algorithms can handle a sequence of different k-adaptive adversaries instead of learning a single flexible adversary strategy. A single k-adaptive strategy is flexible, but a sequence of k-adaptive adversaries is not.

3 Preliminaries

Bounded-memory games are a sub-class of stochastic games, in which outcomes and states satisfy certain properties. Formally, a two-player stochastic game between an attacker A and a defender D is given by $(\mathcal{X}_D, \mathcal{X}_A, \Sigma, P, \tau)$, where $\mathcal{X}_A$ and $\mathcal{X}_D$ are the actions spaces for players A and D, respectively, Σ is the state space, $P : \Sigma \times \mathcal{X}_D \times \mathcal{X}_A \to [0, 1]$ is the payoff function and $\tau : \Sigma \times \mathcal{X}_D \times \mathcal{X}_A \times \{0,1\}^* \to \Sigma$ is the randomized transition function linking the different states. Thus, the payoff during round t depends on the current state (denoted σ^t) in addition to the actions of the defender (d^t) and the adversary (a^t). We use $n = |\Sigma|$ to denote the number of states.

A *bounded-memory game with memory* m ($m \in \mathbb{N}$) is a stochastic game with the following properties: (1) The game satisfies independent outcomes, and (2) The states $\Sigma = \mathcal{O}^m$ encode the last m outcomes, i.e., $\sigma^i = (O^{i-1}, \ldots, O^{i-m})$. An outcome of a given round of play is a signal observed by both players (called "public signal" in games [18]). Outcomes depend probabilistically on the actions taken by the players. We use $\mathcal{O}$ to denote the outcome space and $O^t \in \mathcal{O}$ to denote the outcome during round t. We say that a game satisfies *independent outcomes* if O^t is conditionally independent of $(O^1, \ldots, O^{t-1})$ given d^t and a^t. Notice that the defender and the adversary in a game with independent outcomes may still select their actions based on history. However, once those actions have been selected, the outcome is independent of the game history. Note that a repeated game is a bounded-memory-0 game (a bounded-memory game with memory $m = 0$).

A game in which players only observe the outcome O^t after round t but not the actions taken during a round is called an *imperfect information* game. If both players also observe the actions then the game is a *perfect information* game.

The *history* of a game $H = (O^1, O^2, \ldots, O^i, \ldots, O^t)$, is the sequence of outcomes. We use H_k to denote the k most recent outcomes in the game (i.e., $H_k = (O^{t-k+1}; \ldots; O^t)$), and $t = |H|$ to denote the total number of rounds played. We use H^i to denote the first i outcomes in a history (i.e., $H^i = (O^1, \ldots, O^i)$), and $H; H'$ to denote concatenation of histories H and H'.

A *fixed strategy* for the defender in a stochastic game is a function $f : \Sigma \to \mathcal{X}_D$ mapping each state to a fixed action. F denotes the set of all fixed strategies.

4   Definition of Regret

As discussed earlier, regret minimization in repeated games has received a lot of attention [19]. Unfortunately, the standard definition of regret in repeated games does not directly apply to stochastic games. In a repeated game, regret is computed by comparing the performance of the defender strategy D with the performance of a fixed strategy f. However, in a stochastic game, the actions of the defender and the adversary in round i influence payoffs in each round for the rest of the game. Thus, it is unclear how to choose a meaningful fixed strategy f as a reference. We solve this conundrum by introducing an adversary-based definition of regret.

4.1   Adversary Model

We define a parameterized class of adversaries called k-adaptive adversaries, where the parameter k denotes the level of adaptivity of the adversary. Formally, we say that an agent is k-*adaptive* if its strategy $A(H)$ is defined by a function $f : \mathcal{O}^* \times \mathbb{N} \to \mathcal{X}_A$ such that $A(H) = f(H_i, t)$, where $i = t \mod (k+1)$. Recall that H_i is the i most recent outcomes, and $t = |H|$.

As special cases we define an *oblivious adversary* ($k = 0$) and a *fully adaptive adversary* ($k = \infty$). Oblivious adversaries essentially play without any memory of the previous outcomes. Fully adaptive adversaries, on the other hand, choose their actions based on the entire outcome history since the start of the game. k-adaptive adversaries lie somewhere in between. At the start of the game, they act as fully adaptive adversaries, playing with the entire outcome history in mind. But, different from fully adaptive adversaries, every k rounds, they "forget" about the entire history of the game and act as if the whole game was starting afresh. As discussed earlier, there are numerous practical instances where k-adaptive adversaries are an appropriate model; for instance, in games in which one player (e.g., a firm) has a much longer length of play than the adversary (e.g., a temporary employee), it may be judicious to model the adversary as k-adaptive. In particular, k-adaptive adversaries are similar to the notion of "patient" players in long-run games discussed by [20]. Their notion of "fully patient" players correspond to fully adaptive adversaries, "myopic" players correspond to oblivious adversaries, and "not myopic but less patient" players correspond to k-adaptive adversaries.

Another possible adversary definition could be to consider a sliding window of size k as the adversary memory. But, because such an adversary can play actions to remind herself of events in the arbitrary past, her memory is not actually bounded by k, and regret minimization is not possible. See the full version [3] of this paper for details.

$\mathcal{A}_D^K$ and $\mathcal{A}_A^K$ denote all possible K-adaptive strategies for the defender and adversary, respectively.

4.2 k-adaptive Regret

Suppose that the defender D and the adversary A have produced history H in a game G lasting T rounds. Let $a^1, ..., a^T$ denote the sequence of actions played by the adversary. In hindsight we can construct a hypothetical k-adaptive adversary A_k as follows:

$$A_k\left(H'\right) = A\left(H^{t-i}; H_i'\right) ,$$

where $t = |H'|$ and $i = t \mod (k+1)$. In other words, the hypothetical k-adaptive adversary replicates the plays the real adversary made in the actual game regardless of the strategy of the defender he is playing against, *except* for the last i rounds under consideration where he adapts his strategy to the defender's actions in the same manner the real adversary would.

Abusing notation slightly, we write $P\left(f, A, G, \sigma_0, T\right)$ to denote the expected payoff the defender would receive over T rounds of G given that the defender plays strategy f, the adversary uses strategy A and the initial state of the bounded-memory game G is σ_0. We use $\bar{P}\left(f, A, G, T\right) = P\left(f, A, G, \sigma_0, T\right)/T$ to denote the average per-round payoff. We use

$$\bar{R}_k\left(D, A, G, T, S\right) = \max_{f \in S} \bar{P}\left(f, A_k, G, T\right) - \bar{P}\left(D, A_k, G, T\right) ,$$

to denote the *k-adaptive regret* of the defender strategy D using a fixed set S of experts against an adversary strategy A for T rounds of the game G.

Definition 1. *A defender strategy D using a fixed set S of experts is a γ-approximate k-adaptive regret minimization algorithm for the class of games $\mathcal{G}$ if and only if for every adversary strategy A, every $\epsilon > 0$ and every game $G \in \mathcal{G}$ there exists $T' > 0$ such that $\forall T > T'$*

$$\bar{R}_k\left(D, A, G, T, S\right) < \epsilon + \gamma .$$

If $\gamma = 0$ then we simply refer to D as a k-adaptive regret minimization algorithm. If D runs in time $poly\left(n, 1/\epsilon\right)$ we call D efficient.

k-adaptive regret considers a k-adaptive hypothetical adversary who can adapt within each window of size (at most) $k + 1$. Intuitively, as k increases this measure of regret is more meaningful (as the hypothetical adversary increasingly resembles the real adversary), albeit harder to minimize.

There are two important special cases to consider: $k = 0$ (oblivious regret) and $k = \infty$ (adaptive regret). Adaptive regret is the strongest measure of regret. Observe that if the actual adversary is k-adaptive then the hypothetical adversary A_∞ is same as

Table 2. Speeding game utilities

Actions	S	DS
HI	.19	0.7
LI	0.2	1

(a) Defender utility P

Actions	S	DS
HI	0	0.8
LI	1	0.8

(b) Adversary utility

the hypothetical adversary A_k, and hence $\bar{R}_\infty = \bar{R}_k$. Also, if the actual adversary is oblivious then $\bar{R}_\infty = \bar{R}_0 = \bar{R}_k$.

In this paper $\mathcal{G}$ will typically denote the class of perfect/imperfect information bounded-memory games with memory m. We are interested in expert sets S which contain all of the fixed strategies $F \subseteq S$.

5   Audit Examples

As an example, consider the interaction between a police chief (defender) and drivers (adversary) at a popular tourist destination. The police chief is given the task of enforcing speed limits on local roads. Each day the police chief may deploy resources (e.g., radar, policemen) to monitor local roads, and drivers decide whether or not to speed or not.

Repeated Game. We first model the interaction above using a repeated game. We will consider a simple version of this interaction in which the defender has two actions

$$\mathcal{X}_D = \{\mathbf{HI}, \mathbf{LI}\} \, ,$$

and the adversary has two actions

$$\mathcal{X}_A = \{\mathbf{S}, \mathbf{DS}\} \, .$$

Here, **HI/LI** stands for high/low inspection and **S/DS** stands for speed and don't speed. We consider the defender utilities in Table 2(a).

In this example, the costs of a higher inspection outweigh the benefits of enforcing the policy. In *any* Nash Equilibria the defender will play his dominant strategy – "always play **LI**." Similarly, *any* algorithm that minimizes regret in the standard sense (0-adaptive) – like the regret minimizing audit mechanism from [4] – must eventually converge to the dominant defender strategy **LI**. While this is the best that the defender can do against a byzantine adversary, this may not always be the best result for the defender when playing against a rational adversary. Consider the adversary's utility defined in Table 2(b).

If the defender plays his dominant strategy then the adversary will always play the action **S**, speed. This action profile results in average utility 0.2 for the defender and 1 for the adversary. However, if the defender can commit to his strategy in advance then he can play his Stackelberg equilibrium [6] strategy "play **HI** with probability 0.2 and **LI** with probability 0.8." A rational adversary will respond by playing her best response – the action that maximizes her utility given the defenders commitment. In this case the adversary's best response is to play **DS**. The resulting utility for the defender is now 0.94.

There are two practical challenges with adopting this approach: (1) If the utility of the adversary is unknown then the defender cannot compute the Stackelberg equilibrium. (2) Even if the defender commits to playing a Stackelberg equilibrium it is unlikely that many drivers will respond in purely rational manner for the simple reason that they are uniformed (e.g., a tourist may not know whether or not speed limits are aggressively enforce in an unfamiliar area). If the adversary can learn the Stackelberg Equilibrium from a history of the defender's actions, then she might adapt her play to the best response strategy over time. However, each tourist has a limited time window in which she can make these observations and adjust her behavior (e.g., the tourist leaves after at most k days).

Bounded-memory Game Model with k-adaptive Regret. We model the interaction above using bounded-memory games with k-adaptive adversary model. In each round of our bounded-memory game the defender and the adversary play an action profile, and observe an outcome – a public signal. The action space in our bounded-memory game is identical to the repeated game, and the outcome $\mathcal{O} = \{\textbf{HI}, \textbf{LI}\}$ is simply the defender's action. That is we assume that our tourist driver can observe the defender's inspection level in each round (e.g., by counting the number of police cars by the side of the road). The defender's payoff function is identical to Table 2(a) – the defender's payoff is independent of the current state (e.g., rewards in this particular bounded-memory game are not history-dependent). A k-adaptive regret minimization algorithm could be run without a priori knowledge of the adversary's utility, and will converge to the optimal fixed strategy against any k-adaptive adversary (e.g., any sequence of k-adaptive tourist strategies).

It is reasonable to use a k-adaptive strategy to model the behavior of our tourist drivers. Each tourist initially has no history of the defender's actions – during the first day of her visit a tourist must make the decision about whether or not to speed without any history of the defender's actions. After the first day the tourist may adapt his behavior based on previous outcomes. For example, a tourist might adopt the following k-adaptive strategy: $\mathcal{A}_1$ = "Play **DS** on the first day, and on the remaining $(k - 1)$ days play **S** if the defender has never played **HI** previously, otherwise play **DS**." After k days the tourist leaves and a new tourist arrives. This new tourist may adopt a different k-adaptive strategy (e.g., $\mathcal{A}_2$ = "Play **S** on the first day, and on the remaining $(k - 1)$ days play **S** if the defender has never played **HI** previously, otherwise play **DS**.").

We set the memory of our bounded-memory game to be $m = k$. Now the fixed defender strategies F in our bounded-memory game include strategies like f = "play **HI** every k'th round". Suppose for example that $k = 7$ and the defender plays f. In this case the sequence of rewards that the defender would see against the first k-adaptive adversary $\mathcal{A}_1$ would be $(0.7, 1, 1, 1, 1, 1, 1)$. The sequence of rewards that the defender would see against the second k-adaptive adversary $\mathcal{A}_2$ would be $(0.19, 1, 1, 1, 1, 1, 1)$. It is easy to verify that this is the optimal result for the defender – if the defender does not play **HI** on the first day then the 7-adaptive adversary will speed on day 2. A k-adaptive regret minimization algorithm could be run without a priori knowledge of the adversary's utility, and will converge to the optimal fixed strategy against any k-adaptive adversary (e.g., any sequence of k-adaptive tourist strategies).

Another Example: Hospital Employees. A k-adaptive adversary is also an appropriate model for employees working in organizations. Indeed, most organizations are generally active for much longer than any employee's duration of employment. This is for instance true in the case of hospitals, where employees could be tempted to violate privacy policies supposed to protect patients' sensitive data, for convenience, out of curiosity, or even in the worst case, for financial gain. In this example, we would consider the interaction between the hospital, playing the role of the defender, and the employee playing the role of the adversary. We could naturally transpose the example discussed above by simply replacing the actions **S** and **DS** (e.g., "speed" and "don't speed") by the corresponding actions **B** and **V** (e.g., "behave" and "violate").

Unfortunately, we are able to prove that there is no efficient k-adaptive regret minimization algorithm for general bounded-memory games. However, our results do not rule out the possibility of an efficient γ-approximate k-adaptive regret minimization algorithm. Finding an efficient γ-approximate k-adaptive regret minimization algorithms is an important open problem.

6   Hardness Results

In this section, we show that unless $\mathsf{NP} = \mathsf{RP}$ no oblivious regret minimization algorithm which uses the fixed strategies F as experts can be efficient in the imperfect information setting. In the full version [3] of this paper we explain how our hardness reduction can be adapted to prove that there is no efficient k-adaptive regret minimization algorithm in the perfect information setting for $k \geq 1$.

Specifically, we consider the subclass of bounded-memory games $\mathcal{G}$ with the following properties: $|\mathcal{O}| = O(1)$, $m = O(\log n)$, $|\mathcal{X}_A| = O(1)$, $|\mathcal{X}_D| = O(1)$ and imperfect information. Any $G \in \mathcal{G}$ is a game of imperfect information (on round t the defender observes O^t, but not a^t) with $O(n)$ states. Our goal is to prove the following theorem:

Theorem 1. *For any $\beta > 0$ and $\gamma < 1/8n^\beta$ there is no efficient γ-approximate oblivious regret minimization algorithm which uses the fixed strategies F as experts against oblivious adversaries for the class of imperfect information bounded-memory-m games unless $\mathsf{NP} = \mathsf{RP}$.*

Given a slightly stronger complexity-theoretic assumption called the randomized exponential time hypothesis [21] we can prove a slightly stronger hardness result. The randomized exponential time hypothesis says that no randomized algorithm running in time $2^{o(n)}$ can solve SAT.

Theorem 2. *Assume that the randomized exponential time hypothesis is true. Then for any $\gamma < 1/\left(8\log^2 n\right)$ there is no efficient γ-approximate oblivious regret minimization algorithm which uses the fixed strategies F as experts against oblivious adversaries for the class of imperfect information bounded-memory-m games.*

The proofs of Theorems 1 and 2 use the fact that it is hard to approximate MAX3SAT within any factor better than $\frac{7}{8}$ [22]. This means that unless $\mathsf{NP} = \mathsf{RP}$ then for every constant $\beta > 0$ and every randomized algorithm S in RP, there exists a MAX3SAT instance ϕ such that the expected number of clauses in ϕ unsatisfied by $S(\phi)$ is $\geq \frac{1}{8} - \beta$ even though there exists an assignment satisfying $(1 - \beta)$ fraction of the clauses in ϕ.

We reduce a MAX3SAT formula ϕ with variables $x_1, ..., x_n$ and clauses $C_1, ..., C_\ell$ to a bounded-memory game G described formally below. We provide a high level overview of the game G before describing the details. The main idea is to construct G so that the rewards in G are related to the fraction of clauses of ϕ that are satisfied.

In G, for each variable x there is a state σ_x associated with that variable. The oblivious adversary controls the transitions between variables. This allows the oblivious adversary A_R to partition the game into stages of length n, such that during each stage the adversary causes the game to visit each variable exactly once (each state is associated with a variable). During each stage the adversary picks a clause C at random. In G we have $0, 1 \in \mathcal{X}_D$. Intuitively, the defender chooses assignment $x = 1$ by playing the action 1 while visiting the variable x. The defender receives a reward if and only if he succeeds in satisfying the clause C.

The game G is defined as follows:

Defender actions: $\mathcal{X}_D = \{0, 1, 2\}$

Adversary actions: $\mathcal{X}_A = \{0, 1\} \times \{0, 1, 2, 3\}$

Outcomes and states: Each round i produces two outcomes

$$\tilde{O}^i = \vec{a}^i[1] \quad \text{and} \quad \hat{O}^i = \begin{cases} 1 & \text{if } d^i = 2 \text{ or } d^i = a^i[2]; \\ 0 & \text{otherwise.} \end{cases}$$

Observe that these outcomes satisfy the independent outcomes requirement for bounded-memory games. There are $n = 2^{m+1}$ states, where σ^i is the state at round i, where

$$\sigma^i = \left(\langle \tilde{O}^{i-1}, ..., \tilde{O}^{i-m} \rangle, \hat{O}^{i-1} \right).$$

Observe that each state encodes the last m outcomes $\tilde{O}$ and the last outcome $\hat{O}^i$. Intuitively, the last m outcomes $\tilde{O}^i$ are used to denote the variable x_i, while $\hat{O}^i$ is 1 if the defender has already received a reward during the current phase.

The defender actions $0, 1$ correspond to the truth assignments $0, 1$. The defender receives a reward for the correct assignment. The defender is punished if he attempts to obtain a reward in any phase after he has already received a reward in that phase. Once the defender has already received a reward he can play the special action 2 to avoid getting punished. The intuitive meaning of the adversary's actions is explained below.

If we ignore the outcome $\hat{O}$ then the states form a De Bruijn graph [23] where each node corresponds to a variable of ϕ. Notice that the adversary completely controls the outcomes $\tilde{O}$ with the first component of his action $\vec{a}[1]$. By playing a De Bruijn sequence $S = s_1...s_n$ the adversary can guarantee that we repeatedly take a Hamiltonian cycle over states(for an example see Figure 1).

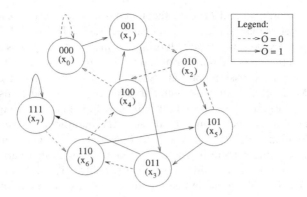

Figure 1. De Bruijn example

Rewards:[2]

$$P\left(\sigma^i, d^i, a^i\right) = \begin{cases} -1 & \text{if } \hat{O}^{i-1} = 1 \text{ and } d^i \neq 2 \text{ and } \vec{a}^i[2] \neq 3; \\ 1 & \text{if } d^i \neq 2 \text{ and } d^i = \vec{a}^i[2] \text{ and } \hat{O}^{i-1} = 0; \\ 0 & \text{otherwise.} \end{cases}$$

An intuitive interpretation of the reward function is presented in parallel with the adversary strategy.

Adversary Strategy: The first component of the adversary's action ($\vec{a}[1]$) controls the transitions between variables. The adversary will play the action $\vec{a}^i[2] = 1$ (resp. $\vec{a}^i[2] = 0$) whenever the corresponding variable assignment $x_i = 1$ (resp. $x_i = 0$) satisfies the clause that the adversary chose for the current phase.

If neither variable assignment satisfies the clause (if $x_i \notin C$ and $\bar{x}_i \notin C$) then the adversary plays $\vec{a}^i[2] = 2$. This ensures that a defender can only be rewarded during a round if he satisfies the clause C, which happens when $d^i = \vec{a}^i[2] = 0$ or 1.

Notice that whenever $\hat{O} = 1$ there is no way to receive a positive reward. The defender may want the game G to return to a state where $\hat{O} = 0$, but unless the adversary plays the special action $\vec{a}^i[2] = 3$ he is penalized when this happens. The adversary action $\vec{a}^i[2] = 3$ is a special 'reset phase' action. By playing $\vec{a}^i[2] = 3$ once at the end of each phase the adversary can ensure that the maximum payoff the defender receives during any phase is 1. See Figure 1 for a formal description of the adversary strategy.

Analysis. At a high level, our hardness argument proceeds as follows:

1. If there is an assignment that satisfies $(1 - \beta)$ fraction of the clauses in ϕ, then there is a fixed strategy that performs well in expectation (see Claim 1).
2. If there a fixed strategy that performs well in expectation, then any γ-approximate oblivious regret minimization algorithm will perform well in expectation (see Claim 2).

---

[2] We use payoffs in the range $[-1, 1]$ for ease of presentation. These payoffs can easily be rescaled to lie in $[0, 1]$.

- **Input:** MAX3SAT instance ϕ, with variables $x_1, \ldots, x_{n-1}$, and clauses $C_1, \ldots, C_\ell$. Random string $R \in \{0,1\}^*$
- **De Bruijn sequence:** $s_0, \ldots, s_{n-1}$
- **Round t:** Set $i \leftarrow t \mod n$.
  1. **Select clause:** If $i = 0$ then select a clause C uniformly at random from $C_1, \ldots, C_\ell$ using R.
  2. **Select move:**

$$a^i = \begin{cases} (s_i, 3) & \text{if } i = 0; \\ (s_i, 1) & \text{if } x_i \in C; \\ (s_i, 0) & \text{if } \bar{x}_i \in C; \\ (s_i, 2) & \text{otherwise.} \end{cases}$$

Figure 2. Oblivious adversary: A_R

3. If an efficiently computable strategy D performs well in expectation, then there is an efficiently computable randomized algorithm S to approximate MAX3SAT (see Claim 3). This would imply that $\mathsf{NP} = \mathsf{RP}$.

Claim 1. *Suppose that there is a variable assignment that satisfies $(1 - \beta) \cdot \ell$ of the clauses in ϕ. Then there is a fixed strategy f such that $E_R\left[\bar{P}\left(f, A_R, G, n\right)\right] \geq (1 - \beta)/n$, where R is used to denote the random coin tosses of the oblivious adversary.*

Claim 2. *Suppose that D is an $\left(\frac{1}{8n} - \frac{3\beta}{n}\right)$-approximate oblivious regret minimization algorithm against the class of oblivious adversaries and there is a variable assignment that satisfies $(1 - \beta)$ fraction of the clauses in ϕ. Then for $T = poly(n)$*

$$E_R\left[\bar{P}\left(D, A_R, G, T\right)\right] \geq \frac{7}{8n} + \frac{\beta}{n},$$

where R is used to denote the random coin tosses of the oblivious adversary.

Claim 3. *Fix a polynomial $p(\cdot)$ and let $\alpha = n \cdot E_R\left[\bar{P}\left(D, A_R, G, T\right)\right]$, where $T = p(n)$ and D is any polynomial time computable strategy. There is a polynomial time randomized algorithm S which satisfies α fraction of the clauses from ϕ in expectation.*

The proofs of these claims can be found in the full version [3] of this paper.

Proof of Theorem 1. The key point is that if an algorithm S runs in time $O\left(p(n)\right)$ on instances of size n^β for some polynomial $p(n)$ then on instances of size n S runs in time $O\left(p\left(n^{1/\beta}\right)\right)$ which is still polynomial time. Unless $\mathsf{NP} = \mathsf{RP}$ $\forall \epsilon, \beta > 0$ and every algorithm S running in time poly(n), there exists an integer n and a MAX3SAT formula ϕ with n^β variables such that

1. There is an assignment satisfying at least $(1 - \epsilon)$ of the clauses in ϕ.
2. The expected fraction of clauses in ϕ satisfied by S is $\leq \frac{7}{8} + \epsilon$.

If we reduce from a MAX3SAT instance with n^β variables we can construct a game with $O(n)$ states ($n^{1-\beta}$ copies of each variable). One Hamiltonian cycle would now corresponds to $n^{1-\beta}$ phases of the game. This means that the expected average reward of the optimal fixed strategy is at least

$$\max_{f \in F} E_R\left[\bar{P}\left(f, A_R, G, T\right)\right] \geq \frac{n^{1-\beta}\left(1 - \epsilon\right)}{n},$$

while the expected average reward of an efficient defender strategy D is at most

$$E_R\left[\bar{P}\left(D, A_R, G, T\right)\right] \leq \frac{n^{1-\beta}\left(\frac{7}{8} + \epsilon\right)}{n}.$$

Therefore, the expected average regret is at least

$$\bar{R}_0\left(D, A_R, G, T, F\right) \geq \left(\frac{1}{8} - 2\epsilon\right)n^{-\beta}.$$

$\square$

The proof of theorem 2 is similar to the proof of theorem 1. It can be found in the full version [3] of this paper.

Our hardness reduction is similar to a result from Even-Dar et al., [12]. They consider regret minimization in a Markov Decision Process where the adversary controls the transition model. Their game is not a bounded-memory game; in particular it does not satisfy our *independent outcomes* condition. The current state in their game can depend on the last n actions. In contrast, we consider bounded-memory games with $m = O\left(\log n\right)$, so that the current state only depends on the last m actions. This makes it much more challenging to enforce guarantees such as "the defender can only receive a reward once in each window of n rounds"—a property that is used in the hardness proof. The adversary is oblivious so she will not remember this fact, and the game itself cannot record whether a reward was given $m + 1$ rounds ago. We circumvented this problem by designing a payoff function in which the defender is penalized for allowing the game to "forget" when the last reward was given, thus effectively enforcing the desired property.

7   Regret Minimization Algorithms

In section 7.1 we present a reduction from bounded-memory games to repeated games. This reduction can be used to create a k-adaptive regret minimizing algorithm (Theorem 3). This is significant because there is no k-adaptive regret minimization algorithm for the general class of stochastic games. A consequence of Theorem 1 is that when the expert set includes all fixed strategies F we cannot hope for an efficient algorithm unless NP = RP. In section 7.2 we present an efficient *approximate* 0-adaptive regret minimization algorithm for bounded-memory games of perfect information. The algorithm uses an implicit weight representation to efficiently sample the experts and update their weights. Finally, we show how this algorithm can be adapted to obtain an efficient approximate 0-adaptive regret minimization algorithm for bounded-memory games of *imperfect* information.

7.1 Reduction to Repeated Games

All of our regret minimization algorithms work by first reducing the bounded-memory game G to a repeated game $\rho(G, K)$. One round of the repeated game $\rho(G, K)$ corresponds to K rounds of G. Before each round of $\rho(G, K)$ both players commit to an adaptive strategy. In $\rho(G, K)$ the reward that the defender gets for playing a strategy $f \in \mathcal{A}_D^K$ is the reward that the defender would have received for using the strategy f for the next K rounds of the actual game G if the initial state were σ_0: $P(f, g, \rho(G, K)) = P(f, g, G, \sigma_0, K)$.

The rewards in $\rho(G, K)$ may be different from the actual rewards in G because the initial state before each K rounds might not be σ_0. Claim 4 bounds the difference between the hypothetical losses from $\rho(G, K)$ and actual losses in G using the bounded-memory property. The proof of Claim 4 is in the full version of this paper [3].

Claim 4. *For any adaptive defender strategy $f \in \mathcal{A}_D^K$ and any adaptive adversary strategy $g \in \mathcal{A}_A^K$ and any state σ of G we have $|P(f, g, G, \sigma, K) - P(f, g, G, \sigma_0, K)| \leq m$.*

The key idea behind our k-adaptive regret minimization algorithm BW is to reduce the original bounded-memory game to a repeated game $\rho(G, K)$ of imperfect information ($K \equiv 0 \mod k$). In particular we obtain the regret bound in Theorem 3. Details and proofs can be found in the full version of this paper [3].

Theorem 3. *Let G be any bounded-memory-m game with n states and let A be any adversary strategy. After playing T rounds of G against A, BW (G, K) achieves regret bound*

$$\bar{R}_k(\text{BW}, A, G, T, S) < \frac{m}{T^{1/4}} + 4\frac{\sqrt{N \log N}}{T^{1/4}},$$

where $N = |S|$ is the number of experts, A is the adversary strategy and K has been chosen so that $K = T^{1/4}$ and $K \equiv 0 \mod k$.

Intuitively, the $m/T^{1/4} = m/K$ term is due to modeling loss from Claim 4 and the other term comes from the standard regret bound of [24].

7.2 Efficient Approximate Regret Minimization Algorithms

In this section we present EXBW (Efficient approXimate Bounded Memory Weighted Majority), an efficient algorithm to approximately minimize regret against an oblivious adversary in bounded-memory games with perfect information. The set of experts $\mathcal{E}$ used by our algorithms contains the fixed strategies F as well as all K-adaptive strategies $\mathcal{A}_D^K$ ($K = m/\gamma$). We prove the following theorem

Theorem 4. *Let G be any bounded-memory-m game of perfect information with n states and let A be any adversary strategy. Playing T rounds of G against A, EXBW runs in total time $Tn^{O(1/\gamma)}$ and achieves regret bound*

$$\bar{R}_0 \left(\text{EXBW}, A, G, T, \mathcal{E} \right) \le \gamma + O \left(\frac{m}{\gamma} \sqrt{\frac{\frac{m}{\gamma} n \log\left(N\right)}{T}} \right),$$

where K has been set to m/γ and $N = \left| \mathcal{A}_D^K \right| = \left(|\mathcal{X}_D| \right)^{n^{1/\gamma}}$ is the number of K-adaptive strategies.

In particular, for any constant γ there is an efficient γ-approximate 0-adaptive regret minimization algorithm for bounded-memory games of perfect information. We can adapt this algorithm to get EXBWII (Efficient approXimate Bounded Memory Weighted Majority for Imperfect Information Games), an efficient approximate 0-adaptive regret minimization algorithm for games of imperfect information using a sampling strategy described in the full version of this paper [3].

Theorem 5. *Let G be any bounded-memory-m game of imperfect information with n states and let A be any adversary strategy. There is an algorithm EXBWII that runs in total time $Tn^{O(1/\gamma)}$ playing T rounds of G against A, and achieves regret bound*

$$\bar{R}_0 \left(\text{EXBWII}, A, G, T, \mathcal{E} \right) \le 2\gamma + O \left(\frac{mn^{1/\gamma}}{\gamma^2} \sqrt{\frac{\frac{mn^{1/\gamma}}{\gamma} n \log\left(N\right)}{T}} \right).$$

where K has been set to m/γ and $N = \left| \mathcal{A}_D^K \right| = \left(|\mathcal{X}_D| \right)^{n^{1/\gamma}}$ is the number of K-adaptive strategies.

The regret bound of Theorem 4 is simply the regret bound achieved by the standard weighted majority algorithm [25] plus the modeling loss term from Claim 4. The main challenge is to provide an efficient simulation of the weighted majority algorithm. There are an exponential number of experts so no efficient algorithm can explicitly maintain weights for each of these experts. To simulate the weighted majority algorithm EXBW implicitly maintains the weight of each expert.

To simulate the weighted majority algorithm we must be able to *efficiently sample* from our weighted set of experts (see **Sample** $(\mathcal{E})$) and efficiently update the weights of each expert in the set after each round of $\rho\left(G, K\right)$ (see update weight stage of EXBW).

Meet the Experts. Instead of using F as the set of experts, EXBW uses a larger set of experts $\mathcal{E}$ $(F \subset \mathcal{E})$. Recall that a K-adaptive strategy is a function f mapping the K most recent outcomes H_K to actions. We use a set of K-adaptive strategies $E = \{ f_\sigma : \sigma \in \Sigma \} \subset \mathcal{A}_D^K$ to define an expert E in $\rho\left(G, K\right)$: if the current state of the real bounded-memory game G is σ then E uses the K-adaptive strategy f_σ in the next round of $\rho\left(G, K\right)$ (i.e., the next K rounds of G). $\mathcal{E}$ denotes the set of all such experts.

Maintaining Weights for Experts Implicitly. To implicitly maintain the weights of each expert $E \in \mathcal{E}$ we use the concept of a game trace. We say that a game trace $p = \sigma, d^1, O^1, ..., d^{i-1}, O^{i-1}, d^i$ is consistent with an expert E if $f_\sigma \left(O^1, ..., O^{j-1} \right) = d^j$ for each j. We define the set $\mathcal{C}\left(E\right)$ to be the set of all such consistent traces of maximum length K and $C = \bigcup_{E \in \mathcal{E}} C(E)$ denotes the set of all traces consistent with some expert $E \in \mathcal{E}$. EXBW maintains a weight w_p on each trace $p \in C$. The weight of an expert E is then defined to be $W_E = \prod_{p \in C(E)} w_p$.

Given adversary actions $\vec{a} = a_1, ..., a_K$ and a trace $p = \sigma, d^1, O^1, ..., d^{i-1}, O^{i-1}, d^i$ we define $\mathcal{R}(\vec{a}, \sigma', p)$:

$$\mathcal{R}(\vec{a}, \sigma', p) = \begin{cases} 0 & \text{if } \sigma \neq \sigma'; \\ \prod_{j<i} \Pr\left[O^j \mid a^j, d^j\right] & \text{otherwise;} \end{cases}$$

Intuitively, $\mathcal{R}(\vec{a}, \sigma', p)$ is the probability that each outcome of p would have occurred given the adversary actions were $\vec{a}$ and the initial state was σ'. We use $\ell(p, \vec{a}, \sigma')$ to denote the payment that the defender received for playing d^i (the last action in p). Formally $\ell(p, \vec{a}, \sigma') = P(\sigma_p^f, d^i, a^i) \mathcal{R}(\vec{a}, \sigma', p)$, where σ_p^f denotes the state reached following the trace p (after observing outcomes $O^1, ..., O^{i-1}$ starting from σ_0) and d^i is the final defender action in the trace. Notice that in the imperfect information setting the defender could not compute ℓ because he would not observe the adversary's actions $\vec{a}$.

Updating Weights Efficiently. While updating weights EXBW maintains the invariant that $w_p = \beta^{\sum_{j=1}^{T/K} \ell\left(p, \vec{a}^j, \sigma^{jK}\right)}$, where σ^{jK} is the state of G after jK rounds and $\vec{a}^t$ is the actions the adversary played during the j'th round of $\rho(G, K)$. The standard weighted majority algorithm maintains the invariant that $W_E = \beta^{\sum_{j=1}^{T/K} P\left(E, \vec{a}^t, \rho(G, K)\right)}$. Claim 5 implies that EXBW also maintains this invariant with its implicit weight representation; the proof of Claim 5 is in the full version of this paper [3].

Claim 5

$$\prod_{p \in C(E)} \beta^{\sum_{j=1}^{T/K} \ell\left(p, \vec{a}^j, \sigma^{jK}\right)} = \beta^{\sum_{j=1}^{T/K} P\left(E, \vec{a}^j, \rho(G, K)\right)}.$$

Sampling Experts Efficiently. We can also efficiently sample from $\mathcal{E}$ using dynamic programming (see **Sample** $(\mathcal{E})$). Using the notation $p \sqsubseteq p'$ for p' extends p we can define $\hat{w}_p$:

$$\hat{w}_p = \sum_{E : p \in C(E)} \prod_{p' \in C(E) \wedge p \sqsubseteq p'} w_{p'}$$

Intuitively, $\hat{w}_{p;O;d}$ represents the weight of the action d from history $p; O$.

Using dynamic programming we can efficiently compute $\hat{w}_p$ for each trace p because there are only $n^{O(1/\gamma)}$ such traces. Using the weights $\hat{w}_p$ we can efficiently sample from $\mathcal{E}$. We use $p; O; d$ to denote a new game trace which contains all of the outcomes/actions in p appended with O and d.

Algorithm: EXBW (γ, G)	**Algorithm: Sample $(\mathcal{E})$**
• **Initialize:** $K = m/\gamma$	• For each trace $p \in \mathcal{C}$ recursively compute
• **Construct:** $\rho(G, K)$	$\hat{w}_p$ using the formula:
• **Each Round:**	
1. $\sigma \leftarrow G.CurrentState$	$\hat{w}_p = \displaystyle\sum_{O \in \mathcal{O}} \sum_{d \in \mathcal{X}_D} \beta^{\sum_{t=1}^{T} \ell\left(p; O; d, \vec{a}^t, \sigma^{Kt}\right)} \hat{w}_{p;O;d}$.
2. $E \leftarrow$ **Sample** $(\mathcal{E})$	
3. Play E	
4. Observe adversary actions	• **Build Strategy** E: For each $p \in \mathcal{C}$ and $O \in$
$\qquad \vec{a} = a^1, ..., a^K$.	$\mathcal{O}$, randomly select $d \in \mathcal{X}_D$
5. **Update Weights:** For each $p \in \mathcal{C}$	$\Pr[d \,|\, p,\, O] = \dfrac{\hat{w}_{p;O;d}}{\sum_{d' \in \mathcal{X}_D} \hat{w}_{p;O;d'}}$.
A. Compute $\ell(p, \vec{a}, \sigma)$	
B. Set $w_p \leftarrow w_p \times \beta^{\ell(p,\vec{a},\sigma)}$.	• E play d any time it observes history $p; O$.

Claim 6 says that **Sample** $(\mathcal{E})$ outputs each expert E with probability proportional to W_E.

Claim 6. *For each expert $E \in \mathcal{E}$, Algorithm* **Sample** $(\mathcal{E})$ *outputs E with probability*

$$\Pr[E] \propto W_E .$$

Given **Sample** $(\mathcal{E})$ it is straightforward to simulate the standard weighted majority algorithm. To update weights EXBW simply loops through all traces $p \in \mathcal{C}$ applying the update rule $w_p = w_p \times \beta^{\ell\left(p, \vec{a}^t, \sigma^{tK}\right)}$, where β is a learning parameter we tune later. The formal proof of Theorem 4 can be found in the full version along with the proof of claim 6.

At a high level our algorithm is similar to the online shortest path algorithm developed by Takimoto and Warmuth [15]. In their work, they consider the set of all source-destination paths in a graph as experts. Since there are exponentially many paths they also maintain the weights of the experts implicitly. In their setting, the defender completely controls the chosen path. In contrast, our experts adapt to adversary actions. The challenge was constructing a new implicit weight representation which works for K-adaptive strategies.

Using this implicit weight representation we could have also used the general barycentric spanner approach to online linear optimization developed by Awerbuch and Kleinberg [16] to design a γ-approximate 0-adaptive regret minimization algorithm running in time $n^{O(1/\gamma)}$. However, we are able to achieve better regret bounds in theorem 4 by simulating the weighted majority algorithm. Awerbuch and Kleinberg [16, Theorem 2.8] achieve the average regret bound $O\left(Md^{5/3}/T^{1/3}\right)$, where d is the dimension of the problem space and M is a bound on the cost vectors. By comparison our regret bounds in Theorems 4 and 5 tend to 0 with $1/\sqrt{T}$. In our setting, the dimension of the problem space is $d = O\left(n^{(1/\gamma)}\right)$ (the number of nodes in the decision tree), and $M = K = m/\gamma$ is the upper bound on the cost vector in each round of $\rho(G, K)$. The average regret bound would be $O\left(\frac{m}{\gamma} n^{5/(3\gamma)}/T^{1/3}\right)$. the regret bound is proportional to $\sqrt{n^{1/\gamma}/T}$. By comparison Theorem 4 has a $\sqrt{n^{1/\gamma}}$ in the numerator.

The standard regret minimization trick for dealing with imperfect information in a repeated game is to break the game up into phases and perform random sampling in each round to estimate the cost of each expert and update weights. The challenge in adapting EXBW is that there are exponentially many experts in $\mathcal{E}$. Our key idea was to estimate $\ell(p, \vec{a}, \sigma)$ for each $p \in \mathcal{C}$ so there are only $n^{O(1/\gamma)}$ samples to take in each phase. We can then update the implicit weight representation using the estimated values $\ell(p, \vec{a}, \sigma)$.

8 Open Questions

In this paper, we defined a new class of games called bounded-memory games, introduced several new notions of regret, and presented hardness results and algorithms for regret minimization in this subclass of stochastic games. Because both the games and the notions of regret we study in this paper rely on novel definitions, they raise a number of interesting open problems: (1) To what extent can the hardness results of Theorems 1 and 2 be further improved? ($\gamma = 1/\log n$?) Could similar hardness results apply to games with perfect information? (2) Is there an efficient *non-approximate* oblivious regret minimization algorithm for bounded-memory games with perfect information? (3) Is there a γ-approximate oblivious regret minimization algorithm with running time $n^{o(1/\gamma)}$? For example, could one design a γ-approximate oblivious regret minimization algorithm with running time $n^{-\log \gamma}$? (4) For repeated games ($m = 0$) is there an efficient γ-approximate k-adaptive regret minimization algorithm if we use $\mathcal{A}_D^K$ as our set of experts ($K = \log n$)?

References

1. Blum, A., Mansour, Y.: Learning, regret minimization, and equilibria. Algorithmic Game Theory, 79–102 (2007)
2. Shapley, L.: Stochastic games. Proceedings of the National Academy of Sciences of the United States of America 39(10), 1095 (1953)
3. Blocki, J., Christin, N., Datta, A., Sinha, A.: Adaptive regret minimization in bounded-memory games. CoRR abs/1111.2888 (2011)
4. Blocki, J., Christin, N., Datta, A., Sinha, A.: Regret minimizing audits: A learning-theoretic basis for privacy protection. In: 24th IEEE Computer Security Foundations Symposium, CSF 2011, pp. 312–327. IEEE (2011)
5. Blocki, J., Christin, N., Datta, A., Sinha, A.: Audit mechanisms for provable risk management and accountable data governance. In: Grossklags, J., Walrand, J. (eds.) GameSec 2012. LNCS, vol. 7638, pp. 38–59. Springer, Heidelberg (2012)
6. Von Stackelberg, H.: Market structure and equilibrium. Springer (2011)
7. Blocki, J., Christin, N., Datta, A., Procaccia, A.D., Sinha, A.: Audit games. In: IJCAI (2013)
8. Tambe, M.: Security and Game Theory: Algorithms, Deployed Systems, Lessons Learned. Cambridge University Press (2011)
9. Mertens, J., Neyman, A.: Stochastic games. International Journal of Game Theory 10(2), 53–66 (1981)
10. Papadimitriou, C., Tsitsiklis, J.: The complexity of optimal queueing network control (1999)
11. Golovin, D., Krause, A.: Adaptive submodularity: A new approach to active learning and stochastic optimization. CoRR abs/1003.3967 (2010)

12. Even-Dar, E., Kakade, S., Mansour, Y.: Experts in a Markov decision process. In: Advances in Neural Information Processing Systems 17: Proceedings of the 2004 Conference, p. 401. The MIT Press (2005)
13. Mannor, S., Shimkin, N.: The empirical bayes envelope and regret minimization in competitive markov decision processes. Mathematics of Operations Research, 327–345 (2003)
14. Even-Dar, E., Mannor, S., Mansour, Y.: Learning with global cost in stochastic environments. In: COLT: Proceedings of the Workshop on Computational Learning Theory (2010)
15. Takimoto, E., Warmuth, M.: Path kernels and multiplicative updates. The Journal of Machine Learning Research 4, 773–818 (2003)
16. Awerbuch, B., Kleinberg, R.: Online linear optimization and adaptive routing. Journal of Computer and System Sciences 74(1), 97–114 (2008)
17. Farias, D.P.D., Megiddo, N.: Combining expert advice in reactive environments. J. ACM 53, 762–799 (2006)
18. Fudenberg, D., Tirole, J.: Game theory. MIT Press (1991)
19. Blum, A., Mansour, Y.: From external to internal regret. Learning Theory, 621–636 (2005)
20. Celentani, M., Fudenberg, D., Levine, D., Pesendorfer, W.: Maintaining a reputation against a patient opponent. Econometrica 64, 691–704 (1996)
21. Impagliazzo, R., Paturi, R.: On the complexity of k-sat. Journal of Computer and System Sciences 62(2), 367–375 (2001)
22. Hastad, J.: Some optimal inapproximability results. Journal of the ACM (JACM) 48(4), 798–859 (2001)
23. Good, I.J.: Normal recurring decimals. Journal of the London Mathematical Society 1(3), 167 (1946)
24. Auer, P., Cesa-Bianchi, N., Freund, Y., Schapire, R.: Gambling in a rigged casino: The adversarial multi-armed bandit problem. In: FOCS, p. 322. IEEE Computer Society (1995)
25. Littlestone, N., Warmuth, M.: The weighted majority algorithm. In: Proceedings of FOCS, pp. 256–261 (1989)

The Cooperative Ballistic Missile Defence Game

Lanah Evers[1,2,3], Ana Isabel Barros[1,2], and Herman Monsuur[2,*]

[1] TNO - Defense, Safety and Security
[2] Netherlands Defence Academy
[3] Econometric Institute, Erasmus University Rotterdam
H.Monsuur@nlda.nl

Abstract. The increasing proliferation of ballistic missiles and weapons of mass destruction poses new risks worldwide. For a threatened nation and given the characteristics of this threat a layered ballistic missile defence system strategy appears to be the preferred solution. However, such a strategy involves negotiations with other nations concerning the use of their defence systems as part of the layered defence system. This paper introduces the Cooperative Ballistic Missile Defense Game, $\mathcal{CBMDG}$, to support the strategic negotiations between a threatened nation and the possible coalition nations. The model determines the assignment of ballistic missile interceptors to the coalition nations that minimizes the expected number of interceptors required to achieve the desired defence level in case of an attack. Simultaneously, it identifies the bargaining strength of each coalition of nations, in order to determine the compensation for participating in the layered defence system to protect the threatened nation.

1 Introduction

As stated by the US Missile Defence Agency and the Intelligence Community, one of the greatest threats facing the world today remains the increasing proliferation of ballistic missiles and weapons of mass destruction [6]. Ballistic Missiles (BMs) follow a ballistic trajectory, see Fig. 1, and can have long range (above 5500km). Moreover, they can carry explosive, nuclear, biological or chemical warheads, see [13]. BMs provide therefore the capability to launch an attack from a large distance (even inter-continental) and enable the projection of power both in a regional and strategic context. Currently, sophisticated ballistic missile technology is available on a wider scale to rogue nations. Therefore, developing an effective and efficient Ballistic Missile Defense strategy has become more relevant.

As shown in Fig. 1 the ballistic trajectory of a BM begins with an ascent phase till the apogee, followed by a descent phase. The ascent phase starts with the boost phase. The boost phase itself ends with the burn-out. During the boost phase detection and tracking of the BM is possible if the Ballistic Missile Defence (BMD) sensors are in close proximity to the BM launch location. After the boost phase, the BM starts with the mid-course phase in space towards its target, following a ballistic trajectory that is determined by the angle and velocity

---

* Corresponding author.

S.K. Das, C. Nita-Rotaru, and M. Kantarcioglu (Eds.): GameSec 2013, LNCS 8252, pp. 85–98, 2013.
© Springer International Publishing Switzerland 2013

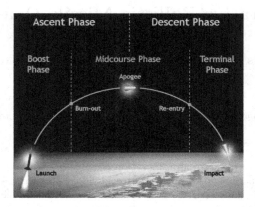

Fig. 1. BM trajectory

at burnout and gravity. Since the mid-course phase is the longest phase (about 75 %), several opportunities to destroy the incoming BM outside the Earth's atmosphere are possible. The last phase, the terminal phase is very short and begins once the missile reenters the atmosphere. It is the last opportunity to make an intercept before the BM reaches its target. An intercept during this phase is difficult due to the short intercept window which should have a little margin for error. Given the nature of the ballistic threat, a ballistic missile defence should not rely on a single defence barrier but on multiple defence layers placed at different sites forming a layered defence system. Such a layered defence system allows for more engagement opportunities and as such increases the probability that incoming BMs are intercepted. Moreover, as we will show, such a strategy yields a reduction on the average number of interceptors required to ensure the desired defence level. Therefore, for a threatened nation, it is important to identify among the nations that have ballistic missile defence capabilities, the coalition of nations that guarantees the desired defence level and at the same time requires the smallest average number of interceptors. In this setting, it is logical to assume that the threatened nation will make the required interceptors available to the coalition nations. On the other hand, from a coalition nation point a view, the fact that they are willing to counter attacks to the threatened nation might pose political and strategic risks for these nations. As such they might require some compensation during the negotiation phase. Such compensation can be, for instance, based on a fair share of the interceptor cost savings for the threatened nation. In order to support the negotiation process, this paper presents a game theoretical model consisting of two phases.

The first phase is an optimization phase used to determine the allocation of interceptor stockpiles to the different candidate cooperating nations. In this optimization phase we assume that the threatened nation has set a desired defence level. Since interceptors are costly munitions, when the intercept window allows for it, a shoot-look-shoot engagement option is usually preferred. A shoot-look-shoot consists of first engaging an incoming BM by an interceptor, followed by an assessment whether or not the BM has been successfully destroyed, and finally followed by a subsequent engagement only if the first engagement failed to destroy the incoming BM. Obviously such an engagement option yields a smaller expected number of interceptors usage then a salvo option where interceptors are simultaneously or sequentially fired in order to ensure kill or BM destruction, compare Washburn and Kress [14]. At the same time this engagement option may satisfy the desired defence level. However, for a single nation, the engagement time-window often does not allow to execute a shoot-look-shoot option on its own. Therefore, the threatened nation has to put in place a cooperative shoot-look-shoot strategy, allowing for more engagement opportunities. Next, the resulting optimal allocation of interceptors is input to the second phase of the model. The second phase is focused on the negotiation problem faced by the nation under threat of attack: finding the compensation to be provided to the cooperating nations given the increase in risk that they will face for supporting the layered defence of the threatened nation. The model computes the bargaining strength of each coalition of nations, in order to determine the compensation for joining the layered defence for the nation that is threatened by an attack with BM. The model falls in the class of *OR games*, see Borm et al. [2]. Related problems are addressed, among others, by Bloemen et al [1] that present an approach to determine a robust defence strategy for the location of BMD systems against ballistic missile threat, and Nguyen and Redding [9] who analyze the effectiveness of layered defense systems. In Menq et al [7] a multi-layered BMD system is modelled as a discrete Markov model, while in the area of air defence, Karasakal [4] addresses the problem of allocating air defense missiles to incoming air targets in order to maximize the air defense effectiveness of a naval task group.

The remainder of this paper is structured as follows. The optimization phase of the $CBMDG$ is defined and illustrated in Sect. 2. Next, in Sect. 3, the cooperative phase of the $CBMDG$ is defined as well as the modeling issues concerning how the nations could be compensated for their cooperation. In Sect. 4 some final remarks are drawn.

2   The Optimization Phase

2.1   Problem Setting

Consider a nation that is defining its ballistic defence strategy. The scale of the long range BM launchers provides an advantage in identifying the possible attack strategies of rogue nations using intelligence sources. Each of these different attack strategies, scenarios, define a specific shot line from a given launch location to a particular High Value Asset (HVA) of this nation. In order to counter this threat, the nation aims at achieving a predefined defence level L, which represents the overall probability of successfully intercepting a BM. Without loss of generality we will assume that nation 1 is the threatened nation. Without cooperation, nation 1 has to deal alone with these threat scenarios. On the other hand, by cooperating with other nations with similar defence capabilities,

$N \setminus \{1\} = \{n, n - 1, \ldots, 2\}$, a layered defence can be set in place. In this case, nation 1 will provide its coalition partners the required interceptors to be used in case of an attack. Of course, this layered defence concept relies on a network architecture to combine the different nations sensors and launchers into one missile defence system. This also makes possible the timely detection of incoming BM attacks.

The nations are indexed from n down to 1 to represent the order in which they are able to engage the BM, if necessary. In a *cooperative* shoot-look-shoot option, after the engagement of nation i, an assessment will be performed in order to determine if the next nation, $i - 1$, needs to take over and perform another engagement, $i = n, n - 1, \ldots, 2$. The probability of successfully intercepting the incoming BM, the so-called Probabilities of Interception (PIs), can be estimated for each cooperating nation and are based on several geographical and physical factors, like the maximum operational range of the BMD system and the trajectory, speed and altitude of the BM. A feasible solution to the optimization phase of the $\mathcal{CBMDG}$ consists of an assignment of M interceptors to the nations in N, $\pi = (m_n, m_{n-1}, \ldots, m_1)$ with $m_n + m_{n-1} + \ldots + m_1 = M$, that ensures that the defence level L is achieved.

2.2   Formal Description

The problem of the optimization phase of the $\mathcal{CBMDG}$ contains the following elements:

- A set N of nations including nation 1, $|N| \geq 2$. These are the nations that are able to engage a BM attacking nation 1.
- A set of PIs p_i, the probability that an interceptor launched from nation i will successfully intercept a BM attacking nation 1.
- The required minimum defence level L (set by nation 1) to be achieved.

Nations $N \setminus \{1\} = \{n, n - 1, \ldots, 2\}$ consider assisting nation 1 in its defence. A feasible solution to the optimization phase of the $\mathcal{CBMDG}$ consists of an assignment of interceptors to the nations in N, $\pi = (m_n, m_{n-1}, \ldots, m_1)$ that ensures the fulfillment of the required defence level L. In order to increase the probability of intercept and due to the trajectory characteristics of a BM, such an assignment will obey the following: nation n first launches a salvo of m_n interceptors. Only if this engagement appears to be unsuccessful, nation $n - 1$ launches a salvo of m_{n-1} interceptors, followed by nation $n - 2$ launching its salvo of m_{n-2} interceptors in case the engagement of nation $n - 1$ also appears to be unsuccessful, etc.

For the sake of simplicity we will consider the situation with one attack scenario. Denote the probability of one interceptor launched from nation i not being successful in intercepting the BM by $q_i = 1 - p_i$. The incoming BM attacking nation 1 will penetrate the layered defence if each of the nations intercept salvos $m_n, m_{n-1}, \ldots, m_1$ fail to intercept the BM. Since the probabilities are independent, the success of the assignment $\pi = (m_n, m_{n-1}, \ldots, m_1)$ therefore is given by:

$$L(\pi) = 1 - \prod_{i \in N} (q_i)^{m_i}. \tag{1}$$

In order for the assignment to be *feasible*, it has to satisfy the predefined defence level:

$$L(\pi) \geq L. \tag{2}$$

The expected number of interceptors that will be launched to intercept a BM attacking nation 1, using an assignment $\pi = (m_n, m_{n-1}, \ldots, m_1)$, with $m_i \geq 0$, is given by:

$$d(\pi) = m_n + (q_n)^{m_n} m_{n-1} + (q_n)^{m_n} (q_{n-1})^{m_{n-1}} m_{n-2} + \ldots + \left\{ \prod_{n \geq i > 1} (q_i)^{m_i} \right\} m_1. \tag{3}$$

Since nation 1 has provided interceptors to its coalition nations, $n, \ldots, 2$, these nations are authorized (by nation 1) to launch the $m_n, m_{n-1}, \ldots, m_2$ interceptors, if necessary. In this way, nation 1 ensures that its defence level is satisfied while the average number of interceptors used to intercept the incoming BM will be smaller than what would be needed if nation 1 would act alone (one defence layer).

For example, consider scenario 1 in Fig. 2, where the estimated PIs are given in Table 1. For scenario 1, we have $N = \{3, 2, 1\}$, for scenario 2 we have $N = \{2, 1\}$. For scenario 1, the assignment $(1, 1, 1)$ results in a defence level of $L(1, 1, 1) = 1 - 0.5 \cdot 0.4 \cdot 0.3 = 0.94$ while the expected number of interceptors launched in case of an attack is $d(1, 1, 1) = 1 + 0.5 \cdot 1 + 0.5 \cdot 0.4 \cdot 1 = 1.7$. The assignment $(0, 0, 3)$ has a defence level of $L(0, 0, 3) = 1 - 0.3^3 = 0.973$ but requires the use of 3 interceptors missiles: $d(0, 0, 3) = 3$. If the defence level would be set at 0.94, obviously the assignment $(1, 1, 1)$ requires in expectation less interceptors than would be needed if nation 1 is not assisted by the other nations.

Table 1. PIs per scenario

nation	PI scenario 1	PI scenario 2
3	0.5	0
2	0.6	0.75
1	0.7	0.8

2.3   The Cost Function and Constraints

The optimization phase of the $\mathcal{CBMDG}$ consists of determining the interceptor assignment π that minimizes the number of expected interceptors used, such that the required defence level L is satisfied:

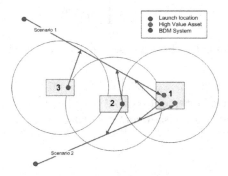

Fig. 2. Example of problem situation

$$\min \quad d(\pi), \tag{4}$$

$$\text{s.t.} \quad L(\pi) \geq L \tag{5}$$

$$\text{where } m_n, m_{n-1}, \ldots, m_1 \in \mathbb{N} \tag{6}$$

The optimal assignment $\pi^* = (m_n, m_{n-1}, \ldots, m_1)$ must be interpreted as follows: nation 1 provides m_i interceptors to nation i and authorizes to launch, if necessary, a salvo of m_i interceptors by nation i to intercept the BM, $i = n, \ldots, 1$. Whenever cooperation would not be possible, nation 1 would have to launch interceptors on its own in order to try to achieve the required defence level. Note that, in some cases, it might be too risky or even impossible to timely intercept the incoming BM in the last stage of the trajectory. In order to fulfill the required defence level with only one defence layer, the required number of interceptors $\overline{M}_L$ is such that:

$$1 - (q_1)^{\overline{M}_L} \geq L \tag{7}$$

where $d(0, \ldots, 0, \overline{M}_L) = \overline{M}_L$. So, for the optimal assignment π^*, we have $d(\pi^*) \leq \overline{M}_L$. Depending on the relative values of the interception probabilities, the sum of the assignments in π^* can be larger or smaller than $\overline{M}_L$.

As an aside, we mention that the cost function just described, but with fixed M and no defence level to be satisfied, is also known in the literature as the salvo size problem [3]. There it is proved that, in case the success probabilities of consecutive salvos are non-decreasing, the policy that minimizes the expected number of shots expended has a non-decreasing sequence of salvos. In our case, this means that for $\pi = (m_n, m_{n-1}, \ldots, m_1)$ with $m_i \leq m_{i+1}$ for $i = 1, \ldots, n-1$, it holds that $d(\pi) \leq d(\tau)$ for any permutation τ of π. In other words, most of the interceptors are located near nation 1.

2.4   Computing the Optimal Assignment

Let π_N be an assignment for any set of cooperating nations N and let $c(N, L)$ be defined by

$$c(N, L) = \min d(\pi_N) \tag{8}$$
$$\text{s.t. } L(\pi_N) \geq L. \tag{9}$$

The resulting optimal assignment π^* minimizes the expected number of interceptors used to successfully intercept an incoming BM, while at the same time it satisfies the desired defence level. It is now possible to relate the optimal layered defence if all nation in N do cooperate, to the optimal layered defence in case nation n, the first nation that is able to engage the BM attacking nation 1, does not cooperate:

Theorem 1. $c(N, L) = \min_{x \in \{0, 1, \dots, \overline{M}_L\}} \{x + q_n^x c(N \setminus n, 1 - \frac{1-L}{q_n^x})\}.$

Proof. ($\geq$) Let $\pi_N^* = (m_n, m_{n-1}, \dots, m_1)$ with $L(\pi_N^*) \geq L$ minimize $c(N, L)$. Let $\pi_{N \setminus n}^* = (m_{n-1}, \dots, m_1)$. Then $L(\pi_{N \setminus n}^*) \geq 1 - \frac{1-L}{q_n^x}$. So, clearly, $d(\pi_{N \setminus n}^*) \geq c(N \setminus n, 1 - \frac{1-L}{q_n^x})$. This implies that $c(N, L) = d(\pi_N^*) = m_n + q_n^{m_n} d(\pi_{N \setminus n}^*) \geq m_n + q_n^{m_n} c(N \setminus n, 1 - \frac{1-L}{q_n^{m_n}})$. ($\leq$) Take some x. Then $c(N \setminus n, 1 - \frac{1-L}{q_n^x}) = d(0, m_{n-1}, \dots, m_1)$ with $m_{n-1}, \dots, m_1$ such that $L(0, m_{n-1}, \dots, m_1) \geq 1 - \frac{1-L}{q_n^x}$. As $L(x, m_{n-1}, \dots, m_1) \geq L$, we have that $c(N, L) \leq x + q_n^x d(0, m_{n-1}, \dots, m_1) = x + q_n^x c(N \setminus n, 1 - \frac{1-L}{q_n^x})$. □

In case all interception probabilities are equal, this result may be used as an efficient dynamic programming approach: one has to keep record of $c(S, \frac{L}{q^k})$ for $k = 0, 1, \dots, \overline{M}$ and $S = \{1\}, \{2, 1\}, \{3, 2, 1, \}, \dots, \{n-1, n-2, \dots, 1\}, N$. Unfortunately, it also indicates that for distinct interception probabilities a dynamic programming approach will not work: for example, if all probabilities differ, one has to compute $c(\{1\}, L)$ for as many defence levels L as there are assignments. In that case we will have to use a suitable heuristic, or, in case $|N|$ is not large, complete enumeration. A good starting solution for a heuristic can be found using the following function $e(N, M)$ that does not take the defence level into account, but fixes the number of interceptors used:

Let $e(N, M)$ be defined by

$$e(N, M) = \min d(\pi_N) \tag{10}$$
$$\text{s.t. } \sum_{i \in N} m_i = M. \tag{11}$$

We then have the following result, the proof of which is similar to that of Th. 1:

Theorem 2. $e(N, M) = \min_{x \in \{0, 1, \dots, M\}} \{x + q_n^x e(N \setminus n, M - x)\}.$

Th. 2 can be used to fill a table T with triple-entries $T_{N,M} = (e(N, M) /$ assignment π / realized defence level), row by row, starting with row 1, for $N = \{1\}, \{2, 1\}, \{3, 2, 1\}, \ldots \{n, n - 1, \ldots, 1\}$ and $M = 1, 2, \ldots$. Regarding the entries $e(N, M)$ in T, we note that this value decreases when more nations join the coalition, and increases with increasing M. The first assertion follows directly from Th. 2 (take $x = 0$); for the second one, we note that $e(N, M) = m_n + q_n^{m_n} m_{n-1} + \ldots + \left\{ \prod_{n \geq i > 1} (q_i)^{m_i} \right\} m_1 > m_n + q_n^{m_n} m_{n-1} + \ldots + \left\{ \prod_{n \geq i > 1} (q_i)^{m_i} \right\} (m_1 - 1) \geq e(N, M - 1)$, where we assume (w.l.o.g.) that $m_1 > 0$.

Using Th. 2 we compute Table 2. Then take the assignment π that satisfies the defence level and has the lowest value of $d(\pi)$. In case there are multiple optima, we choose the one with the highest defence level. This assignment then may serve as a starting solution for a heuristic. For example, if L is set at 0.95, the starting assignment would be $\pi = (0, 1, 2)$, needing $M = 3$ interceptors, while for $L = 0.97$, the starting assignment is $\pi = (1, 1, 2)$ with $M = 4$. Note that in both cases $\overline{M}_L = 3$.

Table 2. Computation of $(e(N, M)$ / assignment π / realized defence level) for Scenario 1 of Figure 2, using the PIs from Table 1

	$M = 1$	$M = 2$	$M = 3$	$M = 4$
$N = \{1\}$	$1/(0, 0, 1)/0.70$	$2.0/(0, 0, 2)/0.91$	$3.0/(0, 0, 3)/0.97$	$4.0/(0, 0, 4)/0.99$
$N = \{2, 1\}$	$1/(0, 0, 1)/0.70$	$1.4/(0, 1, 1)/0.88$	$1.8/(0, 1, 2)/0.96$	$2.2/(0, 1, 3)/0.99$
$N = \{3, 2, 1\}$	$1/(0, 0, 1)/0.70$	$1.4/(0, 1, 1)/0.88$	$1.7/(1, 1, 1)/0.94$	$1.9/(1, 1, 2)/0.98$

As a heuristic, we propose a simple local improvement procedure. First of all, we start with a feasible assignment. At each step, construct 3 possible neighbors for the current assignment $\pi = (m_n, m_{n-1}, \ldots, m_1)$: (1) Choose i, j with $i \neq j$ and $m_j > 0$. Define $m'_i = m_i + 1, m'_j = m_j - 1$; (2) Choose i. Define $m'_i = m_i + 1$; (3) Choose j with $m_j > 0$. Define $m'_j = m_j - 1$. From the set of neighbors and the current assignment, choose the assignment π that satisfies the defence level and has lowest value of $d(\pi)$. In our example, the starting assignments already are optimal assignments. So, if we consider the case $L = 0.97$, the expected total number of interceptors required to intercept an incoming BM, decreases from 3 in a non-cooperative setting to 1.9 in the optimal cooperative strategy. In the next section we will show how to derive a fair compensation for the coalition nations.

3   The Cooperative Phase

In the previous section we presented an approach for the threatened nation to determine the best interceptor assignment strategies. Given a defence level L, it shows how to reduce the expected number of interceptors needed in case of an attack. Clearly, the threatened nation profits from this cooperation. As this nation provides the interceptors to its partners, it may even decide not to share this profit (intercept cost savings) with them. Other nations may reason that the threatened nation needs their cooperation to implement the layered defence. During the negotiations, these nations may therefore claim a fair share of the profit, which is the interceptor cost savings. By comparing the different coalitions in terms of interceptor cost savings, we are able to derive a fair allocation of the total savings of the threatened nation. For this, we will use the notion of a transferable utility (TU) game: A transferable utility (TU) game is a pair (N, v), where $v : 2^N \to \mathbb{R}$ with $v(\emptyset) = 0$. The function v is called the characteristic function: for $S \subset N$, $v(S)$ is the value of the coalition S. For detailed information regarding the various game theoretic concepts that we will use, we refer to Maschler et al. [5].

3.1   The Interceptor Savings Game

We showed how, given a defence level L, cooperation between nations in missile defence reduces the expected number of interceptors needed in case of an attack. To obtain the optimal solution to this OR-problem, all nations $i \in N$ that were able to engage the BM were taken into consideration. In order to define the cooperative phase of the $\mathcal{CBMDG}$, the minimum expected number of interceptors needed has to be defined for every coalition $S \subseteq N$ that contains nation 1. For this, we can use the results of the previous section. In Fig. 2 we may, for example, consider $S = \{3, 1\}$. We may derive that (with $L = 0.97$) we have two optimal assignments $(1, 3)$ and $(2, 2)$, both with expected value (or costs in terms of interceptors) of 2.5. As the defence level of the first assignment is higher, we take $\pi^* = (1, 3)$. For coalition $\{3, 1\}$, we therefore define $c(\{3, 1\}, 0.97) = 2.5$. This may be done for any coalition S, where the optimal assignment will be denoted by π_S^*. This gives the following definition of $c(S, L)$, generalizing the definition of $c(N, L)$ of the previous section:

$$c(S, L) = \begin{cases} d(\pi_S^*) & \text{if } S \ni 1, \\ 0 & \text{elsewhere.} \end{cases} \tag{12}$$

Since we assume only nation 1 to be subjected to a BM attack, we have strictly positive costs if and only if nation 1 is in the coalition. This gives rise to the cooperative phase of the $\mathcal{CBMDG}$, which is stated in terms of, what we will call, an interceptor savings game.

The interceptor savings game is defined by

$$v_{MD}^L(S) = \sum_{i \in S} c(\{i\}, L) - c(S, L), \text{for each } \emptyset \neq S \subset N \tag{13}$$

and

$$v_{MD}^L(\emptyset) = 0. \tag{14}$$

This equation shows that the savings by coalition S is the difference between the sum of savings each member of the coalition S can achieve on its own, minus the savings the coalition S, acting as one, can achieve. As without nation 1 involved in the coalition, no interceptor savings can be established, the savings boil down to $v_{MD}^L(S) = \overline{M}_L - c(S, L)$, for each $\emptyset \neq S \subset N$.

For the example described in the previous section we thus have the following game:

Table 3. Cooperative game of $\mathcal{CBMDG}$, with $|N| = 3, L = 0.97$ ($\overline{M}_L = 3$)

S	{1}	{2}	{3}	{2,1}	{3,1}	{3,2}	{3,2,1}
π_S^*	(0,0,3)	N.A.	N.A.	(0,1,3)	(1,0,3)	N.A.	(1,1,2)
$c(S, L)$	3	0	0	2.2	2.5	0	1.9
$v_{MD}^L(S)$	0	0	0	0.8	0.5	0	1.1

This game has some interesting properties; it is monotone and super additive: An arbitrary game v is **monotone** if $v(S) \leq v(T)$ if $S \subset T$. For our interceptor savings game with $S \subset T$ we clearly have $min_{\pi \in \pi_S} d(\pi) \geq min_{\pi \in \pi_T} d(\pi)$. We thus have $v_{MD}^L(S) = \overline{M} - c(S) \leq \overline{M} - c(T) = v_{MD}^L(T)$, which proves monotonicity of v_{MD}^L. A game v is **super additive** if $v(S \cup T) \geq v(S) + v(T)$ if $S \cap T = \emptyset$. In the $\mathcal{CBMDG}$ game with $S \cap T = \emptyset$ at least one of the coalitions S or T does not contain player 1. By definition a coalition without player 1 has no interceptor savings. Because of that, and because of monotonicity, the game v_{MD}^L is super additive. A game v is **convex** if $v(S \cup \{i\}) - v(S) \leq v(T \cup \{i\}) - v(T)$ if $S \subset T \subset N \setminus \{i\}$. Our game is not convex. To see this we refer to the example game given in Table 3. For $S = \{1\}, T = \{2, 1\}$ and $i = 3$ we have $v_{MD}^L(S \cup \{i\}) - v_{MD}^L(S) = v_{MD}^L(\{3, 1\}) - v_{MD}^L(\{1\}) = 0.5 - 0 > v_{MD}^L(T \cup \{i\}) - v_{MD}^L(T) = v_{MD}^L(\{3, 2, 1\}) - v_{MD}^L(\{2, 1\}) = 1.1 - 0.8.$

3.2 Allocating the Savings

During the negotiation phase, the risk incurred by the nations assisting nation 1 plays a role. Here, a compensation scheme might be used to mitigate these risks. This compensation can be based on a fair share of the value $v_{BM}^L(N)$, the reduction in the expected number of interceptors required to defend nation 1 established by the grand coalition N. Such an allocation, $x = (x)_i$ with $\sum_i x_i = v_{BM}^L(N)$, will have to take into account the intercept savings $v_{BM}^L(S)$ of all possible coalitions S that are able to assist nation 1 in its defence against a BM attack: Each $v_{BM}^L(S)$ can be seen as the claim of coalition S on part of the total value $v_{MD}^L(N)$. As each interceptor is equivalent to a monetary value, say K, in order to compensate the nations for cooperating in the layered defence, a fair compensation would be that nation 1 would provide $x_i K$ to nation i, $i = n, n - 1, \ldots, 2$.

Basic Requirements. Two generally accepted requirements for an allocation $x = (x)_i$ for an arbitrary game v are

- *Efficiency*: $\sum_{i \in N} x_i = v(N)$;
- *Individual rationality*: $x_i \geq v(\{i\})$ for all $i \in N$.

An additional requirement, if possible, is that x is element of the so-called core of the game. The **core** of a game v is defined as

$$C(v) = \left\{ x \in \mathbb{R}^N \mid \sum_{i \in N} x_i = v(N), \sum_{i \in S} x_i \geq v(S) \text{ for all } S \subseteq N \right\}. \quad (15)$$

These allocations are stable with respect to coalitional deviations. This means that no coalition S can rightfully object to x, as the total value allocated to S, $\sum_{i \in S} x_i$ is at least what it can obtain by splitting off, $v(S)$. In general, we have the important result that the core of the game v_{MD}^L is non-empty:

Theorem 3. *The core of the interceptor savings game v_{MD}^L is non-empty.*

Proof. A possible imputation is $(x_n, x_{n-1}, \ldots, x_1) = (0, \ldots, 0, v_{MD}^L(N))$. For any coalition S that includes nation 1, we have $\sum_{i \in S} x_i = v_{MD}^L(N)$, which is, by monotonicity of v_{MD}^L, at least $v_{MD}^L(S)$. Moreover, if S does not include nation 1, $v_{MD}^L(S) = 0$, which concludes the proof. $\square$

To illustrate how this property applies to the $\mathcal{CBMDP}$ game, we will describe the core of the game given in Table 3. For the example we have a core consisting of all $x \in \mathbb{R}^3$ that satisfy $\{x_1 + x_2 + x_3 = 1.1, x_1 + x_2 \geq 0.8, x_1 + x_3 \geq 0.5\}$, which implies $x_3 \leq 0.30$ and $x_2 \leq 0.60$. The core can also be written as the convex hull of the vectors $(x_3, x_2, x_1) = (0, 0, 1.1), (x_3, x_2, x_1) = (0, 0.6, 0.5), (x_3, x_2, x_1) = (0.3, 0.6, 0.2)$ and $(x_3, x_2, x_1) = (0.3, 0, 0.8)$. The center of this set is the allocation $(0.15, 0.30, 0.65)$ for nations 3, 2 and 1. This allocation means that nation 1 has to give nations 2 and 3 a share 0.45 of the total savings 1.1.

The Shapley Value. Instead of presenting a set of allocations from which to choose, one may also apply solutions of cooperative game theory to find a point solution, hopefully in the core of the game. A well-known, classic and often used allocation mechanisms is the Shapley value. The Shapley value of a game v is defined by $\phi_i(v) = \sum_{S \subseteq N \setminus \{i\}} \frac{|S|!(|N|-|S|-1)!}{|N|!} (v(S \cup \{i\}) - v(S))$. For our example, this results in the following Shapley value: $\phi(v_{MD}^L) = (0.183, 0.333, 0.583)$ for nations 3,2 and 1. This allocation is in the core of v_{MD}^L. Unfortunately, in general the Shapley value $\phi(v_{MD}^L)$ is not always in the core of v_{MD}^L. Take, for example, three nations with all interception probabilities equal to 0.7 with $L = 0.99$.

The τ-value. The τ-value, introduced by Tijs [12], is defined for so-called *compromise admissible* games: Define for a game

- $M_i(v) = v(N) - v(N \setminus \{i\})$, and
- $m_i(v) = \max_{S:S \ni i} \{v(S) - \sum_{j \in S \setminus \{i\}} M_j(v)\}$.

The value $M_i(v)$ indicates an upper bound a player reasonable can demand within negotiations regarding the allocation of $v(N)$. Likewise, $m_i(v)$ indicates a kind of minimal value nation i can achieve by satisfying all other nations in a coalition by giving them their utopia demands $M_j(v)$.

A game v is called compromise admissible if

$$m(v) \leq M(v) \text{ and } \sum_{i \in N} m_i(v) \leq v(N) \leq \sum_{i \in N} M_i(v). \tag{16}$$

For our game, one may easily verify that $M_1(v_{MD}^L) = v_{MD}^L(N)$. Because of monotonicity, we have $M_i(v_{MD}^L) \geq 0$ for $i \neq 1$ and also $m_1(v_{MD}^L) \leq v_{MD}^L(N)$, while $m_i(v_{MD}^L) = 0$ for $i \neq 1$. So our game is compromise admissible, and we may define the following allocation rule:

The τ-value is defined by

$$\tau(v_{MD}^L) = \alpha M(v_{MD}^L) + (1 - \alpha)m(v_{MD}^L), \tag{17}$$

with α such that $\sum_{i \in N} \tau_i(v) = v_{MD}^L(N)$. It balances between the minimal rights and the (utopia) vector $M(v_{MD}^L)$. For our example, this results in $\alpha = 0.5$ and a τ-value of $\alpha(0.3, 0.6, 1, 1) + (1 - \alpha)(0, 0, 0.2) = (0.15, 0.30, 0.65)$ for nations 3, 2 and 1. It is at the center of the core of our game.

The Nucleolus. As a final allocation mechanism, we consider an allocation rule that is defined for games for which there exists allocations that are individual rational and efficient, as is the case for our game v_{MD}^L. The rationale behind this allocation rule is based on the excess $E(S, x)$ of a coalition with respect to x: $E(S, x) = v_{MD}^L(S) - \sum_{i \in S} x_i$. The excess is a measure for the dissatisfaction of the coalition S with respect to the proposed allocation x. Then $\theta(x)$ is the excess vector, in (weakly) decreasing order, of all possible coalitions. The nucleolus $n(v_{MD}^L)$, introduced by Schmeidler [11], is the unique allocation, element of the core, that minimizes the maximum dissatisfaction, meaning that the nucleolus is the lexicographic minimum of the set $\theta(x)$.

For example, take $x = (0.15, 0.30, 0.65)$ for the game of Table 3, we obtain as excess vector $\theta(x) = (0.0, 0.0, -0.15, -0.15, -0.30, -0.30, -0.45, -0.65)$. In our case, $n(v_{MD}^L) = \tau(v_{MD}^L) = (0.15, 0.30, 0.65)$ for the nations 3, 2 and 1, again the allocation at the center of the core. This shows that both the τ-value and the nucleolus are suitable allocation mechanisms for our problem.

3.3   Big Boss Game

An arbitrary monotonic game v is called a big boss game if there is one player, denoted by i^*, satisfying the following two conditions (see Muto et al. [8]):

- $A : v(S) = 0$ if $i^* \notin S$
- $B : v(N) - v(S) \geq \sum_{i \in N \setminus S} M_i(v)$ if $i^* \in S$.

A implies that one player i^* is very powerful, i.e., coalitions not containing i^* cannot get anything. Condition B implies that for every coalition $N \setminus S$ not containing i^*, its contribution to the grand coalition (which is $v(N) - v(S)$) is at least as large as the sum of the individual utopia demands. Hence, the weak players may increase their influence by forming coalitions. Economic applications of big boss games include indivisible good market with one seller and many buyers, and bankruptcy problems with one big claimant.

In big boss games the core of the game is equal to

$$C(v) = \left\{ x \in \mathbb{R}^N \mid \sum_{i \in N} x_i = v(N), m_i(v) \leq x_i \leq M_i(v) \text{ for all } i \in N \right\}. \quad (18)$$

We note that for any game with $|N| \leq 3$ and non-empty core, this same expression holds, Quant et al. [10]. For general big boss games, the τ-value coincides with the nucleolus and both are at the center of the core, while the Shapley value differs from the τ-value and nucleolus in case the game is not convex.

For an arbitrary game v, coalition S, $j \in S$ and utopia values $M_i(v, S) = v(S) - v(S \setminus \{i\})$, we have:

Theorem 4. *Suppose that for each coalition $S \ni i^*$ and each $i \in S$ we have that $M_i(v, S) \geq M_i(v, S \cup j)$. Then, if also condition A holds, v is a big boss game.*

Proof. Note that $v(N) - v(S) = (v(N) - v(N \setminus j_n)) + (v(N \setminus j_n) - v(N \setminus \{j_n, j_{n-1}\})) + \ldots + (v(S \cup j_{n-s}) - v(S))$, for $\{j_n, j_{n-1}, \ldots j_{n-s}\} = N \setminus S$. Each term in this sum is equal to $M_j(v, T)$ for some $j \in N \setminus S$ and $S \subset T \subset N$. By assumption, $M_j(v, T) \geq M_j(v, N) = M_j(v)$, proving our claim. $\qquad \square$

For our game v_{MD}^L, the condition in Theorem 4 seems very plausible: the effect of nation i leaving the grand coalition N can more easily be compensated by the remaining nations than if nation i leaves a smaller coalition S of nations. We therefore conjecture that the game v_{MD}^L of $\mathcal{CBMDG}$ belongs to the class of big boss games. For the $\mathcal{CBMDP}$ game, player 1 fulfills the role of i^* used in this definition.

4 Conclusions

The Cooperative Ballistic Missile Defense Game ($\mathcal{CBMDG}$) aims at identifies the bargaining strength of each coalition of nations, based on the optimal assignment of interceptors to the coalition nations. This is used to determine the compensation for each coalition nation for supporting the layered defence system that ensures the required defence level in case of an attack. As such, first the model determines the optimal assignment of the interceptor stocks to nations that minimizes the expected number of interceptors required to achieve a predetermined defense level in case of an attack. Next, the model identifies the benefits of all possible coalitions in a cooperative game and provides possible divisions of the benefits among the cooperating nations. Our approach can easily be adapted to the situation where we have more than one threat scenario and more than just one nation facing the possibility of a BM attack. We believe that the approach presented in this paper provides insights for strategic negotiations when considering a layered defence system, as it identifies the bargaining strengths of each of the participating nations.

References

1. Bloemen, A.A.F., Evers, L., Barros, A.I., Monsuur, H., Wagelmans, A.: A robust approach to the missile defense location problem. The International Journal of Intelligent Defence Support Systems 4(2), 128–147 (2011)
2. Borm, P., Hamers, H., Hendrickx, R.: Operations research games: A survey. TOP 9, 139–216 (2001)
3. Gould, J.: On effient salvo ploicies. Naval Research Logistics Quarterly 31, 159–162 (1984)
4. Karasakal, O.: Air defense missile-target allocation models for a naval task group. Computers and Operations Research 35(6), 1759–1770 (2008)
5. Maschler, M., Solan, E., Zamir, S.: Game theory. Cambridge University Press (2013)
6. MDA: The treat, http://www.mda.mil/system/threat.html (last visited July 18, 2013); Missile Defence Agency
7. Menq, J., Tuan, P., Liu, T.: Discrete Markov ballistic missile defense system modeling. European Journal of Operational Research 178(2), 560–578 (2007)
8. Muto, S., Nakayama, M., Potters, J., Tijs, S.: On big boss games. The Economic Studies Quarterly 39, 303–321 (1988)
9. Nguyen, B., An, D. R.: analytical optimal multi-layer engagement model with incomplete damage assessmet. Technical report, CORA Research Note 9814, 1 CAD/CANR HQs, Winnipeg, Canada (1998)
10. Quant, M., Borm, P., Reijnierse, J., van Velzen, S.: Compromise stable tu-games. Discussion Paper Center 2003-55 (2003)
11. Schmeidler, D.: The nucleolus of a characteristic function game. SIAM Journal of Applied Mathematics 17, 1163–1170 (1969)
12. Tijs, S.: Bound on the core and the τ-value. Game Theory and Mathematical Economics 39, 123–132 (1981)
13. TNO: Missile defence, an overview. Report TNO (2012)
14. Wasburn, A., Kress, M.: Combat modeling. Springer (2009)

Security Games for Virtual Machine Allocation
in Cloud Computing

Yi Han[1], Tansu Alpcan[2], Jeffrey Chan[1], and Christopher Leckie[1]

[1] Department of Computing and Information Systems
[2] Department of Electrical and Electronic Engineering
University of Melbourne, Melbourne, Australia
`andrew.hanyi@gmail.com`,
`{tansu.alpcan,jeffrey.chan,caleckie}@unimelb.edu.au`

Abstract. While cloud computing provides many advantages in accessibility, scalability and cost efficiency, it also introduces a number of new security risks. This paper concentrates on the co-resident attack, where malicious users aim to co-locate their virtual machines (VMs) with target VMs on the same physical server, and then exploit side channels to extract private information from the victim. Most of the previous work has discussed how to eliminate or mitigate the threat of side channels. However, the presented solutions are impractical for the current commercial cloud platforms. We approach the problem from a different perspective, and study how to minimise the attacker's possibility of co-locating their VMs with the targets, while maintaining a satisfactory workload balance and low power consumption for the system. Specifically, we introduce a security game model to compare different VM allocation policies. Our analysis shows that rather than deploying one single policy, the cloud provider decreases the attacker's possibility of achieving co-location by having a policy pool, where each policy is selected with a certain probability. Our solution does not require any changes to the underlying infrastructure. Hence, it can be easily implemented in existing cloud computing platforms.

Keywords: Cloud computing, co-resident attack, game theory, virtual machine allocation policy.

1   Introduction

In cloud computing environments, when a user requests to start a new machine, in most cases the allocated machine is not an entire physical server, but only a virtual machine (VM) running on a specific host. This is enabled by hardware virtualisation technologies [1] such as Hyper-V, VMWare, and Xen, so that multiple VMs of different users can run on the same physical server and share the same underlying hardware resources. While this increases the utilisation rate of hardware platforms, it also introduces a new threat: although in theory, VMs running on the same server (i.e., co-resident VMs) should be logically isolated from each other, malicious users can still circumvent the logical isolation, and obtain sensitive information from co-resident VMs [2].

S.K. Das, C. Nita-Rotaru, and M. Kantarcioglu   (Eds.): GameSec 2013, LNCS 8252, pp. 99–118, 2013.
© Springer International Publishing Switzerland 2013

It has been shown that this new co-resident attack (also known as a co-residence attack, or co-location attack) is indeed feasible in real cloud platforms. By building different kinds of side channels, the attacker can extract a range of private statistics, from the coarse-grained [2], like the victim's workload and traffic rate, to the fine-grained [3], such as cryptographic keys.

Most of the previous work has focused on the side channels, and proposed to solve the problem either by mitigating the threat of side channels [4-7], or designing a new architecture for the cloud system to eliminate side channels [8, 9]. However, few of these are practical for current commercial cloud platforms as they require significant changes to be made. In this paper, we address this issue with a novel decision and game-theoretic approach.

The co-resident attack that we are discussing comprises two steps. Before the attacker can extract any useful information from the victim, they first need to co-locate their own VMs with the target VM. Experiments in [2] show that the attacker can achieve a surprisingly high efficiency rate of 40% (i.e., 4 out of 10 malicious VMs launched by the attacker will be co-resident with the target(s)). This observation motivates us to study practical methods for decreasing this efficiency rate of co-resident attacks. For cloud providers, one important factor they can control that will influence the efficiency rate is the VM allocation policy. Hence, we compare different VM allocation policies in cloud computing, and investigate the impact of these policies on the efficiency of achieving co-residence.

When cloud providers decide on their VM allocation policies, workload balance and power consumption are used as additional important criteria. Therefore, we carry out a comparative study of four basic VM allocation policies using a game theoretic approach, namely: choosing the server (1) with the least number of VMs, (2) with the most number of VMs, (3) randomly, and (4) on a round robin basis. These policies form the basis of most policies used in real cloud systems. Specifically, we model this as a two-player security game, where the attacker's goal is to maximize the attack efficiency, while the defender (cloud provider) aims to minimize it on the premise of balancing the workload and maintaining low power consumption.

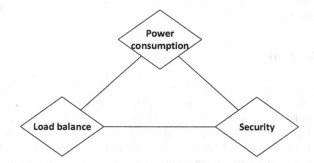

Fig. 1. Different focuses of VM allocation policies

Our **contributions** in this paper include: (1) we introduce a novel game theoretic approach to the problem of defending against co-resident attacks; (2) we model different VM allocation policies using zero- and non-zero-sum security games; (3) we perform extensive numerical simulations to develop and evaluate a practical solution for mitigating the threat of co-resident attacks; and (4) we show that in terms of minimising attack efficiency, the deterministic VM allocation policy behaves the worst, while a mixed policy outperforms any single policy.

The rest of the paper is organized as follows. In Section 2, we further introduce the focus of our study – the co-resident attack, and give our problem statement. In Section 3, we propose our game model. A detailed analysis and comparison of the different VM allocation policies is presented in Section 4, while Section 5 concludes the paper and gives directions for our future work.

2   Background and Problem Statement

In this section, we first introduce the co-resident attack in detail. We discuss how to achieve co-residence, the security risks, and potential countermeasures. We then define the problem that we aim to solve in this paper.

2.1   Methods to Achieve Co-residence

In order to achieve co-residence, i.e., locate their own VM and the victim on the same host, the attackers have several options.

1. The most straightforward approach is to use a brute-force strategy: start as many VMs as possible until co-residence is achieved.
2. Experiments in [2] show that in the popular Amazon EC2 cloud, there is strong sequential and parallel locality in VM placement, which means if one VM is terminated right before another one is started, or if two VMs are launched almost at the same time, then these two VMs are often assigned to the same server. As a result, the attacker can increase the possibility of co-locating their VM with the targets if they are able to trigger the victim to start new VMs, and then launch their own VMs after that.

2.2   Potential Security Risks

After co-residence is achieved, there are a number of potential security risks:

1. *VM workload estimation* – In [2], the authors adopt the Prime+Probe technique [10, 11] to measure cache utilisation. The basic idea is that the execution time of the cache read operation is heavily influenced by the cache utilisation. Hence, by performing intensive read operations and then measuring the execution time, the attacker can infer the cache usage, which also indicates the target VM's workload.
2. *Web traffic rate estimation* – Similarly, the attacker performs cache load measurements on the co-resident VM, and at the same time, they send HTTP requests from

non-co-resident VM(s) to the victim. Experimental results show that there is a strong correlation between the execution time of the cache operation and the HTTP traffic rate. In other words, the attacker is able to obtain information about the web traffic rate on the co-resident VM. This can be useful information if the victim is a corporate competitor.

3. *Private key extraction* – In [3], the authors demonstrate that it is possible to extract cryptographic keys by using cross-VM side channels. In particular, they show how to overcome the following challenges: regaining the control of the physical CPU with sufficient frequency to monitor the instruction cache, filtering out hardware and software noise, and determining if an observation is from the target virtual CPU or not due to the core migration.

In addition, there are a number of papers discussing how to build side channels between co-resident VMs in cloud computing environments [12-17].

Note that there are other types of denial-of-service attacks where the attacker does not care who the victim is, and only aims to obtain an unfair share of resources from the physical server, so that co-resident VMs will experience a degradation of quality of service [18-21]. This type of attack is outside the scope of our research.

2.3   Possible Countermeasures

Previous studies have proposed a number of possible defence methods, which can be broadly classified into the following three categories:

1. *Preventing the attacker from verifying co-residence* – In current cloud computing platforms, it is relatively easy to check if two VMs are on the same host. For example, by performing a *TCP traceroute* the attacker can obtain a VM's Dom0 IP address (where Dom0 is a privileged VM that manages other VMs on the host). If two Dom0 IP addresses are the same, the corresponding VMs are on the same server. If we can prevent the attacker from verifying whether their own VM and the target victim's VM are on the same physical machine, then they will not be able to launch further attacks. However, there are a number of alternative methods to verify co-residence that do not rely on network measurement [2], even though they are more time-consuming. Therefore, it is difficult, if not impossible, to prevent all these methods.

2. *Securing the system to prevent sensitive information of a VM from being leaked to co-resident VMs* – Countermeasures against side channels have already been extensively studied, including (1) mitigating the threat of timing channels by eliminating high resolution clocks [5], or adding latency to potentially malicious operations [6], and (2) redesigning the architecture for cloud computing systems [8, 9]. Nevertheless, these methods are usually impractical for current commercial cloud platforms due to the substantial changes required.

3. *Periodically migrating VMs* – The authors in [22, 23] propose to solve the problem by periodically migrating VMs. The number of chosen VMs and hosts are decided based on game theory. In addition, they also discuss how to place VMs in order to minimize the security risk. However, frequently migrating VMs may increase

power usage and lead to load imbalances, which are undesirable from the cloud provider's perspective.

2.4   Problem Statement

In this paper, we aim to find a solution for defending against the co-resident attack. In order to make our proposed method practical, we assume that the cloud providers (1) do not have any prior knowledge of the attacker; (2) will not apply any additional security patches, and (3) will not have access to an effective detection mechanism. Therefore, the question is under these assumptions, how can they mitigate the threat of co-resident attacks, while maintaining a reasonably high workload balance and low power consumption for the system?

3   Proposed Game Model

We consider this problem as a static game between the attacker and the defender (cloud provider). In this section, we first define the attack and defence scenarios, and then propose our game model.

3.1   Attack Scenarios and Metrics

Before giving the formal description of the game model, we first define the attack scenario: in a system of N servers, there are k (separate) attackers $\{A_1, A_2, ..., A_k\}$, each controlling one single account. No limit on the number of VMs is enforced for an account, which means the attackers can start as many VMs as frequently as they want (in practice, the attackers maybe restricted by costs and other factors). The target for attacker A_i is the set of VMs started by legitimate user L_i, i.e., $Target(A_i) = \sum_t VM(L_i,t) = \{VM_{i1}, VM_{i2}, ..., VM_{iT_i}\}$, where $VM(L_i,t)$ is the set of VMs started by L_i at time t. During one attack started at time t, A_i will launch a number of VMs, $VM(A_i,t)$. Let $SuccVM(A_i,t)$ denote the VMs of attacker A_i that co-locate with at least one of the targets, i.e., $SuccVM(A_i,t) = \{v \mid v \in VM(A_i,t), Servers(\{v\}) \subseteq Servers(Target(A_i))\}$, where $Servers(\{a\ set\ of\ VMs\})$ is the set of servers that host the set of VMs. Similarly, let $SuccTarget(A_i,t)$ denote the VMs of the target user L_i that are co-located with at least one VM of the attacker A_i, i.e., $SuccTarget(A_i,t) = \{u \mid u \in Target(A_i), Servers(\{u\}) \subseteq Servers(VM(A_i,t))\}$. Then an attack is considered as successful if $SuccVM(A_i,t)$ and $SuccTarget(A_i,t)$ are non-empty, i.e., at least one of the attacker's VMs is co-located with at least one of the target VMs.

In order to further measure the success for one attack, two definitions are introduced:

(1) *Efficiency*, which is defined as the number of malicious VMs that are successfully co-located with at least one of the T_i targets, divided by the total number of VMs launched during this attack, i.e.,

$$Efficiency(VM(A_i,t)) = \frac{|SuccVM(A_i,t)|}{|VM(A_i,t)|} \qquad (1)$$

(2) *Coverage*, which is defined as the number of target VMs co-located with malicious VMs started in this attack, divided by the number of targets T_i, i.e.,

$$Coverage(VM(A_i,t)) = \frac{|SuccTarget(A_i,t)|}{|Target(A_i)|} \qquad (2)$$

Table 1. Definitions of symbols used

Name	Definition
Target(A_i)	The target set of VMs that A_i intends to co-locate with. $|Target(A_i)| = T_i$
VM(L_i,t)	The set of VMs started by user L_i at time t
SuccTarget(A_i,t)	A subset of *Target(A_i)* that co-locates with at least one VM from *VM(A_i,t)*
SuccVM(A_i,t)	A subset of *VM(A_i,t)* that co-locates with at least one of the T_i targets
Servers({a set of VMs})	Servers that host the set of VMs

3.2   Defence Policies

Recall that the attack we consider comprises two steps. First, the attacker has a clear set of targets, and they will try different methods to co-locate their own VMs with the targets. Second, after co-residence is achieved, the attacker will use various techniques to obtain sensitive information from the victim.

Because of the assumptions we made in Section 2.4, any solution that focuses on the second step, and any attempt to identify the attacker or their VM requests are infeasible. Therefore, one of the remaining options for the defender is to find an allocation policy that minimizes the overall possibility of achieving co-residence.

For simplicity reasons, we only consider four policies, namely: choosing the server (1) with the least number of VMs ("Least VM"), (2) with the most number of VMs ("Most VM"), (3) randomly ("Random"), and (4) on a round robin basis ("Round Robin"). The reason why we choose these policies is that the first two are two extremes in the policy spectrum, with one spreading the workload and the other one concentrating the workload, while the other two are the most straightforward policies. In addition, most real cloud VM allocation policies are based on these four policies.

We can classify these policies into two main categories: deterministic (Policy 4), and stochastic (Policies 1, 2, 3).

Deterministic VM Allocation Policies

Round Robin: suppose that all the servers form a queue. When a new VM request arrives, all the servers in the queue will be checked sequentially from the beginning, until a server Sr is found with enough remaining resources. Server Sr will be selected to host the new VM, and all the servers that have been checked will be moved to the end of the queue, keeping the original order.

We classify the Round Robin policy as deterministic because the servers will be chosen with the same order, if the cloud system and the workload are the same.

Stochastic VM Allocation Policies
1. *Least VM/Most VM*: for every new VM request, the policy will select the server that hosts the least/most number of VMs, among those with enough resources left (n.b., if multiple servers meet the criterion, the policy will choose one randomly). This kind of policy spreads/concentrates the workload within the system for better workload balance/lower energy consumption.
2. *Random*: for every new VM request, the policy will randomly select one server from those having enough resources.

We classify these three policies as stochastic because in contrast to the deterministic policy, even if the same workload is submitted to the system, the order in which the servers are selected may still be different.

3.3   Game Model

Given the attack and defence scenarios, we define the two-player security game model [24] as follows.

Players
In this strategic game, there are two players, the attacker A, and the defender D: $P = \{A, D\}$.

Action Set
According to our earlier analysis, the defender treats every customer's request in the same way, and their action set is to choose a specific VM allocation policy: $AS^D = \{$Least VM, Most VM, Random, Round Robin$\}$. On the other hand, from the attacker's point of view, they can decide when to start the VMs, and how many VMs to start. In order to simplify the problem, we only consider one single attack (in reality, the attacker may launch the attack periodically). Hence, the action set of the attacker is: $AS^A = \{VM(A,t)\}$.

Utility Functions
The attacker's goal is to maximise the efficiency/coverage rate, and their utility function is:

$$U^A\left(VM(A,t), Policy\right) = w_A \cdot Efficiency(VM(A,t), Policy) + \\ \left(1 - w_A\right) \cdot Coverage(VM(A,t), Policy) \tag{3}$$

where w_A is a weight that specifies the relative importance of efficiency vs. coverage, and $0 \leq w_A \leq 1$.

Note that compared with (2), the efficiency/coverage rate in (3) takes another parameter into consideration – *Policy* – since these two rates are different under the four allocation policies. In addition, the attacker's cost is also implicitly included, because the efficiency rate will be low if the attacker starts a large number of VMs.

In contrast, the defender's goal is to find a policy that achieves an optimal balance between minimising the attacker's efficiency/coverage rate, decreasing the overall power consumption, and balancing the workload. Suppose that P_i and B_i, $i = 1, 2, 3, 4$, represent the system's normalised power consumption and workload balance under the four policies respectively, then the defender's utility function is:

$$U^D \left(Policy \right) = -w_{D_1} \cdot U_i^A - w_{D_2} \cdot P_i + \left(1 - w_{D_1} - w_{D_2} \right) \cdot B_i \qquad (4)$$

where U_i^A is the attacker's utility under policy i, $i = 1, 2, 3, 4$, and w_{D_1} and w_{D_2} are weights such that $0 \leq w_{D_1}$, w_{D_2}, $w_{D_1} + w_{D_2} \leq 1$.

Therefore, the security game G is written as $G = \{P, AS^i, U^i, i \in \{A, D\}\}$. In the next section, we discuss $Efficiency(A, VM(A,t), Policy)$, $Coverage(A, VM(A,t), Policy)$, P_i and B_i in detail.

4   Analysis of VM Allocation Policies Using the Game Model

In this section, we present a simulation-based analysis of the different VM allocation policies. First, we introduce the simulation platform for our experiments. Then we give the detailed results of the efficiency rate, coverage rate, power consumption and workload balance under the four policies, which will help us develop the appropriate parameters in the game model. Finally, we calculate the numerical solution for the game, and discuss the implications of our findings.

4.1   Simulation Environment

We conducted our experiments on the platform CloudSim [25], which has been widely used in previous studies [26, 27]. The settings for our experiments are as follows.

Physical Servers and Virtual Machines
All the configurations of servers and VMs used in our simulations are similar to those of certain real-world models. Note that in CloudSim the CPU speed is measured in MIPS instead of MHz (a higher value of MIPS indicates faster processing speed).

Table 2. Configurations of servers and VMs

	Type	Quantity	CPU speed (MIPS)	No. of CPU cores	RAM (MB)
Servers	1	150	2600	16	24576
VMs	1	random*	2500	1	870
	2	random*	2000	1	1740
	3	random*	1000	1	1740
	4	random*	500	1	613

* Each VM request randomly decides the type of VM it requires.

Background Workload

Our earlier study [28] shows that the VM request arrival and departure processes in cloud computing follow a power law distribution, and exhibit self-similarity. In order to make the background workload more realistic in our simulation, we implement this finding using the program developed in [29]. More specifically, we use this program to generate two self-similar time series, indicating the number of VM requests that arrive/departure in each minute. In addition, we assume that every new request needs only one VM, whose type and CPU utilization for each minute are both randomly selected.

Experimental Settings

In each experiment, a legal user L starts 20 VMs at the 18000th second (note that the system reaches the steady state in terms of the number of started VMs around the 4800th second, so our results are very unlikely to be affected by simulation boot-up behaviours), and a certain time later (we call this time difference the *lag*) at the tth ($t = 18000 + $ lag) second, an attacker A starts $VM(A,t)$ VMs. The simulation stops a few minutes after that. Both the lag and $VM(A,t)$ range from 1 to 100 (note that we use "lag" and "t" interchangeably in the rest of this paper).

For every VM allocation policy/lag combination, we carry out the above experiment 50 times, and the final results presented below are the average values.

4.2   Attack Efficiency under Different VM Allocation Policies

In this subsection, we summarise the attack efficiency under the four policies.

Least VM Allocation Policy

Fig. 2 shows the impact on the efficiency rate of varying the lag and the number of VMs started by the attacker ($VM(A,t)$) under the Least VM policy. The following observations can be made from the experiment:

(1) The number of started VMs, $VM(A,t)$, has little impact on the attack efficiency.

(2) When the lag is small, it is difficult to achieve co-residence. This is consistent with the aim of balancing the workload, which means it is unlikely that a server will be chosen twice within a short period of time.

(3) After the lag reaches 10 minutes, the efficiency rate remains stable. The only exception is when $VM(A,t)$ equals one: the efficiency rate is volatile, but the average value in this case is still similar to the overall average value.

(4) It can be seen from Fig. 2(b) that when the lag is longer than 10 minutes and $VM(A,t)$ is larger than 5, the attack efficiency stays at approximately the same value.

Most VM Allocation Policy

Under the Most VM policy, our simulation shows that:

(1) In most cases, the efficiency rate first grows with $VM(A,t)$, but then the trend reverses. A closer inspection of the trace files shows that only the first $m(t)$ of the $VM(A,t)$ VMs are assigned to different servers, while the rest are all allocated together. $m(t)$ is the number of servers that are already turned on, and have sufficient remaining resources at time t.

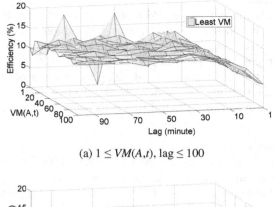

(a) $1 \leq VM(A,t)$, lag ≤ 100

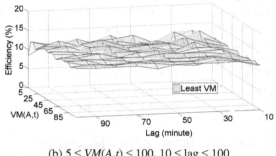

(b) $5 \leq VM(A,t) \leq 100$, $10 \leq$ lag ≤ 100

Fig. 2. The impact of the lag and the number of VMs started by the attacker ($VM(A,t)$) on attack efficiency under the Least VM policy. Fig. 2(a): the overall case, where $1 \leq VM(A,t)$, lag ≤ 100. Fig. 2(b): the stable region, where $5 \leq VM(A,t) \leq 100$, $10 \leq$ lag ≤ 100.

(2) The Most VM policy allocates new VMs to the same server until its remaining resources are less than required. Hence, the efficiency rate is relatively high with small lags, and decreases as the lag increases. However, similar to the situation with the Least VM policy, once the lag is larger than a certain value, the efficiency remains approximately the same.

A clever attacker would learn from the first observation that in order to achieve a higher efficiency with a large $VM(A,t)$, instead of starting all the VMs at the same time, they should start S VMs ($0 < S < VM(A,t)$) at a time, and repeat that $VM(A,t)/S$ times at certain intervals.

We re-ran the experiment with S set to five, and the interval set to one minute. As can be seen from Fig. 3(b), when $VM(A,t)$ is large, the efficiency rate is much higher if the attacker starts their VMs using the staggered approach described above. We use this set of results as the input to our game model.

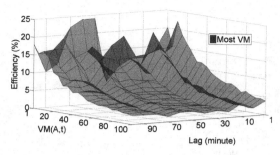

(a) Starting *VM(A,t)* VMs at once

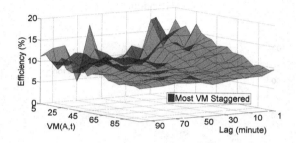

(b) Starting *VM(A,t)* VMs in a staggered way

Fig. 3. The impact of the lag and the number of VMs started by the attacker (*VM(A,t)*) on attack efficiency under the Most VM policy. Fig. 3(a): starting all *VM(A,t)* at once. Fig. 3(b): starting *VM(A,t)* in a staggered way (in batches of *S*).

Random Allocation Policy

The attack efficiency under the Random policy is similar to that of Least VM. It stays at almost the same value regardless of the lag and *VM(A,t)*.

Round Robin Allocation Policy

Under the Round Robin policy, the servers are selected sequentially. As a result, the attacker can only achieve a high efficiency rate if the time when they launch their VMs happens to be close to the time when the target server is chosen. As we can see from Fig. 4(b), there are only a few spikes along the Lag-axis.

However, this is not difficult to implement: because the servers are selected in a fixed order, the attacker can keep starting one VM every a few minutes, and tracking the chosen servers. When they find the target server will be selected again, they can then start their own VMs.

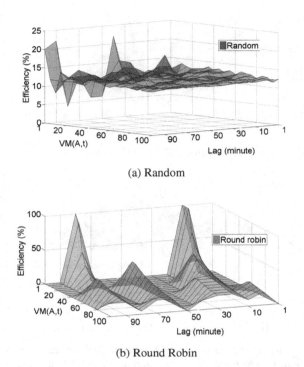

(a) Random

(b) Round Robin

Fig. 4. The impact of the lag and the number of VMs started by the attacker (*VM(A,t)*) on attack efficiency under the Random and Round Robin VM allocation policies

In other words, due to its deterministic behaviour, the Round Robin policy is the least secure. Therefore, in our game model, we set the attack efficiency under the Round Robin policy to 100%.

4.3   Coverage Rate under Different VM Allocation Policies

Under the three stochastic policies, the general trend of the coverage rate is similar: it increases almost linearly with *VM(A,t)*, and the lag has little impact after it reaches 10-20 minutes. The only difference is that when the lag is small, the coverage rate under the Most VM policy is much higher than under the Least VM policy.

As for the Round Robin policy, the situation is similar to that of the attack efficiency, where the attacker can achieve a high rate periodically. Hence, we also set the coverage to 100% for the Round Robin policy in our game model.

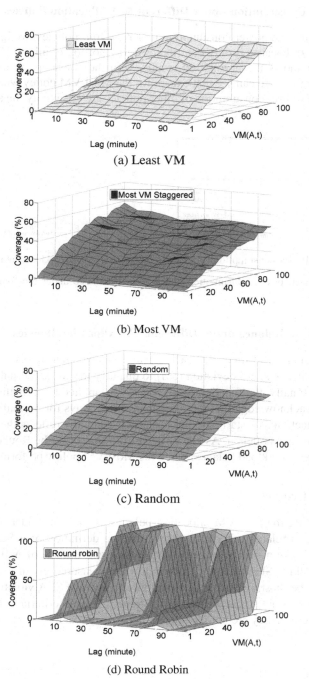

(a) Least VM

(b) Most VM

(c) Random

(d) Round Robin

Fig. 5. The impact of lag and the number of VMs started by the attacker ($VM(A,t)$) on the coverage rate under the four VM allocation policies

4.4   Power Consumption under Different VM Allocation Policies

When comparing the power consumption under the four policies, we ignore the influence of $VM(A,t)$, because it only contributes a tiny portion of all the VMs in the system. Fig. 6(a) shows the normalised results, with the consumption under the Least
VM policy set to 1. We can see that except for the Most VM policy where the value is
around 0.5, the power consumption of other policies are essentially the same.

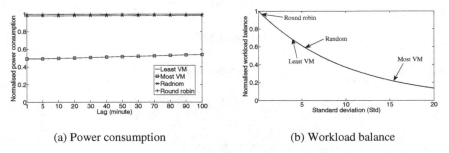

(a) Power consumption   (b) Workload balance

Fig. 6. Normalised power consumption and workload balance under the four VM allocation
policies

4.5   Workload Balance under Different VM Allocation Policies

We count the number of times that each server is selected during one experiment, and
then calculate the standard deviation (*Std*) to quantify the workload balance under the
four policies. Finally, the function $f(Std) = e^{-Std/10}$ is applied to normalise the standard
deviation (we acknowledge that there are many other ways for normalisation, and we
choose this function as a starting point because it generally reflects the degree of balance under the four policies). As can be seen from Fig. 6(b), the Round Robin policy
achieves the best workload balance, while the Most VM policy performs the worst.

4.6   Other Criteria

When comparing the four VM allocation policies, we also considered SLA (service
level agreement) related criteria. Here, we use the definition of a SLA violation in
[30]: "SLA violation occurs when a VM cannot get amount of MIPS that are re-
quested". The three SLA related criteria below are measured in our experiment: SLA
violation time per host, overall SLA violation, and average SLA violation. Our results
show that there is no major difference in terms of these criteria between the four
policies. Therefore, they are not included in our game model.

Table 3. Definitions of SLA related criteria

Name	Definition
SLA violation time per host (%)	$\sum$ SLA violation time of each host / $\sum$ Active time of each host
Overall SLA violation (%)	($\sum$ Requested MIPS - $\sum$ Allocated MIPS) / $\sum$ Requested MIPS
Average SLA violation (%)	Only consider the SLA violation incidents, ($\sum$ Requested MIPS - $\sum$ Allocated MIPS) / $\sum$ Requested MIPS

4.7   Numerical Solutions and Discussion

In the previous subsections, we have presented the attack efficiency, coverage, power consumption and workload balance under the four VM allocation policies. These are used to build the game matrices for the attacker and the defender. In this subsection, we compute the numerical solution of the game using Gambit [31], a tool for constructing and analysing finite, non-cooperative games, and interpret the results.

Zero-Sum Game

We begin with the simplest scenario where $w_{D_1} = w_{D_2} = 0$, which becomes a zero-sum game. We consider the following two situations: (1) $w_A = 1$, $U^A = Efficiency(VM(A,t),Policy)$, $U^D = -U^A$; (2) $w_A = 0$, $U^A = Coverage(VM(A,t),Policy)$, $U^D = -U^A$.

As can be seen from the following figures, both the solutions are mixed strategies. For the attacker, the solution is straightforward: they should start a small number of VMs each time if they aim to maximise the efficiency, but if the goal is to co-locate with as many target VMs as possible, they should start a large number of VMs at a time. For the defender, the result indicates that instead of deploying a single VM allocation policy, it is better to use a set of policies, and when a VM request arrives, each policy will be selected with a pre-set probability.

The following points should be noted. (1) As stated in our previous analysis, the Round Robin policy is the least secure, and is selected in neither case. (2) Even though, generally speaking, the attack efficiency under the Most VM policy is the lowest (especially when the lag is larger than five minutes), the peak value in this case is higher than that under the other two policies. This is the reason why the percentage of choosing Most VM is the smallest, if the defender intends to minimise the attack efficiency. However, if we only consider the situation where $VM(A,t) > 1$ and lag > 1 minute (which is closer to the real case), then the Most VM policy contributes a much larger percentage of the solution. (3) Under the Least VM policy, if the attacker starts multiple VMs at the same time, it is very likely that all of these VMs will be allocated to different servers. In contrast, under the other two policies there is a greater chance that some of these VMs will be located on the same server, which has a negative influence on the coverage rate. As a result, if the defender aims to minimise the coverage rate, they should only use the Most VM and Random policies.

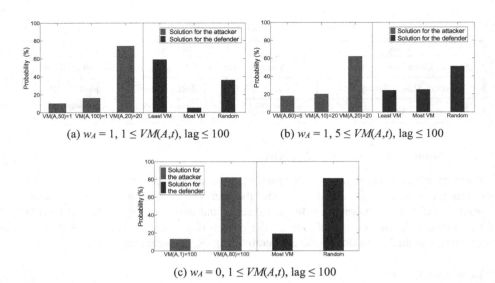

(a) $w_A = 1$, $1 \leq VM(A,t)$, lag ≤ 100 (b) $w_A = 1$, $5 \leq VM(A,t)$, lag ≤ 100

(c) $w_A = 0$, $1 \leq VM(A,t)$, lag ≤ 100

Fig. 7. Nash equilibrium in zero-sum game. Fig. 7(a): in the case where $w_A = 1$, $1 \leq VM(A,t)$, lag ≤ 100, the best strategy for the attacker is to start 1 VM when the lag is 50 minutes, i.e., $VM(A,50)=1$, with a probability of 10%, start 1 VM when the lag is 100 minutes, i.e., $VM(A,100)=1$, with a probability of 16%, and start 20 VMs when the lag is 20 minutes, i.e., $VM(A,20)=20$, with a probability of 74%. The best strategy for the attacker is to choose the Least VM, Most VM, and Random policies with a probability of 59%, 5%, 36%, respectively. The definitions of the symbols in the other two figures are the same.

We re-ran the experiment with the following two sets of configurations: (1) for the attacker, $VM(A,t) = 20$, and the lag ranges from 5 to 100 minutes, while the defender uses the second mixed policy (Least VM, 24%, Most VM, 25%, and Random, 51%); (2) for the attacker, $VM(A,t) = 100$, $1 \leq t \leq 100$, and the defender uses the third mixed policy (Most VM, 19%, and Random, 81%). The result shows that in overall terms, the average efficiency/coverage rate is the lowest under the mixed policies.

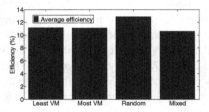   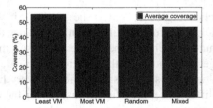

(a) Mixed policy 2: minimising the efficiency   (b) Mixed policy 3: minimising the coverage

Fig. 8. Comparison between the mixed policies and the stochastic policies in terms of the efficiency/coverage rate

Non-zero-sum Game

Different policies have their own advantages/disadvantages. For instance, the power consumption under the Most VM policy is the lowest, while the other policies achieve better workload balance. In practice, the defender can adjust the weights of security, power consumption, and workload balance, according to their different requirements.

Here we consider the situation where the three aspects are considered as equally important, i.e., $w_{D_1} = w_{D_2} = 1/3$. From Fig. 9, we can see that the results are similar to those shown in Fig. 7, which further demonstrates that a mixed policy may outperform any single policy.

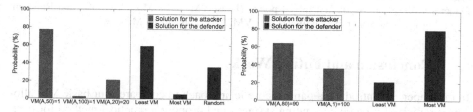

Fig. 9. Nash equilibrium in non-zero-sum game (all the definitions of symbols are the same as those in Fig. 7)

Similarly, we re-ran the experiment with the following configurations to compare the mixed policy with the four pure policies: (1) for the attacker, $VM(A,t) = 1$, and $1 \le t \le 100$, while the defender uses the first mixed policy (Least VM, 59%, Most VM, 5%, and Random, 36%); (2) for the attacker, $VM(A,t) = 90$, $1 \le t \le 100$, and the defender uses the second mixed policy (Least VM, 21%, and Most VM, 79%).

However, in this case, the defender cannot simply mix the policies with the specified percentages. Otherwise, on the one hand, an excessive number of servers will be turned on because the mixed policy contains Least VM and Random policies. On the other hand, the workload will not be balanced due to the Most VM policy. In other words, the mixed policy integrates the disadvantages instead of the advantages of each policy.

Therefore, we make the following changes and the allocation process comprises two rounds. In the first round, only the servers that are already being used and have sufficient remaining resources will be considered. If such a kind of server does not exist, then in the second round all servers are taken into consideration. In both rounds, each policy is still selected with the specified probability. As we can see from Fig. 10, the defender's utility is highest under mixed policies in both cases.

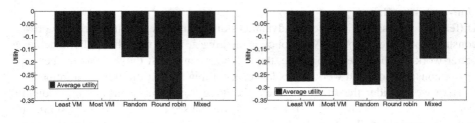

(a) First mixed policy   (b) Second mixed policy

Fig. 10. Comparison between the mixed policies and the pure policies in terms of the defender's utility. Fig. 10(a): first mixed policy (Least VM, 59%, Most VM, 5%, and Random, 36%). Fig 10(b): second mixed policy (Least VM, 21%, and Most VM, 79%).

5   Conclusion and Future Work

In this paper, we introduce a game theoretic approach to compare four basic VM allocation policies for cloud computing systems, and propose a practical method for mitigating the threat of the co-resident attack. Our results show that in order to minimise the efficiency and coverage rates for the attacker, the cloud provider should use a policy pool, such that for each VM request, a policy is chosen at random from the pool according to their predefined probabilities.

In the future, we intend to test our findings in larger scale systems. In addition, we will also study what the differences are between the behaviours of the attacker and normal users under the mixed policy, and how to identify them.

References

1. Barham, P., Dragovic, B., Fraser, K., Hand, S., Harris, T., Ho, A., Neugebauer, R., Pratt, I., Warfield, A.: Xen and the Art of Virtualization. Operating Systems Review 37, 164–177 (2003)
2. Ristenpart, T., Tromer, E., Shacham, H., Savage, S.: Hey, You, Get Off of My Cloud: Exploring Information Leakage in Third-Party Compute Clouds. In: 16th ACM Conference on Computer and Communications Security, CCS 2009, pp. 199–212 (2009)
3. Zhang, Y., Juels, A., Reiter, M., Ristenpart, T.: Cross-VM Side Channels and Their Use to Extract Private Keys. In: 2012 ACM Conference on Computer and Communications Security, CCS 2012, pp. 305–316 (2012)
4. Aviram, A., Hu, S., Ford, B., Gummadi, R.: Determining Timing Channels in Compute Clouds. In: 2010 ACM Workshop on Cloud Computing Security Workshop, CCSW 2010, pp. 103–108 (2010)
5. Vattikonda, B., Das, S., Shacham, H.: Eliminating Fine Grained Timers in Xen. In: 3rd ACM Workshop on Cloud Computing Security Workshop, CCSW 2011, pp. 41–46 (2011)
6. Wu, J., Ding, L., Lin, Y., Min Allah, N., Wang, Y.: XenPump: A New Method to Mitigate Timing Channel in Cloud Computing. In: 2012 IEEE Fifth International Conference on Cloud Computing, pp. 678–685 (2012)

7. Shi, J., Shi, J., Song, X., Chen, H., Zang, B.: Limiting Cache-based Side-channel in Multi-tenant Cloud using Dynamic Page Coloring. In: 2011 IEEE/IFIP 41st International Conference on Dependable Systems and Networks Workshops (DSN-W), pp. 194–199 (2011)
8. Jin, S., Ahn, J., Cha, S., Huh, J.: Architectural Support for Secure Virtualization under a Vulnerable Hypervisor. In: 44th Annual IEEE/ACM International Symposium on Micro-architecture, MICRO-44 2011, pp. 272–283 (2011)
9. Szefer, J., Keller, E., Lee, R., Rexford, J.: Eliminating the Hypervisor Attack Surface for a More Secure Cloud. In: 18th ACM Conference on Computer and Communications Security, CCS 2011, pp. 401–412 (2011)
10. Osvik, D.A., Shamir, A., Tromer, E.: Cache Attacks and Countermeasures: The case of AES. In: Pointcheval, D. (ed.) CT-RSA 2006. LNCS, vol. 3860, pp. 1–20. Springer, Heidelberg (2006)
11. Tromer, E., Osvik, D.A., Shamir, A.: Efficient Cache Attacks on AES, and Countermeasures. Journal of Cryptology 23, 37–71 (2010)
12. Hlavacs, H., Treutner, T., Gelas, J.-P., Lefevre, L., Orgerie, A.-C.: Energy Consumption Side-Channel Attack at Virtual Machines in a Cloud. In: 2011 IEEE Ninth International Conference on Dependable, Autonomic and Secure Computing, pp. 605–612 (2011)
13. Xu, Y., Bailey, M., Jahanian, F., Joshi, K., Hiltunen, M., Schlichting, R.: An Exploration of L2 Cache Covert Channels in Virtualized Environments. In: 3rd ACM Workshop on Cloud Computing Security, CCSW 2011, pp. 29–39 (2011)
14. Okamura, K., Okamura, K., Oyama, Y.: Load-based Covert Channels between Xen Virtual Machines. In: 2010 ACM Symposium on Applied Computing, SAC 2010, pp. 173–180 (2010)
15. Wu, J., Ding, L., Wang, Y., Han, W.: Identification and Evaluation of Sharing Memory Covert Timing Channel in Xen Virtual Machines. In: 2011 IEEE 4th International Conference on Cloud Computing, pp. 283–291 (2011)
16. Kadloor, S., Kadloor, S., Kiyavash, N., Venkitasubramaniam, P.: Scheduling with Privacy Constraints. In: 2012 IEEE Information Theory Workshop, pp. 40–44 (2012)
17. Xia, Y., Yetian, X., Xiaochao, Z., Lihong, Y., Li, P., Jianhua, L.: Constructing the On/Off Covert Channel on Xen. In: 2012 Eighth International Conference on Computational Intelligence and Security, pp. 568–572 (2012)
18. Bedi, H., Shiva, S.: Securing Cloud Infrastructure Against Co-Resident DoS Attacks Using Game Theoretic Defense Mechanisms. In: International Conference on Advances in Computing, Communications and Informatics, ICACCI 2012, pp. 463–469 (2012)
19. Varadarajan, V., Kooburat, T., Farley, B., Ristenpart, T., Swift, M.: Resource-Freeing Attacks: Improve Your Cloud Performance (at Your Neighbor's Expense). In: 2012 ACM Conference on Computer and Communications Security, CCS 2012, pp. 281–292 (2012)
20. Yang, Z., Yang, Z., Fang, H., Wu, Y., Li, C., Zhao, B., Huang, H.H.: Understanding the Effects of Hypervisor I/O Scheduling for Virtual Machine Performance Interference. In: 4th IEEE International Conference on Cloud Computing Technology and Science, pp. 34–41 (2012)
21. Zhou, F.F., Goel, M., Desnoyers, P., Sundaram, R.: Scheduler Vulnerabilities and Coordinated Attacks in Cloud Computing. In: 10th IEEE International Symposium on Network Computing and Applications, NCA (2011)
22. Zhang, Y., Li, M., Bai, K., Yu, M., Zang, W.: Incentive Compatible Moving Target Defense against VM-Colocation Attacks in Clouds. In: Gritzalis, D., Furnell, S., Theoharidou, M. (eds.) Information Security and Privacy Research, vol. 376, pp. 388–399. Springer, Heidelberg (2012)

23. Li, M.: Improving cloud survivability through dependency based virtual machine placement. In: The International Conference on Security and Cryptography, SECRYPT 2012, pp. 321–326 (2012)
24. Alpcan, T., Baar, T.: Network Security: A Decision and Game-Theoretic Approach. Cambridge University Press (2010)
25. CloudSim, http://www.cloudbus.org/cloudsim/
26. Calheiros, R., Ranjan, R., Beloglazov, A., De Rose, C.A.F., Buyya, R.: CloudSim: a Toolkit for Modeling and Simulation of Cloud Computing Environments and Evaluation of Resource Provisioning Algorithms. Software, Practice and Experience 41, 23–50 (2011)
27. Beloglazov, A., Abawajy, J., Buyya, R.: Energy-aware Resource Allocation Heuristics for Efficient Management of Data Centers for Cloud Computing. Future Generation Computer Systems 28, 755–768 (2012)
28. Han, Y., Chan, J., Leckie, C.: Analysing Virtual Machine Usage in Cloud Computing. In: IEEE 2013 3rd International Workshop on Performance Aspects of Cloud and Service Virtualization, CloudPerf 2013 (to appear, 2013)
29. Synthetic self-similar traffic generation,
    http://glenkramer.com/ucdavis/trf_research.html
30. Buyya, R., Beloglazov, A., Abawajy, J.: Energy-Efficient Management of Data Center Resources for Cloud Computing: A Vision, Architectural Elements, and Open Challenges. In: 2010 International Conference on Parallel and Distributed Processing Techniques and Applications, PDPTA 2010 (2010)
31. Gambit: Software Tools for Game Theory,
    http://www.gambit-project.org/gambit13/index.html

Monotonic Maximin: A Robust Stackelberg Solution against Boundedly Rational Followers

Albert Xin Jiang[1], Thanh H. Nguyen[1], Milind Tambe[1], and Ariel D. Procaccia[2]

[1] University of Southern California, Los Angeles, USA
{jiangx,thanhhng,tambe}@usc.edu
[2] Carnegie Mellon University, Pittsburgh, USA
arielpro@cs.cmu.edu

Abstract. There has been recent interest in applying Stackelberg games to infrastructure security, in which a defender must protect targets from attack by an adaptive adversary. In real-world security settings the adversaries are humans and are thus boundedly rational. Most existing approaches for computing defender strategies against boundedly rational adversaries try to optimize against specific behavioral models of adversaries, and provide no quality guarantee when the estimated model is inaccurate. We propose a new solution concept, *monotonic maximin*, which provides guarantees against *all* adversary behavior models satisfying *monotonicity*, including all in the family of Regular Quantal Response functions. We propose a mixed-integer linear program formulation for computing monotonic maximin. We also consider top-monotonic maximin, a related solution concept that is more conservative, and propose a polynomial-time algorithm for top-monotonic maximin.

1 Introduction

Stackelberg games have been used to model resource allocation problems in infrastructure security, in which a defender must allocate limited security resources to protect targets from attack by an adversary [1, 2, 10, 16]. Due to surveillance by the adversary, any pure strategy by the defender can be exploited. The defender thus should commit to a mixed strategy as the leader in this Stackelberg game, taking into account the response by the adversary who is the follower. Classical solution concepts such as Strong Stackelberg Equilibrium assume that the follower is perfectly rational. However, in real-world security settings the adversaries are humans and thus this perfect rationality assumption is problematic. There has been much recent progress on optimal defender strategies for Stackelberg security games against boundedly rational adversaries, for various behavior models including epsilon-best response, anchoring bias, prospect theory and logit quantal response models [14, 18].

The quantal response (QR) model is well-supported by the social and behavioral science literature [11–13] and has performed well in laboratory experiments for the Stackelberg game setting [18]. Within the QR framework, there is some freedom in the choice of functional families (logit, probit, etc.) and parameter values, e.g., the parameter λ in the logit QR model which measures the adversary's level of rationality. Once the function form is selected and parameter estimated (e.g., from real-world data

S.K. Das, C. Nita-Rotaru, and M. Kantarcioglu (Eds.): GameSec 2013, LNCS 8252, pp. 119–139, 2013.
© Springer International Publishing Switzerland 2013

or lab experiments), optimal defender strategies can be computed using optimization algorithms such as BRQR and PASAQ [18, 19].

However, there is some uncertainty about the best modeling parameters to use in real-world settings. In particular, real-world data on terrorist attacks are difficult to obtain. One can try to overcome this by running laboratory experiments, but models and parameters that give good fits in laboratory settings might not perform as well in actual security settings, due to factors such as different populations and different environments. And when the parameter estimate is inaccurate, *current algorithms provide no worst-case guarantee with respect to the solution quality*.

At the other extreme, there is the maximin solution: a leader strategy that maximizes leader expected utility when the follower is playing the worst-case strategy, i.e., play as if the follower is trying to minimize the leader's utility, even though the game is generally not zero-sum. The maximin solution provides utility guarantee without making any assumption on the attackers' behavior model. The maximin solution is computationally tractable: it can be solved by linear programming. However, the solution concept may be too conservative; in particular, the leader is disregarding any knowledge she may have about the follower's utilities in the game.

Are there robust solutions that do make use of recent advances in behavioral sciences, but are less sensitive to the choice of modeling parameters? In this paper we propose an approach that, instead of optimizing against a particular QR model, aims to guarantee good defender utility against all "reasonable" QR attackers. We note that QR in its most general form [13] covers all possible player behavior [6], so restriction to some notion of "reasonableness" is necessary. Goeree, Holt and Palfrey [5] proposed four properties that all reasonable QR models should satisfy, and called models satisfying all four properties Regular Quantal Response. In this paper, we impose constraints on attacker strategies that correspond to a relaxed version of Regular QR. Specifically, we assume that the attacker's strategies satisfy one of the four Regular QR properties, namely *monotonicity*, which is the property that actions with lower expected utility are played with smaller probability. (We further discuss the choice of monotonicity in Section 3.2.) We propose the following "monotonic maximin" solution concept to Stackelberg games: a defender plays a mixed strategy that maximizes defender expected utility, against the worst-case monotonic attacker mixed strategy. Since all Regular QR attackers satisfy monotonicity, monotonic maximin provides utility guarantees against all Regular QR attackers. Monotonic maximin is a robust alternative to the optimal Stackelberg strategy against specific QR models: it provides utility guarantees against all "reasonably rational" attackers (as defined by Regular QR) without making assumptions about parameters. This can be thought of as a "model-free" or "non-parametric" approach to Stackelberg games with boundedly rational followers.

The resulting computational problem might appear similar to a standard maximin problem, but is more challenging because the constraints for attacker's monotonicity now depend on the defender strategy. In this paper we propose an algorithm for this problem, based on LP duality and mixed-integer programming.

It is also interesting to consider attackers satisfying relaxations of the monotonicity constraint: the resulting defender strategies are *more* robust, as we are considering a larger set of possible attacker strategies. Specifically, we consider top monotonicity, the property that the follower's probability of playing each best response action is no less that that of any other action. We propose a polynomial-time algorithm for computing the resulting top-monotonic maximin solution concept.

We ran computational experiments to compare monotonic maximin and top-monotonic maximin against previously-proposed solution concepts including strong Stackelberg equilibrium [4], maximin, MATCH [15], as well as logit QR models with various parameter settings. Overall monotonic maximin is significantly more robust against monotonic adversaries compared to the previously-proposed solution concepts.

1.1   Related Work

There has been some recent work on designing defender strategies in security games that are robust against uncertainties, including uncertainties about the opponent as well as about the environment. One line of work is based on probabilistic models of uncertainties, and aims for security strategies that maximize the defender's expected utility under such probabilistic models. These include approaches based on specific models of bounded rationality, such as logit quantal response, prospect theory, and anchoring bias [14, 18]. A drawback of such approaches is the requirement on the availability and accuracy of probabilistic models; if an inaccurate probabilistic model is chosen, there is no quality guarantee with respect to the resulting security solution.

Another line of work, which includes our approach in this paper, adopts the robust optimization framework [3, 17] from Operations Research: define an *uncertainty set* that represents the space of likely models, and compute a security strategy that maximizes defender's utility under the worst case choice of models from that uncertainty set. For example, the BRASS algorithm [14] was designed to be robust against all adversaries playing epsilon-best response. An algorithm that is related to our approach is MATCH [15], which aims to provide a robust approach to Stackelberg security games against human attackers. MATCH is based on a similar intuition as our approach, that places less importance on attacker's actions with worse expected utilities. Specifically, MATCH bounds the defender's potential loss due to attacker's irrational behavior by a β-multiple of the attacker's loss due to his irrational behavior. Thus the robustness guarantee provided by MATCH is relative to the amount of the attacker's loss due to irrational behavior, and gets worse against less rational attackers. In contrast, our approach provides guarantees on defender utility against all Regular QR attackers.

At a high level, one drawback of these previous robust approaches is that they are still dependent on their parameter settings to define the sizes of their uncertainty sets. If the parameters are set so that the uncertainty sets are too small, the resulting solutions will be insufficiently robust. If the parameters are set so that the uncertainty sets are too large, the resulting solutions approach the maximin solution and are thus too conservative. While for certain cases it may be possible to come up with suitable parameters, our monotonic maximin approach avoids the requirement for parameters altogether. On the other hand, one could ask the same question about the uncertainty set defined by monotonic maximin: does the uncertainty set have the "right" shape and size? In particular, one potential criticism against monotonic maximin would be that it may be too conservative, because it uses only one of the four Regular QR conditions. In Section 3.2 we show that the uncertainty set for monotonic maximin is tight for Regular QR attackers, that is, any point in the uncertainty set could be arbitrarily approached by the behavior of a Regular QR attacker.

Finally, we mention work on modeling the game's uncertainties in aspects other than the adversary's behavior. Bayesian games were proposed to model players' probabilistic uncertainty about payoffs of the game [7]. There is also work that uses Bayesian games to model probabilistic uncertainties about defender's ability to execute the strategies as well as attacker's observation of defender strategies [22]. Within the robust optimization framework, The RECON algorithm [20] was designed to be robust against observation and execution uncertainties within a certain (hyperrectangular) error bound. Kiekintveld et al. [8] proposed robust solutions for security games against interval payoff uncertainties. While our paper's focus is on the behavior of the adversary, in Section 3.3 we briefly mention how our approach can be applied to achieve robustness against certain types of payoff uncertainty.

2 Preliminaries

Let $\mathbf{1}$ be a vector of 1's, the dimension of which will be clear from context. Let e_i be the i-th basis vector. Denote by $[n]$ the set $\{1, \ldots, n\}$.

We consider a two-player Stackelberg game between a leader and a follower. Leader's mixed strategy is denoted by $x \in X \subset \mathbb{R}^m$ where $X = \{x \in \mathbb{R}^m | Cx \leq d\}$ is a polytope. This includes the standard case where x is the distribution over m leader actions, when X is the simplex $\{x \in \mathbb{R}^m | x \geq 0, \mathbf{1}^T x = 1\}$; it also includes cases where x is a compact representation of mixed strategy as marginal probabilities (e.g., marginal coverage on targets [9], or marginal flow on a network [21]). Follower has n actions, labeled from 1 to n; i.e., his set of actions is $[n]$. Follower's mixed strategy is denoted by $y \in Y$, where $Y = \{y \in R^n | y \geq 0, \mathbf{1}^T y = 1\}$ is the standard simplex. The game's payoff matrices are $A, B \in \mathbb{R}^{m \times n}$. Expected utilities for the leader and the follower are $x^T A y$ and $x^T B y$ respectively. The game is general-sum: the sum of the players' utilities is not necessarily a constant.

Stackelberg Security Games. Although the solution concepts proposed in this paper apply to two-player Stackelberg games in general, we will frequently consider Stackelberg Security Games (SSGs) [9], a class of games with utility structure corresponding to the real-world problem of infrastructure security. Specifically, an SSG is a two-player Stackberg game between a defender (the leader) and an an adversary/attacker (the follower). There is a set of n targets $T = [n]$. The defender can deploy resources to cover some of the targets. Let $Z \subset \{0, 1\}^n$ be the set of feasible allocations of defender resources to targets, where for each allocation $z \in Z$ and target $j \in T = [n]$, $z_j = 1$ means the target is covered by the defender, and $z_j = 0$ means the target is not covered. Defender's set of mixed strategies X can then be represented by the convex hull of feasible allocations Z: $X = \text{conv}(Z) \subset \mathbb{R}^n$. The attacker chooses one target to attack, i.e., his set of mixed strategies Y is the standard simplex $\{y \in R^n | y \geq 0, \mathbf{1}^T y = 1\}$.

The payoffs to the players depend only on which target is attacked, and whether that target is covered by the defender. In other words, whether the defender covers an unattacked target does not affect the payoffs. Specifically, for each target $t \in T$, we denote by $U_d^u(t)$ the defender's utility for an uncovered attack on t, and $U_d^c(t)$ for a covered attack. Similarly, $U_a^u(t)$ and $U_a^c(t)$ are the attacker's payoffs for uncovered and covered attacks on t, respectively. In terms of the payoff matrices $A, B \in \mathbb{R}^{n \times n}$, this means that A_{ij} is equal to $U_d^c(j)$ if $i = j$, and $U_d^u(j)$ otherwise; while B_{ij} is equal to $U_a^c(j)$ if $i = j$ and $U_a^u(j)$ otherwise. We further assume that $U_d^c(t) > U_d^u(t)$ and $U_a^c(t) < U_a^u(t)$ for all $t \in T$.

In this paper we will focus on SSGs in which the set of feasible defender allocations Z has a simple structure: the defender has r resources, and each resource can protect any single target. Thus any allocation that uses r resources is feasible. The corresponding convex hull X can be described using a small number of constraints: $X = \{x \in \mathbb{R}^n | 0 \le x \le 1, 1^T x = r\}$. We call such a game an *SSG with r resources*.

Table 1. An example 3-target Stackelberg security game

	Target 1	Target 2	Target 3
U_d^c	7	10	2
U_d^u	-10	-8	-10
U_a^u	3	10	4
U_a^c	-10	-4	-10

Example 1. Table 1 shows the payoffs of an example Stackelberg security game with 3 targets. Specifically, the columns represent the targets and for each column the defender's utilities for covered attack (U_d^c) and uncovered attack (U_d^u), and the attacker's utilities for uncovered attack (U_a^u) and covered attack (U_a^c) are given.

Strong Stackelberg Equilibrium (SSE) is one of the standard solution concepts of Stackelberg games. In an SSE, the leader is maximizing her expected utility, assuming that the follower plays a best response. When the follower has multiple best responses, he is assumed to break ties in favor the leader. Formally, the SSE strategy for the leader is $\arg\max_{x \in X, y \in BR(x)} x^T A y$, where $BR(x) = \arg\max_{y \in Y} x^T B y$ is the set of best responses of the follower given leader strategy x.

Quantal Response is in general defined by a function $P : \mathbb{R}^n \to Y$ from the vector of expected payoffs of an agent's actions to a probability distribution over the actions. Denote by $P_j(u)$ the probability of playing action j given the vector $u \in \mathbb{R}^n$ of expected payoffs. For example, the logit quantal response function has the form $P_j(u) = \frac{e^{\lambda u_j}}{\sum_{j'} e^{\lambda u_{j'}}}$ where $\lambda \ge 0$ is a parameter. Other examples of P include probit, and the constant mapping to the uniform distribution.

Quantal Response Equilibrium (QRE) [13] is a solution concept for simultaneous games in which all players are playing quantal response strategies. In security domains, the adversary is human (and therefore not perfectly rational) while the defender can be assumed to be a rational decision maker aided by computers. This "Stackelberg against Quantal Response" model has been studied by Yang *et al* [18], who assumed that the adversary's quantal response function is known to the defender. In this paper we consider the case where the defender knows that the adversary behaves according to some quantal response model but does not know the specific quantal response function P.

Regular QRE. Goeree, Holt and Palfrey [5] proposed constraints that all reasonable QRE models should satisfy. Formally, P is a *regular quantal response function* if it satisfies the following:

1. Interiority: $P_j(u) > 0$ for all j.
2. Continuity: $P_j(u)$ is continuously differentiable.
3. Responsiveness: $\frac{\partial P_j(u)}{\partial u_j} > 0$ for all j.
4. Monotonicity: $u_j > u_k \Rightarrow P_j(u) > P_k(u)$ for all j, k.

They also point out that Continuity and Monotonicity imply $u_j = u_k \Rightarrow P_j(u) = P_k(u)$. The logit and the probit distributions are examples of regular quantal response functions. On the other hand, choosing a best response is not a regular quantal response function because it does not satisfy Interiority and Continuity.

The maximin solution is the optimal defender strategy assuming that the attacker is choosing the strategy that is worst for the defender: $\arg\max_{x \in X} \min_{y \in Y} x^T A y$. This solution concept is extremely conservative: the defender has to take into account an attacker that does completely arbitrary things, and as a result is disregarding his knowledge about the attacker payoff matrix B and treating the game as a zero-sum game.

3   Monotonic Maximin

Our overall approach is to modify maximin by imposing constraints on the attacker strategy. Specifically, we assume that the attacker strategy satisfies monotonicity. Since all Regular QR attackers satisfy monotonicity, our approach is able to provide guarantee against all Regular QR attackers. For computational convenience we will use the following form of monotonicity.

Definition 1. *Given $x \in X, y \in Y$, we say y satisfies* closed monotonicity *if for all $i, j \in [n]$, $x^T B e_i \geq x^T B e_j \Rightarrow y_i \geq y_j$.*

Recall that $x^T B e_i$ is the follower's expected utility of choosing action i, given that the leader plays x. There are strategies that are closely monotonic but not monotonic, for example the uniformly random strategy. It is straightforward to show the following:

Proposition 1. *If attacker is acting according to a regular quantal response function, then his mixed strategy y satisfies closed monotonicity.*

Proof. We need to show that for all i, j, $x^T B e_i \geq x^T B e_j \Rightarrow y_i \geq y_j$. Given i, j, suppose $x^T B e_i > x^T B e_j$. Then by the assumption of Monotonicity we have $y_i > y_j$ which implies $y_i \geq y_j$. Now suppose $x^T B e_i = x^T B e_j$. Then by Continuity and Monotonicity, $y_i = y_j$ which implies $y_i \geq y_j$.

Observe that closed monotonicity is not necessarily a weaker version of Monotonicity; nevertheless it is a consequence of Continuity and Monotonicity.

Let $Q(x) \subseteq Y$ be the set of follower mixed strategies that satisfy closed monotonicity given x. Then $Q(x) = \{y \in Y | \forall (i, j) \in E(x), y_i \geq y_j\}$, where $E(x) = \{(i, j) \in [n] | x^T B e_i \geq x^T B e_j\}$.

Definition 2. *The* monotonic maximin *solution is*

$$\arg\max_{x \in X} \min_{y \in Q(x)} x^T A y. \tag{1}$$

Let the monotonic maximin value be the corresponding objective value: $\max_{x \in X} \min_{y \in Q(x)} x^T A y$. By definition, the monotonic maximin solution provides guaranteed expected utility of at least the monotonic maximin value against all attacker strategies satisfying the monotonicity property.

Example 2. Consider the 3-target Stackelberg security game from Example 1. Suppose the defender has one resource. The defender's strategies generated by monotonic maximin, maximin, and SSE are shown in Table 2. For example, the second row indicates the defender's strategy generated by monotonic maximin, i.e., target 1 will be covered by the defender 37% of the time while there are 53% and 10% that target 2 and 3 will be covered by the defender, respectively.

When the strategy of the defender is generated by monotonic maximin, the defender's expected utility is -3.65 given a worst-case monotonic attacker strategy. Multiple monotonic attacker strategies tied for the worst case, including $(\frac{1}{2}, 0, \frac{1}{2})$ and $(\frac{1}{3}, \frac{1}{3}, \frac{1}{3})$. On the other hand, when maximin is used, the defender's expected utility is -4.38 for any actions of the monotonic attacker. Finally, when SSE is used, the attacker's expected utilities for all targets are the same and equal to 1.05. Thus the only feasible action for the monotonic attacker is $(\frac{1}{3}, \frac{1}{3}, \frac{1}{3})$. The defender's expected utility in this case is -3.8.

Table 2. Defender's strategy

	Target 1	Target 2	Target 3
Monotonic maximin	0.3732	0.5277	0.0991
Maximin	0.3306	0.2011	0.4683
SSE	0.15	0.6393	0.2107

The following proposition shows that the monotonic maximin concept is of most interest when the game is not zero sum.

Proposition 2. *For zero-sum games, the monotonic maximin solution coincides with maximin solution.*

Intuitively, if we consider e.g., a logit QR follower with $\lambda \to \infty$, then his behavior approaches that of a perfectly rational player and the leader can do no better than the maximin solution in a zero-sum game.

3.1 Existence of Monotonic Maximin Solutions

The standard Extreme Value Theorem states that a continuous function on a compact domain has a maximum. Since the set of monotonic follower strategies $Q(x)$ is not continuous in x, the value of the inner minimization $\min_{y \in Q(x)} x^T A y$ is not necessarily continuous in x. A natural question arises: does the monotonic maximin solution always exist? Of course if the maximum does not exist we could take the supremum instead, but the corresponding defender strategy would no longer be guaranteed to be robust.

Proposition 3. *The monotonic maximin solution exists in all Stackelberg games.*

This will be a direct consequence of Proposition 6 in Section 4, which provides an algorithm for monotonic maximin.

3.2 Optimality against Interiority, Continuity and Responsiveness

One potential criticism is that by focusing on monotonicity (and not the other conditions of Regular QR), monotonic maximin may be too conservative as it does not take advantage of all information about the follower behavior provided by the Regular QR model.

Proposition 4. *The monotonic maximin solution is arbitrarily close to optimal against an attacker who chooses the worst (for the defender) strategy satisfying both closed monotonicity and interiority.*

Proof (sketch). It is sufficient to show that given any x, $\min_{y \in Q(x)} x^T A y = \inf_{y \in Q(x) \cap \text{Int}} x^T A y$ where Int is the set of strategies satisfying interiority. Given an attacker strategy y that does not satisfy interiority (say a solution of the LHS), we can construct another strategy y' that satisfies interiority by re-assigning a small amount of probability mass to actions with zero probability in y. It is also straightforward to show that this can be done in a way that preserves closed monotonicity. y and y' achieve almost the same expected payoffs for both players.

Let us now consider continuity and responsiveness. Unlike monotonicity and interiority, which can be expressed as "local constraints" on y, continuity and responsiveness are properties of the response function $P(u)$ and correspond to constraints on the values of P given multiple inputs.

Consider the inner minimization problem of monotonic maximin: $q(x) = \arg\min_{y \in Q(x)} x^T A y$. This defines a response function $P(u)$ for the attacker,[1] which likely violates continuity and responsiveness. But is that necessarily the response function of the attacker we face? In particular does the attacker's response function have to be the same regardless of the defender's mixed strategy? Instead, we allow the attacker to "pick a response function" after observing defender's mixed strategy x, which is consistent with our overall robust optimization approach. It turns out that it is possible to pick the response function in a way that satisfies all conditions of regular quantal response, at the same time outputting the worst-case closedly monotonic strategy given x. This shows that monotonic maximin remains the optimal solution concept even when we consider attackers that satisfy all conditions of regular quantal response.

---

[1] Actually for P to be well-defined, it requires that $q(x) = q(x')$ whenever $x^T B = x'^T B$. This holds for Stackelberg security games, in which the players' utilities depend on the coverage on targets.

Proposition 5. *Given $x \in X$, there exists a regular quantal response function P : $\mathbb{R}^n \to Y$ such that $P(x^T B)$ is arbitrarily close to $\arg\min_{y \in Q(x)} x^T A y$.*

We give a proof in Section 5.2.

3.3   Capturing Other Behavioral and Uncertainty Models

In this section we show that monotonic maximin provides guarantees not only against regular quantal response attackers, but also other models of attacker behavior. Furthermore, if uncertainties in the game model (e.g., in the game's payoffs, attacker's capabilities, defender's execution, etc.) result in attacker behavior that is monotonic, then we can use monotonic maximin as a robust solution concept against such uncertainties.

Behavioral Models. A mixture (i.e., convex combination) of regular quantal response models is also a regular quantal response model, and therefore satisfies closed monotonicity. For example, one can have some probabilistic prior belief over the values of parameter λ in logit QR models, resulting in a mixture of logit QR models. As another example, consider a mixture of a regular QR model with the model that attacker plays a uniformly random mixed strategy. This is also a mixture of Regular QR models because the uniformly random strategy is a special case of Regular QR. Monotonic maximin provides utility guarantees against all such models.

Such guarantees are also applicable to behavior models that are not Regular QR but satisfy closed monotonicity. For example, consider the following "uniform best response" model: the attacker chooses a best response; if there are multiple pure-strategy best responses the attacker uniformly randomizes among those best responses. This is not a Regular QR function since it is not continuously differentiable; but it satisfies closed monotonicity. More generally, consider the *uniform top-K* strategy, in which the top K actions in terms of expected utilities are played, each with equal probability of $1/K$. These are not Regular QR but are nevertheless closedly monotonic. We will see later that these strategies have importance in monotonic maximin solutions.

Payoff Uncertainty. A simple consequence of [13] is that if we add i.i.d. noise with a smooth distribution of zero mean to the entries of the follower's payoff matrix B, and assuming that the follower plays a best response, then the resulting average follower strategy is monotonic. However this kind of noise does not preserve the structure of Stackelberg security games. For Stackelberg security games, consider the following type of payoff noise: i.i.d. with a smooth distribution of zero mean, added for each target to the payoffs of covered and uncovered attacks. For each instantiation of this noise the resulting game is still a Stackelberg security game. Given a defender mixed strategy, this would result in zero-mean i.i.d. noise over the expected attacker utilities of attacking each target. By the same argument as in [13], if the follower plays a best response, then the resulting average follower strategy is monotonic.

However, if the follower has a monotonic response function (not just a best response), the resulting average strategy under i.i.d. payoff noise is not guaranteed to be monotonic. Indeed, our numerical experiments show that when the follower plays the worst-case monotonic strategy with respect to perturbed utilities, the resulting average strategy is not always monotonic. It is possible to show that monotonicity is preserved under such noise if we assume the follower's response function is symmetric with respect to actions; we leave the detailed discussion to a future extended version of the paper.

3.4 Top-Monotonic Maximin

We define *top monotonicity* to be the property that for each best response action of the attacker, the probability of that action is no less than that of any other action. Formally, y satisfies top monotonicity given x if for all $i \in [n]$,

$$x^T B e_i \geq x^T B e_j \; \forall j \Rightarrow y_i \geq y_j \; \forall j.$$

We denote by $\widehat{Q}(x) \subset Y$ the set of top-monotonic follower strategies given x.

Top-monotonicity is a relaxation of closed monotonicity: the inequality $y_i \geq y_j$ only needs to hold between the best response action i and each of the other actions j. In other words, the corresponding *top-monotonic maximin* solution

$$\arg\max_{x \in X} \min_{y \in \widehat{Q}(x)} x^T A y$$

is more conservative than monotonic maximin.

Top-monotonic maximin is interesting partially because there have been extensive studies on various solution concepts that focus on pairwise comparisons between the best-response action against possible deviations.[2] Furthermore, we will show later that top-monotonic maximin can be computed in polynomial time.

4 Computation of Monotonic Maximin

4.1 Multiple-LP Formulation

Unlike the Maximin problem, we cannot directly use linear programming to solve (1). This is because the feasible set $Q(x)$ for y now depends on x. Fortunately, $Q(x)$ depends only on $E(x)$, which is essentially the ordering of attacker actions in terms of attacker utilities. So in theory we could solve an LP for each possible ordering, and return the one with best defender utility.

Since the attacker utilities are real numbers, the binary relation $E(x) \subset [n] \times [n]$ satisfies the constraints of a *total order*, i.e., transitivity: $(i, j), (j, k) \in E(x) \Rightarrow (i, k) \in E(x)$ and totality: $(i, j) \in E(x) \vee (j, i) \in E(x)$. Given a total order $\mathcal{E} \subset [n] \times [n]$, let $E^{-1}(\mathcal{E}) = \{x \in X : \mathcal{E} = E(x)\}$, i.e., the set of leader strategies inducing the order $\mathcal{E}$ on follower expected utilities.

---

[2] For example, in a Strong Stackelberg Equilibrium, any non-best response of the follower receives zero probability; for the epsilon-best responses considered in the BRASS algorithm [14], any follower action that is more than ϵ worse than the best response receives zero probability; In a MATCH solution [15], if an adversary action j is ϵ worse than the best response, the defender's potential loss if the adversary chooses j instead of the best response is bounded by $\beta\epsilon$, where $\beta > 0$ is a parameter of the solution concept.

Thus, for each $\mathcal{E}$ that corresponds to a total order, we solve

$$\max_{\boldsymbol{x} \in E^{-1}(\mathcal{E})} \min_{\boldsymbol{y} \in Q(\boldsymbol{x})} \boldsymbol{x}^T A \boldsymbol{y}, \tag{2}$$

and output the solution that achieves the best objective value. However, the set

$$E^{-1}(\mathcal{E}) = \{\boldsymbol{x} : \forall (i,j) \in \mathcal{E}, (i,j) \in E(\boldsymbol{x}); \; \forall (i,j) \notin \mathcal{E}, (i,j) \notin E(\boldsymbol{x})\}$$
$$= \{\boldsymbol{x} : \forall (i,j) \in \mathcal{E}, \boldsymbol{x}^T B \boldsymbol{e}_i \geq \boldsymbol{x}^T B \boldsymbol{e}_j; \; \forall (i,j) \notin \mathcal{E}, \boldsymbol{x}^T B \boldsymbol{e}_i < \boldsymbol{x}^T B \boldsymbol{e}_j\}$$

is not closed in general, since it involves strict inequalities for pairs not in $\mathcal{E}$. This presents problems such as potential nonexistence of solutions of (2). We instead use the closure of $E^{-1}(\mathcal{E})$, which is

$$\mathrm{cl}E^{-1}(\mathcal{E}) = \{\boldsymbol{x} : \forall (i,j) \in \mathcal{E}, \boldsymbol{x}^T B \boldsymbol{e}_i \geq \boldsymbol{x}^T B \boldsymbol{e}_j; \; \forall (i,j) \notin \mathcal{E}, \boldsymbol{x}^T B \boldsymbol{e}_i \leq \boldsymbol{x}^T B \boldsymbol{e}_j\}.$$

Since $\mathcal{E}$ is a total order, $(i,j) \notin \mathcal{E}$ implies that $(j,i) \in \mathcal{E}$, so the above can be simplified to $\{\boldsymbol{x} : \forall (i,j) \in \mathcal{E}, \boldsymbol{x}^T B \boldsymbol{e}_i \geq \boldsymbol{x}^T B \boldsymbol{e}_j\}$. Given $\mathcal{E}$, define the matrix $F \in \mathbb{R}^{n \times n(n-1)}$ such that its (i,j)-th column $F_{(i,j)}$ is $\boldsymbol{e}_i - \boldsymbol{e}_j$ if $(i,j) \in \mathcal{E}$ and the 0 vector otherwise. Then $\mathrm{cl}E^{-1}(\mathcal{E})$ can be written as $\{\boldsymbol{x} : \boldsymbol{x}^T B F \geq 0\}$. We will show below that replacing $E^{-1}(\mathcal{E})$ with $\mathrm{cl}E^{-1}(\mathcal{E})$ will not introduce incorrect solutions.

The inner minimization problem of (2) can then be written as

$$\min \boldsymbol{x}^T A \boldsymbol{y} \tag{3}$$
$$F^T \boldsymbol{y} \geq 0 \tag{4}$$
$$\mathbf{1}^T \boldsymbol{y} = 1 \tag{5}$$
$$\boldsymbol{y} \geq 0 \tag{6}$$

where $F^T \boldsymbol{y} \geq 0$ is the matrix form for constraints $y_i \geq y_j \; \forall (i,j) \in \mathcal{E}$.

Given $\boldsymbol{x}$, the above is an LP. By LP duality, its optimal solution is equal to that of its dual LP

$$\max t \tag{7}$$
$$F \boldsymbol{\lambda} + t\mathbf{1} \leq A^T \boldsymbol{x} \tag{8}$$
$$\boldsymbol{\lambda} \geq 0. \tag{9}$$

Now that the inner min becomes an max, max-min becomes a max-max problem. Recall that $\boldsymbol{x} \in X$ can be expressed as the linear constraints $C\boldsymbol{x} \leq d$, and $\boldsymbol{x} \in \mathrm{cl}E^{-1}(\mathcal{E})$ can be expressed as the linear constraints $\boldsymbol{x}^T B F \geq 0$. Then (2) can be written as

$$V_F = \max_{\boldsymbol{x}, \boldsymbol{\lambda}, t} t \tag{10}$$
$$C\boldsymbol{x} \leq d \tag{11}$$
$$\boldsymbol{x}^T B F \geq 0 \tag{12}$$
$$F \boldsymbol{\lambda} + t\mathbf{1} \leq A^T \boldsymbol{x} \tag{13}$$
$$\boldsymbol{\lambda} \geq 0 \tag{14}$$

which is an LP.

Proposition 6. *Consider the following Multiple-LP algorithm: given a two-player general sum game, solve the LP* (10) *for each F corresponding to a total order over the set of attacker actions. For the LP achieving the highest objective value V_F, output its solution $\boldsymbol{x}$. Then $\boldsymbol{x}$ is a monotonic maximin solution of the game.*

Proof (sketch). By construction, $\{E^{-1}(\mathcal{E}) : \mathcal{E} \text{ is a total order}\}$ partitions X; however each of the $E^{-1}(\mathcal{E})$ is not necessarily closed. Instead, the feasible sets for $\boldsymbol{x}$ in the LP instances (10) are closures of $E^{-1}(\mathcal{E})$, and thus cover X. These feasible sets have overlaps: consider total orders $\mathcal{E}$ and $\mathcal{E}'$, corresponding to matrices F and F' respectively, such that $\mathcal{E}' \subset \mathcal{E}$. Then $\mathrm{cl}E^{-1}(\mathcal{E}) \subseteq \mathrm{cl}E^{-1}(\mathcal{E}')$. Such overlap presents potential problems if the LP for F' has a solution $\boldsymbol{x}' \in \mathrm{cl}E^{-1}(\mathcal{E})$, with an objective value greater than V_F, the optimal objective of the LP for F; this is because such a solution would mask the correct solution for the region $\mathrm{cl}E^{-1}(\mathcal{E})$. We claim that this masking will never happen. Take this $\boldsymbol{x}'$, which is feasible for the LPs for F and F', and compare the objective values achieved by $\boldsymbol{x}'$ in the two LPs. Given $\boldsymbol{x}'$, the objective for the LP for F will be higher, intuitively because $\mathcal{E}' \subset \mathcal{E}$ and closed monotonicity implies that the follower is subject to more constraints in the case of $\mathcal{E}$, which makes the leader better off. Therefore the objective value of $\boldsymbol{x}'$ can never be higher than V_F, and the output of the Multiple-LP algorithm will be the monotonic maximin solution.

A direct consequence of this result is the existence of monotonic maximin solutions in all Stackelberg games (Proposition 3).

However, we would need to solve one LP for each total order on $[n]$. The following proposition shows that we only need to consider the *strict* orderings on $[n]$, i.e., those $\mathcal{E}$ in which for each pair of actions i, j, exactly one of (i, j) and (j, i) is in $\mathcal{E}$.

Proposition 7. *Consider a "non-strict" total order $\mathcal{E}^t$, with corresponding matrix F^t, i.e., there exists (i, j) such that $(i, j), (j, i) \in \mathcal{E}^t$. We say total order $\mathcal{E}^c$ (with corresponding matrix F^c) is a* sharpening *of $\mathcal{E}^t$ if $\mathcal{E}^c \subset \mathcal{E}^t$ and $\mathcal{E}^c$ is a strict order; i.e., for every pair $(i, j), (j, i) \in \mathcal{E}^t$, either (i, j) or (j, i) belongs to $\mathcal{E}^c$, not both. Let $\mathcal{F}(\mathcal{E}^t)$ be the set of matrices corresponding to sharpenings of $\mathcal{E}^t$. Then $V_{F^t} \leq \max_{F^c \in \mathcal{F}(\mathcal{E}^t)}\{V_{F^c}\}$.*

The proof is given in an online appendix available at http://teamcore.usc.edu/people/jiangx/papers/MMappendix.pdf.

This reduces the number of orderings we need to consider, but there are still $n!$ strict orderings to consider, corresponding to permutations of $[n]$. One approach to overcome this is to formulate the problem as a mixed-integer linear program (MILP).

4.2   MILP Formulation

The main idea is to have a binary integer variable z_{ij} that indicates whether $(i, j) \in E$. Then $F_{(i,j)} = z_{ij}(\boldsymbol{e_i} - \boldsymbol{e_j})$. To ensure that E corresponds to a total order, we can have constraints $z_{ij} + z_{ji} \geq 1$ and $(1 - z_{ij}) + (1 - z_{jk}) + z_{ik} \geq 1$. Then $\boldsymbol{x}^T BF \geq 0$ can be expressed as

$$\boldsymbol{x}^T B\boldsymbol{e_i} + M(1 - z_{ij}) \geq \boldsymbol{x}^T B\boldsymbol{e_j}, \; \forall i, j \tag{15}$$

where M is a sufficiently large positive constant that upper bounds $|\boldsymbol{x}^T B(\boldsymbol{e_i} - \boldsymbol{e_j})|$, e.g. $M = (\max_{i \in [m], j \in [n]} B_{ij} - \min_{i \in [m], j \in [n]} B_{ij}) \max_{\boldsymbol{x} \in X} \|x\|_1$.

One issue is that $F\lambda = \sum_{i,j} \lambda_{ij} F_{(i,j)} = \sum_{i,j} \lambda_{ij} z_{ij}(e_i - e_j)$, which now involves quadratic terms. We can transform this quadratic expression into MILP constraints using standard techniques, by replacing $\lambda_{ij} z_{ij}$ with a new variable w_{ij} satisfying the following constraints:

$$w_{ij} \geq 0 \tag{16}$$

$$w_{ij} \geq \lambda_{ij} - (1 - z_{ij})N \tag{17}$$

$$w_{ij} \leq \lambda_{ij} \tag{18}$$

$$w_{ij} \leq z_{ij}N \tag{19}$$

where N is a large positive constant. In fact we can eliminate λ since it is not used elsewhere, i.e., we do not need to include the constraints (17) and (18). Taking these together, we have a polynomial-sized MILP

$$\max_{x,w,t,z} t \tag{20}$$

$$Cx \leq d \tag{21}$$

$$x^T Be_i + M(1 - z_{ij}) \geq x^T Be_j, \ \forall i, j \tag{22}$$

$$\sum_{i,j} w_{ij}(e_i - e_j) + t\mathbf{1} \leq A^T x \tag{23}$$

$$0 \leq w_{ij} \leq z_{ij}N \tag{24}$$

$$z_{ij} \in \{0, 1\} \tag{25}$$

$$z_{ij} + z_{ji} \geq 1 \tag{26}$$

$$(1 - z_{ij}) + (1 - z_{jk}) + z_{ik} \geq 1. \tag{27}$$

4.3 Computing Top-Monotonic Maximin

Top-monotonic maximin can be efficiently computed by solving a small number of LPs. There are n possible best response actions of the attacker corresponding to n LPs: for each $i \in [n]$, let $E = \{(i, j) | j \in [n], j \neq i\}$ and solve (10). These correspond to partial orders as opposed to the total orders used previously. What about the cases with multiple best responses? The same argument as in Proposition 7 shows that we only need to consider the case with a single best response.

Proposition 8. *Top-monotonic maximin can be computed in time polynomial in n, m, and the number of constraints that define X.*

We can define similar relaxations of closed monotonicity, that focus on the best L actions: y is top-L monotonic for positive integer $L \geq 2$ if the closed monotonicity condition holds for any pair of actions (i, j) in which at least one of i and j is among the top L actions.[3] Top-L monotonic maximin is defined analogously. For the corresponding computational problem, we only need to solve $\frac{n!}{(n-L)!}$ LPs, one for each way of selecting an ordered L-tuple from n actions as the top L actions. This number of LPs is polynomial when L is fixed to be a constant.

---

[3] The notion of top L actions is well-defined: we do not need to consider the case of ties, by the same argument as in Proposition 7.

Proposition 9. *For constant L, top-L monotonic maximin can be computed in time polynomial in n, m, and the number of constraints that define X.*

5   Structure of Monotonic Maximin Solutions

5.1   Extreme Points of the Set of Monotonic Follower Strategies

Given x, the attacker's feasible region $Q(x)$ is a polytope. Consider the inner minimization problem $min_{y \in Q(x)} x^T A y$. Since the objective is linear, it is sufficient to consider only vertices of the polytope $Q(x)$. These vertices correspond to points y for which a sufficient number of inequalities of $Q(x)$ become tight. The inequalities of $Q(x)$ are of the form $y_i \geq 0$ and $y_i \geq y_j$. Therefore, we have the following:

Lemma 1. *Let y be a vertex of the polytope $Q(x)$. Then each action that is played with positive probability in y is played with equal probability.*

Thus a vertex is specified by a support set $R \subseteq [n]$, which is the set of attacker actions played with positive probability. Given R, the vertex y^R has $y_i^R = 0$ for all $i \notin R$, and $y_i^R = 1/|R|$ for $i \in R$.

The support set R of a vertex of $Q(x)$ has the following properties:

1. All best responses of the follower are required to be in R.
2. If $i \in R$ then all pure strategies that are better for the follower than i are also in R.
3. Since x induces a total ordering on $[n]$ of the follower expected utilities, there exists an action $j \in [n]$ such that $i \in R$ iff $x^T B_i \geq x^T B_j$. We call such an action j a *threshold action*.

In other words, given x, threshold action determines the support set. Since there are at most n possible threshold actions, we have the following characterization of the polytope $Q(x)$:

Proposition 10. *$Q(x)$ has at most n vertices, each of which is specified by a threshold action.*

These vertices correspond to *uniform top K* strategies for various K (recall that these are strategies in which the top K actions in terms of expected utilities are played, each with equal probability of $1/K$). This ranges from the "rational" uniform best response strategy, to the completely mixed uniform random strategy.

5.2   Proof of Proposition 5

In this section, we give a proof of Proposition 5, making use of the structure of $Q(x)$ that we showed in Section 5.1.

Proof (of Proposition 5). Recall that given $x \in X$, Proposition 5 asks for a regular quantal response function P such that $P(x^T B)$ is arbitrarily close to $\arg\min_{y \in Q(x)} x^T A y$. Given x, we know that at least one attacker strategy in $\arg\min_{y \in Q(x)} x^T A y$ is a vertex of $Q(x)$, and is therefore a uniform top-K strategy for some K. Now fix this K, and consider an attacker who plays the uniform top-K strategy against any x. Since the top K actions depends only on the attacker utilities $x^T B$, this defines a response function P'. However, P' does not satisfy interiority and continuity. We instead consider P^λ, a smooth version of P', defined as follows: given a vector of expected utilities $u \in \mathbb{R}^n$, P^λ first selects a subset S of K actions according to the probability distribution

$$\Pr(S) = \frac{e^{\lambda \sum_{j \in S} u_j}}{\sum_{S' \subset [n]:|S'|=K} e^{\lambda \sum_{j' \in S'} u_{j'}}},$$

then randomize uniformly over the K selected actions. For any $\lambda > 0$, P^λ is a response function satisfying monotonicity, interiority, continuity and responsiveness. As we take $\lambda \to \infty$, P^λ becomes arbitrarily close to P.

6 Evaluation

We ran computational experiments to compare the performance of monotonic maximin and its variants (top-L monotonic maximin) against other previously-proposed solution concepts, including Strong Stackelberg Equilibrium (SSE), MATCH [15], maximin, and logit Quantal Response. Both the solution quality and the runtime performance are examined, on instances of Stackelberg security games across a wide range of number of targets and number of defender resources.

6.1 Payoff Structures

The performances of solution concepts are affected by payoffs of the game. In particular, it is known that for zero-sum games, SSE, MATCH and maximin solutions coincide [15]; indeed monotonic maximin also coincides with maximin for zero-sum games (Proposition 2). We generated payoff structures for Stackelberg security games with different covariance values, by adapting the covariance game generator of the GAMUT package. The covariance value r, which is chosen within the range $[-1.0, 0.0]$, measures the correlation between the defender's payoff and the adversary's payoff. For example, when $r = -1.0$, the game becomes zero-sum whereas there will be no correlation between the defender and the adversary's payoffs when $r = 0.0$. The rewards for success of both the defender and the adversary are positive integers which lie within the range $[1, 10]$. On the other hand, the penalties for failure are negative integers within the range $[-10, -1]$.

6.2 Solution Quality against Worst-Case Monotonic Attackers

In the first set of experiments, we compared the solution quality (i.e., defender expected utility) of the different solution concepts against the worst-case closely monotonic attacker. That is, given a defender strategy x provided by one of the solution concepts, the attacker chooses $y \in \arg\min_{y \in Q(x)} x^T A y$. We would expect monotonic maximin to achieve the best performance, since it is by definition the optimal solution in this

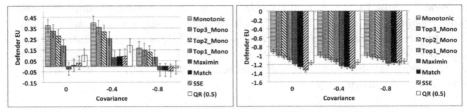

(a) 6 Targets, 3 Defender Resources   (b) 8 Targets, 3 Defender Resources

Fig. 1. The defender expected utility against the monotonic adversary, exact payoff structures

measure. The purpose of these experiments is to observe the magnitudes of differences in performance between monotonic maximin and others, and to check whether our top-L monotonic maximin algorithms provide good approximations to monotonic maximin. Specifically, we compared the performances among monotonic maximin, top-3, top-2, top-1 monotonic maximin, Maximin, MATCH, SSE, and logit Quantal Response. For logit Quantal Response, we tried a number of different values for λ, ranging from 1/32 to 8, but will only present the results for the best-performing value, which is $\lambda = 0.5$. The results for 6-target and 8-target games with 3 defender resources are shown in Figure 1. The x-axis represents the covariance value ranging from 0 to -0.8 with the step size of 0.4 while the y-axis shows the average of the defender expected utility when the adversary chooses the worst monotonic strategy. These computed values are averaged across 200 generated payoff structures for each covariance value, and error bars indicate standard deviations. As shown in Figure 1, monotonic maximin obtains a much higher defender's expected utility than Maximin, MATCH, SSE, and logit Quantal Response. In particular, for logit Quantal Response, even though the key parameter λ is carefully selected, it still performs poorly in comparison with monotonic maximin, implying its non-robustness against a monotonic adversary. For example, in the case of 6 targets and 3 resources (Figure 1a), when the covariance $r = 0$, while the defender achieves an average of expected utility of 0.37 using monotonic maximin, her expected utility is only -0.026, 0.007, 0.029, and 0.099 when using Maximin, MATCH, SSE, and logit Quantal Response, respectively. Furthermore, Figure 1 shows that top-1, top-2, and top-3 also significantly outperform Maximin, MATCH, SSE, and Quantal Response while their performance is in turn closer to monotonic maximin when the number of targets in the top set increases.

As predicted, when the games are zero-sum games (i.e., $r = -1.0$), the defender strategies generated by all algorithms excepts for Quantal Response turn out to be the same. In addition, as shown in Figure 1, when the covariance value r is closer to -1, the differences in defender expected utilities obtained by the compared algorithms tend to be smaller. Indeed, we observed that when the games are close to zero-sum games, the defender strategies generated by the algorithms are similar, thus the difference in their performances becomes small.

The promising performance of our algorithms in the case of small games motivated us to investigate their performance in larger games. In the next experiment, we evaluated the performance of top-1, top-2, and top-3 as well as Maximin, MATCH, SSE, and Quantal Response in 12-target and 14-target games with 6 defender resources. In this

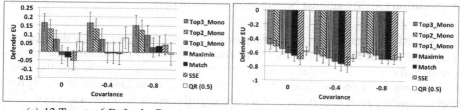

(a) 12 Targets, 6 Defender Resources   (b) 14 Targets, 6 Defender Resources

Fig. 2. The defender expected utility against the monotonic adversary, large games

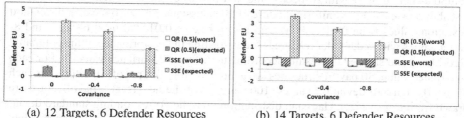

(a) 12 Targets, 6 Defender Resources   (b) 14 Targets, 6 Defender Resources

Fig. 3. Comparison results between the monotonic adversary and the expected adversary

experiment, we did not examine monotonic maximin due to its runtime limitation which we will describe in detail later. The results are shown in Figure 2. For each covariance value, 50 payoff structures are generated.

Figure 2 clearly shows that our top-L monotonic maximin algorithms with $L = 1, 2,$ 3 outperform Maximin, MATCH, SSE, and logit Quantal Response in terms of the obtained defender expected utility. For example, in Figure 2b, when the covariance value $r = 0$, the defender expected utility obtained by top-3, top-2, and top-1 are in turn -0.48, -0.51, and -0.54 while the defender expected utility obtained by Maximin, MATCH, SSE, and logit Quantal Response are -0.61, -0.65, -0.69, and -0.58, respectively. This result demonstrates that our algorithms still perform much better than the other compared algorithms in large game scenarios. It also suggests that our top-L algorithm is a promising approach for handling monotonic adversaries in large games.

In this paper we have argued that previous algorithms such as SSE and logit Quantal Response are not robust because such algorithms only attempt to address a specific type of adversary which could lead to deterioration in their performance when their assumptions are inaccurate. To check whether this is confirmed by our experiments, for both SSE and logit Quantal Response we compared the defender's expected utility of the expected objective (assuming correct model) and the expected utility against the worst case monotonic adversary. As shown in Figure 3, these two algorithms' performance against the monotonic adversary is significantly worse than the performance that they expected. For example, in Figure 3a, when the covariance $r = 0$, Quantal Response obtains the defender expected utility of only 0.05 against the worst case monotonic adversary while its expected objective value is 0.67. Also, SSE obtains only -0.05 against the worst case monotonic adversary while its expected value is 4.09.

6.3   Solution Quality against Non-monotonic Attackers

In the second set of experiments, we compared the solution concepts when the attacker is playing a non-monotonic strategy. The motivation for such experiments is that unlike the setting of our previous experiment, in practice our estimates about the payoffs of the adversary may be inaccurate. Recall from Section 3.3 that monotonicity is generally not preserved under payoff uncertainty, even if the payoff noise is zero-mean and i.i.d.

We added i.i.d. zero-mean noise to the reward and penalty of the adversary at every target, and calculated defender expected utilities given that the adversary responds with the worst-case monotonic strategy in each of the perturbed games. The defender computes her strategy with respect to the non-perturbed game. Specifically, the noise distribution we used is a uniform mixture of 10 Gaussians with zero mean and standard deviation values from 0.01 to 0.10 with a step size of 0.01. We used 6-target and 8-target games with 3 defender resources for evaluating the performance of the compared algorithms, and showed the results according to different covariance values r. For each covariance value, 50 payoff structures are generated; for each payoff structure, for each of the 10 standard deviation values, 100 samples of the adversary's payoff noise are drawn from a zero-mean Gaussian distribution with the corresponding standard deviation. That is, 50 x 10 x 100 = 50000 samples are generated for each covariance value. The result is shown in Figure 4, in which we plot the average defender expected utility across all noisy samples. As shown in Figure 4, monotonic maximin still outperforms the other compared algorithms although the differences are smaller than in the previous experiment with exact payoffs for the adversary. In addition, the top-L monotonic maximin algorithms with L = 1, 2, and 3 also obtained higher defender expected utilities than Maximin, MATCH, and SSE. This result indicates that our algorithms are robust to some amount of noise in the adversary's payoffs, even when such noise induces non-monotonic behavior from the adversary.

On the other hand, as we increase the magnitude (i.e., standard deviation) of noise, monotonic maximin no longer has a significant advantage over the previous solution concepts. Intuitively, as the noise becomes larger, the resulting average strategy of the attacker becomes farther away from the set of monotonic strategies, which are the strategies that monotonic maximin is designed to be robust against.

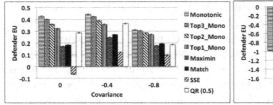
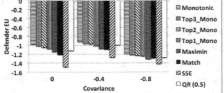

(a) 6 Targets, 3 Defender Resources   (b) 8 Targets, 3 Defender Resources

Fig. 4. The defender expected utility against the monotonic adversary, uncertainty in the adversary's payoffs

6.4   Runtime Performance

Finally, we tested the runtime scaling behavior of the algorithms. The results are shown in Figure 5. Figure 5a shows the runtime comparison between monotonic maximin, top-3, top-2, top-1, Maximin, MATCH, and SSE (as implemented by the ERASER algorithm [9]) in small games, i.e., 5-10 target games with 3 defender resources. The x-axis indicates the number of targets and the y-axis shows the average runtime in seconds for each algorithm to compute the defender's strategy given a payoff structure. The runtime is averaged over 300 different payoff structures. As shown in this figure, monotonic maximin's runtime grows very quickly with regard to the number of targets compared to other algorithms. When the number of targets increases to 10, its runtime reaches 446 seconds while top-3, top-2, and top-1 require only 11.91 seconds, 1.53 seconds, and 0.19 seconds, respectively. In this case, it takes Maximin, MATCH, and SSE only 0.02 seconds, 0.1 seconds, and 0.1 seconds, respectively.

In figure 5b, the runtime of top-3, top-2, top-1, Maximin, MATCH, and SSE in large game scenarios (i.e., 10-70 targets and 6 defender resources) are illustrated. This figure shows that when the number of targets is up to 20 targets, the runtime of top-3 increases to 146.49 seconds. In the case of 70 targets, top-2's runtime reaches 99.68 seconds while the runtime of top-1 is about 1.83 seconds, and the runtime of Maximin, MATCH, and SSE are all less than 1 second.

Overall, monotonic maximin and its variants have been shown to outperform Maximin, MATCH, logit QR and SSE in various game settings, i.e., different number of targets and different number of defender resources, and different groups of payoff structures with corresponding covariance values. In terms of scalability, even though our algorithms are not as fast as algorithms for these existing solution concepts, we have shown that our approach (especially top-L monotonic maximin) is feasible for large game scenarios. Among different variants of monotonic maximin, there is a trade-off between solution quality and runtime performance.

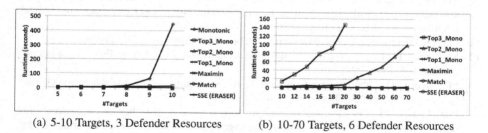

(a) 5-10 Targets, 3 Defender Resources   (b) 10-70 Targets, 6 Defender Resources

Fig. 5. Runtime comparison

7   Conclusion and Future Work

We proposed monotonic maximin, a novel robust solution concept for Stackelberg games with boundedly rational followers. We showed both theoretically and through numerical experiments on security games that monotonic maximin provides defender strategies that are robust against all regular quantal response attackers.

Our work points the way to a variety of new research challenges and potential future directions, including extending our robust optimization approach to other behavior models such as risk averseness, as well as applying the solution concept to games with multiple followers.

Acknowledgments. This research was supported by MURI Grant W911NF-11-1-0332.

References

1. Agmon, N., Kraus, S., Kaminka, G.A.: Multi-robot perimeter patrol in adversarial settings. In: ICRA (2008)
2. Basilico, N., Gatti, N., Amigoni, F.: Leader-follower strategies for robotic patrolling in environments with arbitrary topologies. In: AAMAS (2009)
3. Ben-Tal, A., El Ghaoui, L., Nemirovski, A.: Robust optimization. Princeton University Press (2009)
4. Conitzer, V., Sandholm, T.: Computing the optimal strategy to commit to. In: EC: Proceedings of the ACM Conference on Electronic Commerce (2006)
5. Goeree, J.K., Holt, C.A., Palfrey, T.R.: Regular quantal response equilibrium. Experimental Economics 8(4), 347–367 (2005)
6. Haile, P.A., Hortaçsu, A., Kosenok, G.: On the empirical content of quantal response equilibrium. The American Economic Review, 180–200 (2008)
7. Harsanyi, J.: Games with incomplete information played by "Bayesian" players, i-iii. part i. the basic model. Management Science 14(3), 159–182 (1967)
8. Kiekintveld, C., Islam, T., Kreinovich, V.: Security games with interval uncertainty. In: AAMAS (2013)
9. Kiekintveld, C., Jain, M., Tsai, J., Pita, J., Ordóñez, F., Tambe, M.: Computing optimal randomized resource allocations for massive security games. In: Proceedings of The 8th International Conference on Autonomous Agents and Multiagent Systems, pp. 689–696. International Foundation for Autonomous Agents and Multiagent Systems (2009)
10. Korzhyk, D., Conitzer, V., Parr, R.: Complexity of computing optimal stackelberg strategies in security resource allocation games. In: Proc. of The 24th AAAI Conference on Artificial Intelligence, pp. 805–810 (2010)
11. Luce, R.D.: Individual Choice Behavior: A Theoretical Analysis. Wiley (1959)
12. McFadden, D.: Conditional logit analysis of qualitative choice behavior. Frontiers of Econometrics, 105–142 (1974)
13. McKelvey, R.D., Palfrey, T.R.: Quantal Response Equilibria for Normal Form Games. Games and Economic Behavior 10(1), 6–38 (1995)
14. Pita, J., Jain, M., Ordóñez, F., Tambe, M., Kraus, S., Magori-Cohen, R.: Effective solutions for real-world stackelberg games: When agents must deal with human uncertainties. In: Proc. of The 8th International Conference on Autonomous Agents and Multiagent Systems (AAMAS) (2009)
15. Pita, J., John, R., Maheswaran, R., Tambe, M., Kraus, S.: A robust approach to addressing human adversaries in security games. In: ECAI (2012)
16. Tambe, M.: Security and Game Theory: Algorithms, Deployed Systems, Lessons Learned. Cambridge University Press (2011)
17. Wald, A.: Statistical decision functions which minimize the maximum risk. The Annals of Mathematics 46(2), 265–280 (1945)
18. Yang, R., Kiekintveld, C., Ordonez, F., Tambe, M., John, R.: Improving Resource Allocation Strategy Against Human Adversaries in Security Games. In: IJCAI (2011)

19. Yang, R., Ordonez, F., Tambe, M.: Computing optimal strategy against quantal response in security games. In: AAMAS (2012)
20. Yin, Z., Jain, M., Tambe, M., Ordonez, F.: Risk-Averse Strategies for Security Games with Execution and Observational Uncertainty. In: Proc. of the 25th AAAI Conference on Artificial Intelligence (AAAI), pp. 758–763 (2011)
21. Yin, Z., Jiang, A.X., Johnson, M.P., Tambe, M., Kiekintveld, C., Leyton-Brown, K., Sandholm, T., Sullivan, J.: TRUSTS: Scheduling randomized patrols for fare inspection in transit systems. In: IAAI (2012)
22. Yin, Z., Tambe, M.: A unified method for handling discrete and continuous uncertainty in Bayesian Stackelberg games. In: AAMAS (2012)

Defeating Tyranny of the Masses in Crowdsourcing: Accounting for Low-Skilled and Adversarial Workers

Aditya Kurve[1], David J. Miller[1], and George Kesidis[2]

[1] Department of EE
[2] Department of EE and CSE
The Pennsylvania State University, PA, USA

Abstract. Crowdsourcing has emerged as a useful learning paradigm which allows us to instantly recruit workers on the web to solve large scale problems, such as quick annotation of image, web page, or document databases. Automated inference engines that fuse the answers or opinions from the crowd to make critical decisions are susceptible to unreliable, low-skilled and malicious workers who tend to mislead the system towards inaccurate inferences. We present a probabilistic generative framework to model worker responses for multicategory crowdsourcing tasks based on two novel paradigms. First, we decompose worker reliability into *skill level* and *intention*. Second, we introduce a stochastic model for answer generation that plausibly captures the interplay between worker skills, intentions, and task difficulties. This framework allows us to model and estimate a broad range of worker "types". A generalized Expectation Maximization algorithm is presented to jointly estimate the unknown ground truth answers along with worker and task parameters. As supported experimentally, the proposed scheme de-emphasizes answers from low skilled workers and leverages malicious workers to, in fact, improve crowd aggregation. Moreover, our approach is especially advantageous when there is an (*a priori* unknown) majority of low-skilled and/or malicious workers in the crowd.

Keywords: crowd aggregation, information fusion, malicious workers, probabilistic modeling.

1 Introduction

Crowdsourcing systems leverage the diverse skill sets of a large number of Internet workers to solve problems and execute projects. In fact, the Linux project and Wikipedia can be considered products of crowdsourcing. These systems have recently gained much popularity with web services such as Amazon MTurk [1] and Crowd Flower [2], which provide a systematic, convenient and templatized way for requestors to post problems to a large pool of online workers and get them solved quickly. The success of crowdsourcing has been demonstrated for annotating and labeling images and documents [22], writing and reviewing software code [4], designing products [21], and also raising funds [3]. Here, we focus

S.K. Das, C. Nita-Rotaru, and M. Kantarcioglu (Eds.): GameSec 2013, LNCS 8252, pp. 140–153, 2013.

on crowdsourcing tasks where workers make a choice from a categorical answer space.

Although the crowd expedites annotation, its anonymity allows noisy or even malicious labeling to occur. Due to this, the application of crowdsourcing in mission and life critical tasks (such as identifying a plane crash in one of thousands of satellite images [5] or identifying malicious activity in a collection of video snippets) has been limited. Online reputation systems can help reduce the effect of noisy labels, but are susceptible to Sybil [10] or whitewashing [11] attacks. A second mitigation strategy is to assign each task to multiple workers and aggregate their answers in some way to estimate the ground truth answer. The estimation may use simple voting or more sophisticated aggregation methods. In this work, we present a stochastic model for answer generation that plausibly captures the interplay between worker skills, intentions, and task difficulties. To the best of our knowledge, this is the first model that incorporates the difference between worker skill and task difficulty (measured on the real line) in modeling the accuracy of workers on individual tasks. We also make a clear distinction between worker intention and skill level. Inferring intention allows us to identify malicious workers and to exploit their anticipated behavior to actually improve the accuracy of crowd aggregation. From the parameters of our model, we can infer the presence of several "types" of workers in the crowd, including adversarial workers. We formalize the notion of an adversarial worker and discuss and model different types of adversaries. A simple adversary gives incorrect answers "to the best of his skill level". More "crafty" adversaries can attempt to evade detection by only giving incorrect answers on the more difficult tasks solvable at their skill level. The detection of adversaries and the estimation of both worker skills and task difficulties can be assisted by knowledge of ground-truth answers for some (probe) tasks. Accordingly, we first propose a semisupervised approach[1], invoking a generalized EM algorithm to maximize the joint log likelihood over the (known) true labels for the "probe" tasks and the answers of the crowd for all tasks. Interestingly, our crowd aggregation rule comes precisely from the E-step, since the ground-truth answers are treated as the hidden data [9] in our EM approach. We emphasize that, unlike some approaches, which assume all tasks are drawn from the same classification domain, e.g., [17], our approach is applicable even when the batch of tasks is *heterogeneous*, *i.e.*, not necessarily drawn from the same domain.

Our experimental evaluation of the proposed scheme consisted of three levels. First, we evaluated the robustness of our EM algorithm using synthetic data. We observed that our scheme was especially effective when there is large variation in worker skills and task difficulties. Moreover, the algorithm was able to accurately identify adversarial workers with a high level of accuracy. Second, we evaluated performance using a crowd of "simulated" workers that do not generate answers in a fashion closely matched to our model. Specifically, each worker was a strong learner, formed from an ensemble of weak learners. Each weak learner was a

---

[1] When probe tasks are unavoidable, this method simply specializes to an unsupervised algorithm.

decision tree, with the ensemble (and thus, a strong learner) obtained by multi-class boosting. A strong worker's skill was controlled by varying the number of boosting stages used. We performed experiments on UC Irvine data sets [6] and observed that our method outperforms a (semisupervised) weighted plurality voting method. Our final experiment involved a crowdsourcing task we posted using Amazon Mturk [1]. Our conclusion is that the proposed scheme achieves strong performance and is able to overcome "tyranny of the masses", *i.e.*, when there is a minority of highly skilled workers among a pool of mostly low skilled workers. Such scenarios can occur when the tasks require specialized expertise rarely found in the crowd.

2   Modeling Paradigm

We separately model worker intention and skill. A worker's intention is a binary parameter indicating if he is adversarial or not. An honest worker provides accurate answers to the best of his skill level whereas an adversarial worker may provide incorrect answers to the best of his skill level. In the case of binary crowdsourcing tasks, adversarial workers can be identified by a negative weight [14] given to their answers. Here we extend malicious/adversarial worker models to *multicategory* tasks and hypothesize both "simple" and "crafty" adversaries.

Our approach incorporates task difficulty and worker skill explicitly and, unlike previous approaches [23] [22] [14], characterizes the interplay between them. Task difficulty and worker skill are both represented on the real line, with our generative model for a worker's answer based on their difference. If the task difficulty exceeds a worker's skill level, the worker answers randomly (whether honest or adversarial). For an adversary, if the task difficulty is less than his skill level, he chooses randomly from the set of incorrect answers. We also acknowledge another category of worker type known as "spammers". These are lazy workers who simply answer randomly for all tasks. In our model, they are represented with large negative skill levels.

3   Framework

3.1   Notation

Suppose a crowd of N workers is presented with a set of T_u unlabeled tasks for which the ground truth answers are unknown. There are also T_l probe tasks with known ground truth answers. We assume the crowd is unaware which tasks are probes. Accordingly, a malicious worker cannot alter his answering strategy in a customized way for the probe tasks to "fool" the system. Let $\{1, 2, ..., T_l\}$ be the index set of the probe tasks and $\{T_l + 1, T_l + 2, ..., T_l + T_u\}$ be the index set for non-probe tasks. We assume without loss of generality that each worker is asked to solve all the tasks[2]. The answers are chosen from a set $\mathcal{C} := \{1, 2, ..., K\}$. Let

---

[2] We only make this assumption for notational simplicity. Our methodology in fact applies generally to the setting where each worker solves only a subset of the tasks.

$z_i \in \mathcal{C}$ be the ground truth answer and let $\tilde{d}_i \in (-\infty, \infty)$ represent the difficulty level of task i. The intention of worker j is indicated by $v_j \in \{0, 1\}$, where $v_j = 1$ denotes an honest worker and $v_j = 0$ an adversary. $d_j \in (-\infty, \infty)$ represents the jth worker's skill level. Finally the response provided by the jth worker to the ith task is denoted $r_{ij} \in \mathcal{C}$.

3.2 Stochastic Generation Model

We define our model's *parameter set* $\Lambda = \{\{(v_j, d_j, a_j) \; \forall \; j\}, \{\tilde{d}_i \; \forall \; i\}\}$. We hypothesize the generation of the answers for non-probe tasks in two steps. Independently for each non-probe task $i \in \{T_l + 1, ..., T_l + T_u\}$:

1. Randomly choose the ground truth answer (z_i) from $\mathcal{C}$ according to a uniform pmf[3] $\frac{1}{K}$.
2. For each worker $j \in \{1, ..., N\}$, generate $r_{ij} \in \mathcal{C}$ for task i based on the *parameter-conditional* pmf $\beta(r_{ij}|\Lambda_{ij}, z_i)$, where $\Lambda_{ij} := \{v_j, d_j, a_j, \tilde{d}_i\}$[4].

Also, independently for each probe task $i \in \{1, ..., T_l\}$ and each worker j, generate the answer $r_{ij} \in \mathcal{C}$ based on the parameter-conditional pmf $\beta(r_{ij}|\Lambda_{ij}, z_i)$.

3.3 Worker Types

We model the ability of a worker to solve the task correctly using a sigmoid function based on the difference between the task difficulty and the worker's skill[5], *i.e.*, the probability that worker j can solve task i correctly is $\frac{1}{1+e^{-a_j(d_j - \tilde{d}_i)}}$. Note we have included a degree of freedom a_j which attempts to capture the individuality of workers. It is also possible to tie this parameter, *i.e.*, set $a_j = a, \forall j$.

Honest Workers. For an honest worker $(v_j = 1)$, the pmf β is defined as:

$$\beta(r_{ij} = l|\Lambda_{ij}, v_j = 1, z_i) = \begin{cases} \frac{1}{1+e^{-a_j(d_j - \tilde{d}_i)}} + \left(\frac{1}{K}\right)\left(\frac{e^{-a_j(d_j - \tilde{d}_i)}}{1+e^{-a_j(d_j - \tilde{d}_i)}}\right) & \text{for } l = z_i \\ \left(\frac{1}{K}\right)\left(\frac{e^{-a_j(d_j - \tilde{d}_i)}}{1+e^{-a_j(d_j - \tilde{d}_i)}}\right) & \text{otherwise} \end{cases} \quad (1)$$

Here, the worker essentially answers correctly with high probability if $d_j > \tilde{d}_i$, and with probability $\frac{1}{K}$ otherwise. Next, we discuss two models for adversarial workers.

Simple Adversarial Workers. For the simple adversarial model, β is given by

$$\beta(r_{ij} = l|\Lambda_{ij}, v_j = 0, z_i) = \begin{cases} \left(\frac{1}{K}\right)\left(\frac{e^{-a_j(d_j - \tilde{d}_i)}}{1+e^{-a_j(d_j - \tilde{d}_i)}}\right) & \text{for } l = z_i \\ \left(\frac{1}{K}\right)\left(\frac{e^{-a_j(d_j - \tilde{d}_i)}}{1+e^{-a_j(d_j - \tilde{d}_i)}}\right) + \left(\frac{1}{K-1}\right)\left(\frac{1}{1+e^{-a_j(d_j - \tilde{d}_i)}}\right) & \text{otherwise} \end{cases} \quad (2)$$

---

[3] One can always randomize the indexing of the answers for every task to ensure that the index of the true answer is uniformly distributed across all possible indices.

[4] The specific parametric dependence of β on Λ_{ij} will be introduced shortly.

[5] Alternative (soft) generalized step functions could also in principle be used here.

Here, essentially, the worker only chooses the correct answer (randomly) if the task difficulty defeats his skill level; otherwise he excludes the correct answer.

Complex Adversarial Workers. In this case, the adversarial worker is more evasive. He answers correctly for simpler tasks with difficulty level below a certain value. Assume $\theta_j < d_j$ to be such a threshold for worker j. The pmf β for this (complex) adversarial worker is given by:

$$\beta(r_{ij} = l | \Lambda_{ij}, v_j = 0, z_i) = \begin{cases} (\frac{1}{K})(\frac{e^{-a_j(d_j - \tilde{d}_i)}}{1+e^{-a_j(d_j - \tilde{d}_i)}}) + (\frac{1}{1+e^{-b_j(\theta_j - \tilde{d}_i)}})(\frac{1}{1+e^{-a_j(d_j - \tilde{d}_i)}}) \\ \qquad\qquad\qquad\qquad \text{if } l = z_i \\ (\frac{1}{K})(\frac{e^{-a_j(d_j - \tilde{d}_i)}}{1+e^{-a_j(d_j - \tilde{d}_i)}}) + (\frac{1}{K-1})(\frac{e^{-b_j(\theta_j - \tilde{d}_i)}}{1+e^{-b_j(\theta_j - \tilde{d}_i)}})(\frac{1}{1+e^{-a_j(d_j - \tilde{d}_i)}}) \\ \qquad\qquad\qquad\qquad \text{otherwise} \end{cases} \tag{3}$$

Here, essentially, the worker answers correctly with high probability for easy tasks ($\theta_j > \tilde{d}_i$), he excludes the correct answer for more difficult tasks below his skill level, and for even more difficult tasks that defeat his skill level ($d_j < \tilde{d}_i$), he answers correctly at random ($\frac{1}{K}$). In this work we will only investigate the simple model of adversarial workers.

3.4 Incomplete, Complete and Expected Complete Data Log Likelihood

The observed data $\mathcal{X} = \mathcal{R} \cup \mathcal{Z}_\mathcal{L}$ consists of the set $\mathcal{R}$ of answers given by the workers to all the tasks, i.e., $r_{ij} \ \forall \ i, j$ and the set $\mathcal{Z}$ of ground truth answers to the probe tasks, i.e., z_i, $i \in \{1, 2, ..., T_l\}$. We express $\mathcal{R} = \mathcal{R}_L \cup \mathcal{R}_U$, i.e., the union of answers to probe tasks and non-probe tasks. We choose the hidden data [9] $\mathcal{H}$ to be the ground truth answers to the non-probe tasks, i.e., Z_i, $i \in \{T_l + 1, ..., T_l + T_u\}$. The incomplete data log-likelihood that we seek to maximize in estimating Λ is given by

$$\log \mathcal{L}_{\text{inc}} = \log \mathrm{P}(\mathcal{R}, \mathcal{Z}_\mathrm{L} | \Lambda) = \log \mathrm{P}(\mathcal{R}_\mathrm{L}, \mathcal{R}_\mathrm{U}, \mathcal{Z}_\mathrm{L} | \Lambda) = \log \mathrm{P}(\mathcal{R}_\mathrm{L}, \mathcal{Z}_\mathrm{L} | \Lambda) + \log \mathrm{P}(\mathcal{R}_\mathrm{U} | \Lambda)$$

$$= \sum_{i=1}^{T_l} \sum_{j=1}^{N} \log \frac{1}{K} \beta(r_{ij} | \Lambda_{ij}, z_i) + \sum_{i=T_l+1}^{T_l+T_u} \sum_{j=1}^{N} \log \frac{1}{K} \sum_{k=1}^{K} \beta(r_{ij} | \Lambda_{ij}, z_i = k)$$

$$\propto \sum_{i=1}^{T_l} \sum_{j=1}^{N} \log \beta(r_{ij} | \Lambda_{ij}, z_i) + \sum_{i=T_l+1}^{T_l+T_u} \sum_{j=1}^{N} \log \sum_{k=1}^{K} \beta(r_{ij} | \Lambda_{ij}, z_i = k).$$

The expected complete data log-likelihood, where the expectation is with respect to the pmf $\mathrm{P}(Z_i = k | \mathcal{X}, \Lambda)$, can be written as:

$$\mathbb{E}[\log \mathcal{L}_c | \mathcal{X}, \Lambda] \propto \sum_{i=1}^{T_l} \sum_{j=1}^{N} \log \beta(r_{ij}|\Lambda_{ij}, z_i) + \sum_{i=T_l+1}^{T_u} \sum_{j=1}^{N} \sum_{k=1}^{K} [P(z_i = k | \mathcal{X}, \Lambda) \log \beta(r_{ij}|\Lambda_{ij}, z_i = k)]$$

$$= \sum_{i=1}^{T_l} \sum_{j:r_{ij}=z_i} \left[v_j \log \left(\beta(r_{ij}|\Lambda_{ij}, v_j = 1, z_i = r_{ij}) \right) + (1 - v_j) \log \left(\beta(r_{ij}|\Lambda_{ij}, v_j = 0, z_i = r_{ij}) \right) \right]$$

$$+ \sum_{i=1}^{T_l} \sum_{j:r_{ij}\neq z_i} \left[v_j \log \left(\beta(r_{ij}|\Lambda_{ij}, v_j = 1, z_i \neq r_{ij}) \right) + (1 - v_j) \log \left(\beta(r_{ij}|\Lambda_{ij}, v_j = 0, z_i \neq r_{ij}) \right) \right] \qquad (5)$$

$$+ \sum_{i=T_l+1}^{T_l+T_u} \sum_{k=1}^{K} \sum_{j:r_{ij}=k} P(z_i = k) \left[v_j \log \left(\beta(r_{ij}|\Lambda_{ij}, v_j = 1, z_i = k) \right) + (1 - v_j) \log \left(\beta(r_{ij}|\Lambda_{ij}, v_j = 0, z_i = k) \right) \right]$$

$$+ \sum_{i=T_l+1}^{T_l+T_u} \sum_{k=1}^{K} \sum_{j:r_{ij}\neq k} P(z_i = k) \left[v_j \log \left(\beta(r_{ij}|\Lambda_{ij}, v_j = 1, z_i \neq k) \right) + (1 - v_j) \log \left(\beta(r_{ij}|\Lambda_{ij}, v_j = 0, z_i \neq k) \right) \right]$$

4   The Generalized EM (GEM) Algorithm

We formulate our algorithm using the above defined expected complete data log-likelihood. The EM algorithm ascends monotonically in log $\mathcal{L}_{inc}$ with each iteration of the E and M steps [9]. In the *expectation* step, we calculate the pmf $P(Z_i = k | \mathcal{X}, \Lambda^t)$ using the current parameter values Λ^t, and in the *maximization* step, we compute $\Lambda^{t+1} = \arg\max_\Lambda \mathbb{E}[\log \mathcal{L}_c | \mathcal{X}, \Lambda^t]$.

E step: In the E-step we compute the expected value of $\mathcal{Z}_u$ given the observed data $\mathcal{X}$ and the current parameter estimates Λ^t. Based on our assumed stochastic model (section 3.2), $P(\mathcal{Z}_u | \mathcal{X}, \Lambda^t) = \prod_{i=T_l+1}^{T_l+T_u} P(Z_i = z_i | \mathcal{X}, \Lambda^t)$. Moreover, again based on the assumed stochastic model and applying Bayes' rule, we can derive the closed form expression for the pmf in the E-step as:

$$P_i(Z_i = k | \mathcal{X}, \Lambda^t) = \frac{\prod_{j=1}^{N} \beta(r_{ij}|\Lambda_{ij}^t, Z_i = k)}{\sum_{l=1}^{K} \prod_{j=1}^{N} \beta(r_{ij}|\Lambda_{ij}^t, Z_i = l)}, \quad i \in \{T_l + 1, ..., T_u + T_l\}. \quad (6)$$

Generalized M step: In the M-step of EM, one maximizes the expected complete data log-likelihood with respect to the model parameters:

$$\Lambda^{t+1} = \arg\max_\Lambda \mathbb{E}[\log \mathcal{L}_c | \mathcal{X}, \Lambda^t]. \qquad (7)$$

Since Λ consists of mixed (both continuous and discrete) parameters, with a particular parametric dependence and with 2^N (honest, adversarial) crowd configurations, it is not practically feasible to find a closed form solution to (7) for our model. Instead, we use a generalized M-step approach [16][12] to iteratively maximize over the parameters $\{v_j \ \forall \ j\}$, and $\{\{(d_j, a_j) \ \forall \ j\}, \{d_i \ \forall \ i\}\}$.

M1 Substep: We can find a closed form solution for $v_j \ \forall \ j$ keeping all other parameters fixed:

$$v_j = \arg\max_{\{0,1\}} \mathbb{E}(log \boldsymbol{L}_c | \mathcal{X}_j, \Lambda \backslash \{v_j\}), \qquad (8)$$

where $\mathcal{X}_j$ is the set of answers provided by the jth worker and the ground truth answers for the probe tasks that he answered.

M2 Substep: We maximize $\mathbb{E}[\log \mathcal{L}_c | \mathcal{X}, \Lambda^t]$ with respect to $\Lambda \backslash \{v_j : \forall j\}$ given $\{v_j : \forall j\}$ fixed. For this, we use a gradient ascent algorithm which ensures monotonic increase in $\log \mathcal{L}_{\text{inc}}$, but which may only find a local maximum, rather than a global maximum of $\mathbb{E}[\log \mathcal{L}_c | \mathcal{X}, \Lambda^t]$.

Inference: Note that the E-step (6) estimates the *a posteriori* probabilities of ground-truth answers. Thus, after our GEM learning has converged, a maximum *a posteriori* decision rule applied to (6) gives our crowd-aggregated estimates of the true answers for the non-probe tasks.

4.1   Unsupervised GEM

Note that when probe tasks are not included in the batch of tasks, an unsupervised specialization of the above GEM algorithm is obtained. In particular, we have $T_l = 0$, with the first term in (4) and the first two terms in (5) not present. Our above GEM algorithm is accordingly specialized for this case. In Section 5, we will evaluate the unsupervised GEM based scheme along with all other methods.

5   Experiments

Experiments were performed using synthetic data as well as data generated by a crowdsourced multicategory labeling task on Amazon MTurk [1]. Additionally, for a number of UC Irvine domains, we generated a collection of heterogeneous classifiers to be used as a "simulated" crowd. We generated adversaries of the simple type in all our experiments. We compared our scheme (both with and without probe tasks) with simple (multicategory) plurality voting and weighted plurality voting, which exploits the probe tasks[6].

5.1   Experiments with Synthetic Data

The synthetic data was produced according to the stochastic generation described in section 3.2. The goal here was to evaluate the GEM algorithm by comparing the estimated parameters and the estimated hidden ground truth answers with their actual values used in generation of the synthetic data. We generated a crowd of 100 workers with $d_j \sim \mathcal{N}(1, 400)$, $a_j \sim \mathcal{N}(0.3, 0.2)$; 10% of workers were adversarial. The tasks were generated with $\tilde{d}_i \sim \mathcal{N}(8, \sigma^2)$, where σ^2 was varied. The ground truth answer for each task was chosen randomly from $\{0, 1, ..., 4\}$. We observed that in this regime of high variance in worker skill and

---

[6] The answer from each worker was weighted by the fraction of probe tasks that he answered accurately.

task difficulty, there is a definite advantage in using the proposed scheme over other benchmark schemes, as shown in Table 1. Table 2 shows performance as a function of the number of workers assigned to each task. In each case, a random regular bipartite graph of workers and tasks was generated. We also see in Figure 1 the high level of correlation between the estimated and actual values of worker skills and task difficulties.

5.2   Simulating a Crowd Using an Ensemble of Classifiers

We also leveraged ensemble classification to generate a set of automated workers (each an ensemble classifier) using boosting [19]. Each such classifier (worker) is a strong learner obtained by applying multiclass boosting to boost decision tree-based weak learners. The strength (accuracy) of each worker was varied by controlling the number of boosting stages. Each weak learner was trained using a random subset of the training data to add more heterogeneity across the workers' hypotheses. Note that unlike Section 5.1, this approach to simulated generation of a crowd is *not* matched to our stochastic data generation model in Sections 3.2 and 3.3. Thus, this more complex simulation setting provides a more realistic challenge for our model and learning.

Table 1. Synthetic data: different values of task difficulty variance

Task Variance	Average erroneous tasks			
	Simple Plurality	Weighted Plurality	Proposed With Probes	Proposed Without Probes
4000	23.4	22.1	16.9	19.8
2000	21.2	20	14.6	15.2
1000	19.1	16.2	9.6	9.8
500	11.1	8.1	3.14	4.1
250	5.8	5.8	2.1	2.5

Table 2. Synthetic data: different values of worker assignments

Assignment degree	Average erroneous tasks			
	Simple Plurality	Weighted Plurality	Proposed With Probes	Proposed Without Probes
20	31.2	31.6	27	28
40	22.85	21.2	18.4	18.8
60	19.4	18.6	14.2	15.4
80	16.5	16	13.1	13.4

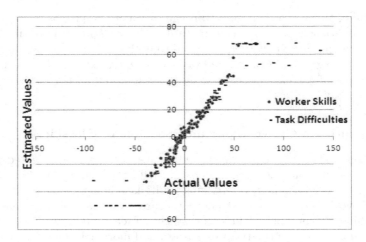

Fig. 1. Comparison of actual and estimated parameters

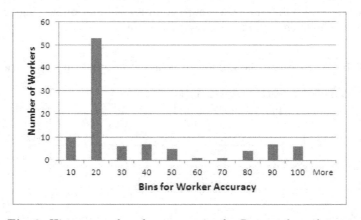

Fig. 2. Histogram of worker accuracies for Dermatology dataset

We ran Multiboost [7] 100 times to create a crowd of 100 workers for four domains that are realistic as crowdsourcing domains: Pen Digits[7], Vowel, Dermatology, and Nominal[8]. For each experimental trial, 100 crowd tasks were created by randomly choosing 100 data samples from a given domain; 10 of them were randomly chosen to be probe. The rest of the data samples from the domain were used for training the strong (ensemble) classifiers/workers. The average of the number of crowd-aggregated erroneous tasks was computed across 5 trials, where each trial consisted of a freshly generated crowd of workers and set of

---

[7] We resampled the dataset to have only odd digits.

[8] Hungarian named entity dataset [20]. Identifying and classifying proper nouns into four categories: not a proper noun, person, place, and organization.

tasks. In Table 3, we give performance evaluation for different worker accuracy means and variances. We did not directly control these values since they were an outcome of the boosting mechanism. However we could control the number of boosting stages used by each worker. We also show the performance when 10% of the workers were replaced by adversarial workers. These synthetic adversaries retained the estimated skill level of the (replaced) workers and generated their answers using the stochastic model described in section 3.2. In Table 3, the worker accuracy mean and variance across all workers is based on the number of correctly answered tasks (both probe and non-probe) for each worker. We can see in Table 3 the gains in performance of our method over the other methods, especially in the presence of adversarial workers. Note also that our method *exploits* simple adversaries, achieving improved crowd aggregation accuracy compared with the case when no adversaries are present, in almost all cases in Table 3. We observed a clear improvement when the mean worker accuracy is low, especially when only a few workers in the crowd are skilled enough to answer most tasks accurately. To illustrate this, we plot the histogram of worker accuracies for a sample run of the Dermatology dataset in Figure 2 (corresponding to the lowest mean accuracy in Table 3). From the results in Table 3, for low mean accuracies, we can see that our proposed scheme is able to identify and leverage the expertise of a small subset of highly skilled workers, thus defeating "tyranny of the masses".

5.3   MTurk Experiment

Fig. 3. The MTurk experiment: a few of the sample images

We designed an image labeling task where workers had to provide the country of origin, choosing from *Afghanistan, India, Iran, Pakistan, Tajikistan*. Some of the regions in these countries look very similar in their culture, geography, and demography and hence only people with domain experience and a deep understanding of the region will likely know the true answer. For instance, the blue rickshaws are typical to Pakistan and the yellow taxis are more common in Kabul. One can also guess *e.g.* from the car models on the street or from the script on street banners. We posted a task consisting of 50 such images on Amazon MTurk [1] and asked all workers to upload a file with their answers on all tasks. We received responses from 62 workers. In order to evaluate under the scenario where workers answer only a subset of the tasks, for each task, we used answers from a sampled set of workers using a randomly generated degree regular bipartite graph consisting of worker and task nodes. A worker's answer

Table 3. Experiments using UC Irvine datasets

Number of erroneous non-probe tasks for Pen Digits dataset

Worker accuracy mean	Worker accuracy variance	Task accuracy variance	Without adversarial workers				With 10% adversarial workers			
			Simple Plurality	Weighted Plurality	Proposed with Probes	Proposed Without Probes	Simple Plurality	Weighted Plurality	Proposed with Probes	Proposed without Probes
24.1	262.8	24.2	23.8	11	8.4	9.4	45.2	13.8	5.2	5.2
26.2	294.2	32.5	22	10.4	8.6	8.6	29	13.8	5.3	5.4
29.5	465.5.3	42.9	17.6	11.8	9.2	9.5	21.4	14.4	5.8	6.4
33.9	485.5.6	88.8	14	7.6	7.8	8.2	16.4	7.8	8	8.2

Number of erroneous non-probe tasks for Dermatology dataset

Worker accuracy mean	Worker accuracy variance	Task accuracy variance	Without adversarial workers				With 10% adversarial workers			
			Simple Plurality	Weighted Plurality	Proposed with Probes	Proposed Without Probes	Simple Plurality	Weighted Plurality	Proposed with Probes	Proposed without Probes
19.7	215.1	44.1	48	23.2	8.6	24.2	62.4	23	7.4	32.4
22.6	399.8	66.3	40.2	12.2	8.2	20.4	50.4	11.2	7	17.4
26.9	480.3	76.6	22.4	8.6	6.8	7	30	8.4	4	4.6
27.6	563.1	88.9	26.8	9.6	5.4	5.6	36.6	9	2.2	2.8
31.7	632.2	110.3	14.4	4	4	4	19	5	3.6	3.8

Number of erroneous non-probe tasks for Nominal dataset

Worker accuracy mean	Worker accuracy variance	Task accuracy variance	Without adversarial workers				With 10% adversarial workers			
			Simple Plurality	Weighted Plurality	Proposed with Probes	Proposed Without Probes	Simple Plurality	Weighted Plurality	Proposed with Probes	Proposed without Probes
45.4	575.4	339.1	19	18.8	18.2	18.4	22.6	18.8	18.1	18.1
49.1	599.2	328.6	10.4	8.4	6.1	7.7	11.6	8.2	5.9	7
51.3	649.9	325.1	9.8	7.8	5.6	7.7	12.2	6.6	5.4	6.1
52.6	678.5	298.5	9.6	6.2	4.7	5.1	11	4.8	2.9	3.2
54.3	878.1	233.1	4.2	3.2	2.6	2.9	5.2	3	2.3	2.4

Number of erroneous non-probe tasks for Vowel dataset

Worker accuracy mean	Worker accuracy variance	Task accuracy variance	Without adversarial workers				With 10% adversarial workers			
			Simple Plurality	Weighted Plurality	Proposed with Probes	Proposed Without Probes	Simple Plurality	Weighted Plurality	Proposed with Probes	Proposed without Probes
24.2	390.1	35.6	25.6	15.8	15.6	16.2	26.2	15.6	10	11.8
26.8	412.4	42.3	23.1	14.2	15.7	15.9	26	14.7	9.4	10.4
29	445.8	57.4	22.5	13.8	15.2	15.6	42.4	14.4	5.2	9.6
32.2	624.6	81.8	17.8	12.2	13.6	13.6	25	13	7.6	7.8
35.9	716.2	118.9	16.4	13.6	15.6	16.4	20	13.4	9.2	12.2

to a task was used only when a link existed in the bipartite graph between the two corresponding nodes. Table 4 shows the average number of erroneous crowd-aggregated answers for the methods under study as we varied the number of tasks assigned to each worker. The average was computed over 5 trials, each time generating a new instance of a random bipartite graph and using 5 randomly chosen probe tasks.

Table 4. MTurk Experiment: Average number of erroneous tasks

Assignment Degree	Simple Plurality Voting	Weighted Plurality Voting	Proposed Scheme with Probes	Proposed Scheme without Probes
10	15.4	18.2	13.4	13.9
20	14.2	13.6	7.2	7.2
30	14.2	11.2	5.2	5.2
40	12.2	7.6	4	4.2
50	13.4	8	3.4	3.4

6   Related Work

This paper develops novel approaches for inference in multicategory crowdsourcing tasks by explicitly modeling worker skill and intention and task difficulty. Our approach is specifically designed to overcome "tyranny of the masses" by de-emphasizing low-skilled workers and identifying malicious workers to improve crowd aggregation. Stochastic models for the generation of workers' answers have been previously considered in [8], [18], [23], [22]. In [8] the parameters of the model are the per-worker confusion matrices that are jointly estimated with the ground truth answers. This method was extended in [24], where a distinct probability distribution was given over the answers for every task-worker pair. But [24] does not consider a per-task parameter representing task difficulty. In our model, the distribution over answers for a task explicitly depends on the task difficulty that is essentially "perceived" by workers attempting to solve it. More specifically, we explicitly model the *interplay* between worker skill and task difficulty. In [23] and [22], task difficulty was considered explicitly, but only for the binary case. [15] considered task difficulty for ordinal data, wherein a higher difficulty level adds more variance to the distribution of workers' elicited answers without affecting the mean (which equals the ground truth value). Our method of incorporating task difficulties is novel, as we use them in a comparative paradigm with worker skill in our soft threshold-based model. Unlike [22], which assumes all tasks are drawn from the same classification domain, our approach is applicable even when the batch of tasks is *heterogeneous, i.e.*, not necessarily all drawn from the same (classification) domain. Adversarial workers in the binary case were accounted for in [23] and [14]. In this work, we characterize adversarial behavior for a more generalized (multicategory) setting. Moreover, we propose several malicious worker models and our approach exploits responses from (simple) adversaries to actually improve overall performance. [14] and [13] consider other statistical methods, such as correlation-based rules. In those works, the worker accuracies and task difficulties are computed as messages over the bipartite graph of worker-to-task assignments in an unsupervised way. These methods have been studied for binary classification tasks and their extension to multiclass problems does not seem very straightforward.

7 Conclusion

In this paper we studied a crowd aggregation method which is robust to the presence of a large number of low skilled and malicious workers in the crowd. In our approach, we proposed a stochastic generative model of workers' responses along with an inference mechanism based on a generalized EM algorithm. We chose our model in order to succinctly represent different worker types such as spammers, low-skilled and malicious workers. We also proposed a complex adversary which is a more evasive variant of a simple adversary. In future, we would like to evaluate the more complex adversarial models experimentally. We would also like to explore the possibility of alternative attacks on crowdsourcing systems, for instance collusion attacks, where a group of adversarial workers collude and submit the same (but incorrect) answer for a task.

References

1. Amazon Mechanical Turk, http://www.mturk.com
2. Flower, C.: http://crowdflower.com
3. Funding, C.: http://www.mid-day.com/news/2013/feb/150213-twos-company-tweets-a-crowdfunder.htm
4. Topcoder, http://topcoder.com
5. 50,000 volunteers join distributed search for Steve Fossett. Wired (September 11, 2007)
6. Bache, K., Lichman, M.: UCI machine learning repository (2013)
7. Benbouzid, D., Busa-Fekete, R., Casagrande, N., Collin, F.D., Kégl, B., et al.: Multiboost: a multi-purpose boosting package. Journal of Machine Learning Research 13, 549–553 (2012)
8. Dawid, A.P., Skene, A.M.: Maximum likelihood estimation of observer error-rates using the EM algorithm. Applied Statistics, 20–28 (1979)
9. Dempster, A.P., Laird, N.M., Rubin, D.B.: Maximum likelihood from incomplete data via the EM algorithm. Journal of the Royal Statistical Society. Series B (Methodological), 1–38 (1977)
10. Douceur, J.R.: The sybil attack. In: Druschel, P., Kaashoek, M.F., Rowstron, A. (eds.) IPTPS 2002. LNCS, vol. 2429, pp. 251–260. Springer, Heidelberg (2002)
11. Feldman, M., Papadimitriou, C., Chuang, J., Stoica, I.: Free-riding and whitewashing in peer-to-peer systems. In: Proceedings of the ACM SIGCOMM Workshop on Practice and Theory of Incentives in Networked Systems, pp. 228–236 (2004)
12. Graham, M.W., Miller, D.J.: Unsupervised learning of parsimonious mixtures on large spaces with integrated feature and component selection. IEEE Transactions on Signal Processing 54(4), 1289–1303 (2006)

13. Karger, D.R., Oh, S., Shah, D.: Budget-optimal crowdsourcing using low-rank matrix approximations. In: 49th IEEE Annual Allerton Conference on Communication, Control, and Computing, pp. 284–291 (2011)
14. Karger, D.R., Oh, S., Shah, D.: Iterative learning for reliable crowdsourcing systems. In: Advances in Neural Information Processing Systems (2011)
15. Lakshminarayanan, B., Teh, Y.W.: Inferring ground truth from multi-annotator ordinal data: a probabilistic approach. arXiv preprint arXiv:1305.0015 (2013)
16. Meng, X.L., Van Dyk, D.: The EM algorithm an old folk-song sung to a fast new tune. Journal of the Royal Statistical Society: Series B (Statistical Methodology) 59(3), 511–567 (1997)
17. Raykar, V.C., Yu, S.: Eliminating spammers and ranking annotators for crowdsourced labeling tasks. Journal of Machine Learning Research 13, 491–518 (2012)
18. Raykar, V.C., Yu, S., Zhao, L.H., Valadez, G.H., Florin, C., Bogoni, L., Moy, L.: Learning from crowds. Journal of Machine Learning Research 11, 1297–1322 (2010)
19. Schapire, R.E.: The strength of weak learnability. Machine Learning 5(2), 197–227 (1990)
20. Szarvas, G., Farkas, R., Kocsor, A.: A multilingual named entity recognition system using boosting and c4. 5 decision tree learning algorithms. In: Todorovski, L., Lavrač, N., Jantke, K.P. (eds.) DS 2006. LNCS (LNAI), vol. 4265, pp. 267–278. Springer, Heidelberg (2006)
21. Vukovic, M.: Crowdsourcing for enterprises. In: IEEE World Conference on Services-I, pp. 686–692 (2009)
22. Welinder, P., Branson, S., Belongie, S., Perona, P.: The multidimensional wisdom of crowds. In: Advances in Neural Information Processing Systems, vol. 6, p. 8 (2010)
23. Whitehill, J., Ruvolo, P., Wu, T., Bergsma, J., Movellan, J.: Whose vote should count more: Optimal integration of labels from labelers of unknown expertise. In: Advances in Neural Information Processing Systems, vol. 22, pp. 2035–2043 (2009)
24. Zhou, D., Platt, J., Basu, S., Mao, Y.: Learning from the wisdom of crowds by minimax entropy. In; Advances in Neural Information Processing Systems, pp. 2204–2212 (2012)

Quantifying Network Topology Robustness under Budget Constraints: General Model and Computational Complexity

Aron Laszka[1] and Assane Gueye[2]

[1] Laboratory of Cryptography and System Security (CrySyS Lab),
Budapest University of Technology and Economics
laszka@crysys.hu
[2] National Institute of Standards and Technology (NIST)
assane.gueye@nist.gov

Abstract. Recently, network blocking game (NBG) models have been introduced and utilized to quantify the vulnerability of network topologies in adversarial environments. In NBG models, the payoff matrix of the game is only *"implicitly"* given. As a consequence, computing a Nash equilibrium in these games is expected to be harder than in more conventional models, where the payoff matrix is *"explicitly"* given.

In this paper, we first show that computing a Nash equilibrium of a NBG is in general NP-hard. Surprisingly, however, there are particular interesting cases for which the game can be solved in polynomial time. We revisit these cases in a framework where the network is to be operated under budget constraints, which previous models did not consider. We generalize previous blocking games by introducing a budget limit on the operator and consider two constraint formulations: the maximum and the expected cost constraints.

For practical applications, the greatest challenge posed by blocking games is their computational complexity. Therefore, we show that the maximum cost constraint leads to NP-hard problems, even for games that were shown to be efficiently solvable in the unconstrained case. On the other hand, we show that the expected cost constraint formulation leads to games that can be solved efficiently.

Keywords: network topology robustness, robustness metrics, game theory, blocking games, computational complexity.

1 Introduction

Designing network topologies that are robust and resilient to attacks has been and continues to be an important and challenging topic in the area of communication networks. One of the main difficulties resides in quantifying the robustness of a network in the presence of an intelligent attacker, who might exploit the structure of the network topology to design harmful attacks. Quantifying the robustness or, equivalently, the vulnerability of topologies has been extensively

S.K. Das, C. Nita-Rotaru, and M. Kantarcioglu (Eds.): GameSec 2013, LNCS 8252, pp. 154–174, 2013.

studied [1–5]; however, the simultaneous and strategic decision making of the defender and the adversary, which is key to the security of information systems, has received only little attention.

To capture the strategic nature of the interactions between a defender and an adversary, game-theoretic models have been gaining a lot of interest in the study of the security of communication networks. In a recent line of research [6–10], *network blocking games* (NBGs) have been introduced and applied to the analysis of the robustness of network topologies. An NBG takes as input the communication model and the topology of a network, and casts the strategic interactions between an adversary and the defender, called the network operator, as a two-player game. The Nash equilibrium strategies are then used to predict the attacker's most likely actions; and the attacker's equilibrium payoff[1] serves as a quantification of the vulnerability (i.e., inverse robustness) of the network.

A particularity of NBG models is that the payoff matrix of the game is not given as an input. In other words, the strategy set of (at least) one player (and hence the payoff matrix) is only *implicitly* defined, and the actual strategy sets need to be computed from the input of the game (here, the communication model and the network topology). Furthermore, in most NBG models, checking whether a given action is a feasible strategy can be done efficiently; however, computing the complete strategy set is inherently difficult. For instance, in the game described in [6], the operator's strategy space is the set of feasible network flows. In general, checking whether a given flow is feasible can be done efficiently. However, computing the set of all feasible network flows (which is required for computing the payoff matrix) is difficult: the number of feasible flows is exponential in the number of nodes and links in the graph, so they cannot be enumerated in polynomial time.

Hence, with respect to the complexity of computing a Nash equilibrium, NBG models present two challenges: first, the game is only implicitly defined; second, the payoff matrix is potentially exponential in size. Thus, solving network blocking games can be expected to be harder than solving games for which the payoff matrix is "explicitly given". Recall that computing a NE for "explicit" two-player games has been shown to be PPAD-complete (*Polynomial Parity Arguments on Directed graphs*), a class of problems that are believed to be hard, but not necessarily NP-hard [11]. In this paper, we show that computing a Nash equilibrium of a network blocking game is NP-hard in general.

Interestingly though, in the series of NBG papers cited above, new algorithms have been developed to *efficiently* compute a Nash equilibrium in a number of communication models: All-to-All (e.g., Ethernet) networks with constant [7] and linear loss [9], All-to-One (e.g., access and sensor) networks [10], and Supply-Demand networks [6]. These algorithms are mostly based on the theory of network flows and, for some models, on the minimization of submodular functions. More precisely, the problem of finding a Nash equilibrium is cast as a network

---

[1] It has been shown that the attacker's payoff is the same in every equilibrium of a network blocking game; thus, it suffices to find a single equilibrium in order to characterize the robustness of a network.

flow problem (or a submodular function minimization problem), which enables bypassing the computation of the payoff matrix. In this paper, we revisit some of these models and discuss the complexity of computing their NE in scenarios where the network operator has access to only a limited budget to operate the network.

Such budget constraints were not considered in previous NBG models, which implicitly assume that the operator can use the network elements at zero cost. However, this assumption is not realistic: indeed, links in a network have positive usage costs (e.g., operation/maintenance costs, protection costs) and these costs might be non-uniform. Since network operators do not have an unlimited budget, they cannot use any combination of network element. In [6], a usage cost model as well as a budget constraint have been introduced for the particular case of Supply-Demand (S-D) networks. This budget constraint means that the network operator can use a set of network elements (links) only if its associated cost does not exceed a given budget.

In the present paper, we extend the budget constraint idea to network blocking games in general, and provide a number of complexity results with regard to the computation of the equilibrium payoff. Recall that the aim of solving these models is to derive a quantification of the network's robustness in the presence of a strategic adversary, and that the equilibrium payoff is used as the vulnerability metric. Thus, computational complexity is of central importance in these models, and analyzing it is the primary goal of this paper.

This paper builds upon the studies in [6] and [12], but considers a more general setting and presents many additional results compared to those papers. [6] is the first study to introduce the idea of a budget limit and usage costs in the context of a NBG. However, it considers only the special case of Supply-Demand networks and (what we call here) the maximum cost constraint. Furthermore, it does not provide a complexity analysis. [12] presents a complexity analysis and introduces a new constraint formulation (the expected cost constraint), but limits the discussion to the special case of the All-to-One communication model with zero attack costs. In the present paper, we consider a unifying framework and provide a thorough complexity analysis for NBGs in general. The main contributions of this paper are the following:

- In Section 3, we show that solving a blocking game is generally NP-hard (Theorem 1).
- In Section 4, we generalize the network blocking game model by introducing a budget limit for the operator. We consider two constraint formulations: the maximum cost constraint (MCC) and the expected cost constraint (ECC).
- In Section 5, we show that the problem of determining the equilibrium payoff is NP-hard under the MCC in the previously proposed models, which can be solved efficiently in the unconstrained game (Theorem 2).
- In Section 6, we show how to solve the game under the ECC in polynomial time given a linear characterization of the operator's mixed strategy space (Theorem 3).

Notational Conventions. We use lower case bold letters (e.g., $\boldsymbol{\alpha}$) and upper case bold letters (e.g., $\boldsymbol{S}$) to denote column vectors and matrices, respectively. We use the prime sign ($'$) to denote transpose, and subindices (e.g., $\boldsymbol{\alpha}_T$) to refer to elements of vectors.

2   Unconstrained Network Blocking Games

In this section, we summarize the previous work on network blocking games. Since these models do not consider a budget constraint, we will refer to them as *unconstrained network blocking games* when the distinction is important.

As it was stated earlier, network blocking games are defined by the communication model and the topology of the network. The topology of the network is represented by a connected simple graph $G = (V, E)$, where V is the set of nodes and E is the set of links. The edges can be undirected or directed depending on the communication models (as we will see later). The network operator wants to guarantee *"some" connectivity* between the nodes of the network. For this, she selects a collection $T \subseteq E$ of the links as the communication infrastructure. The type of *connectivity* and the set of feasible collections (denoted by $\mathcal{T}$) are determined by the communication model (see the next subsection for examples of communication models).

Assume that the operator chooses collection T for her communication and that a given link e in the network fails. In this paper, we only consider failures that are due to the actions of a malicious and strategic adversary. If $e \notin T$, then the communication is not affected at all. If, on the other hand, $e \in T$, then e can longer be used: the operator incurs some *usage loss*, which is how much she would transmit on the link if it were intact. For a given T and e, we let $\boldsymbol{\lambda}(T, e)$ denote this usage loss (or zero if $e \notin T$). Notice that all results presented in this paper also hold if the attacker is allowed to attack nodes as well[2], but we restrict our analysis to link attacks only due to the lack of space.

2.1   Communication Models

The communication model defines the type of "connectivity" that the network operator is trying to achieve, the set of feasible collections which she can use for that, and the usage losses $\boldsymbol{\lambda}(T, e)$ for the network elements. Next, we introduce the three communication models that are of interest in this paper.

All-to-One Model. In an *All-to-One* network [10], the primary goal of the network operator is to enable all nodes to communicate with a designated node r. This models sensor and access networks, where all nodes are trying to reach a gateway or data collection node (or, alternatively, a set of nodes, which can be modeled by a designated super-node).

To get all nodes connected to r, the network operator chooses a collection of links T that forms a spanning tree. Hence, the set of feasible collections $\mathcal{T}$ is the

---

[2] The results for both node and edge attacks can be derived using vertex splitting.

set of all spanning trees. In practice, a spanning tree can be implemented, for example, as the next-hop forwarding table entries for r, which are stored at the individual nodes of the network.

Let the network be connected using a spanning tree T. Then, if a given link $e \in E$ fails, some nodes might no longer be able to communicate with r and can be considered lost for the network operator. Thus, we define the usage (loss) $\lambda(T, e)$ as *the number of those nodes that are disconnected from r*.

All-to-All Model. In an *All-to-All* network [7, 9], the goal of the network operator is to enable each node to communicate with every other node, using the minimum number of links. For example, this is the case for bridged Ethernet LANs, where every node should be able to "logically" communicate with every other node, but the topology has to be loop-free. Assuming that links are undirected, spanning trees are the subgraph structures that (looplessly) connect all nodes with the minimum number of links. Hence, the network operator selects a spanning tree as communication infrastructure. Thus, the set of feasible collections $\mathcal{T}$ corresponds to the set of all spanning trees.

Let the network be connected using a spanning tree T and assume that link e fails. If link e does not belong to T, then the network remains connected and the operator does not lose any connectivity. If, on the other hand, $e \in T$, the network is cut into two separate components that are unable to communicate. Now, if e is a link connecting a leaf to the rest of the spanning tree, only that leaf gets disconnected and all the other nodes can still reach each other. In this case, the operator loses some connectivity, but the loss can be considered *minor*. If, on the other hand, the removal of link e cuts the network into two components of comparable size, then connections between many pairs of nodes are now missing, and the loss to the operator is considerably *larger*. In general, the more fractured the network is, the more severe the loss is. To capture this phenomenon, the usage (loss) $\lambda(T, e)$ is defined as *the size of the smaller connected component of $G(V, T \setminus e)$*, where $G(V, T \setminus e)$ is the subgraph containing only the links in $T \setminus e$.

Supply-Demand Model. In a Supply-Demand (S-D) network [6], the operator wants to carry a fixed amount of goods from a nonempty set $S \subseteq V$ of "source" nodes to a nonempty set $D \subseteq V$ of "destination" nodes using the network links. We assume that $S \cap D = \emptyset$ and that network links are directed. With each node $u \in S$, we associate a nonnegative number $s(u)$, the "supply" at u, and with each node $u \in D$, we associate a nonnegative number $d(u)$, the "demand" at u. We consider *uncapacitated* networks, where each link can carry an unlimited amount of goods[3]. We also assume that links carry only *integer* amounts of goods and that the total amount of goods to be carried from S to D is also a given positive integer.

---

[3] The analysis of *capacitated* network follows from the study in this paper, but it is not considered in this paper due to space limitation.

To transport the goods, the network operator chooses a collection of links that forms a *feasible (integer) flow*. A feasible flow $T \in \mathcal{T}$ is a function that assigns to each link e the amount of goods $T(e)$ (≥ 0) it carries, such that the *conservation of flow* property is satisfied at each node. Hence, the set of collections $\mathcal{T}$ is equal to the set of all feasible flows.

The usage (loss) $\lambda(T, e)$ is defined to be the *amount of goods $T(e)$ that flow T assigns to link e.* This is how much the operator will lose if she uses a feasible flow $T \in \mathcal{T}$ and link e fails.

2.2 Game-Theoretic Measure of Robustness

Given the communication model and the topology of the network, a two-player game is defined between the network operator and a strategic attacker. The network operator wants to guarantee "some" connectivity by choosing a *feasible* collection of links in the network (i.e., her strategy space is the set $\mathcal{T}$ of feasible collections). The type of connectivity and the set of feasible collections are defined by the communication model, as previously discussed. At the same time, a strategic and malicious adversary is trying to disrupt the communication by attacking a link (i.e., her strategy space is the set E of links in the network). We assume that to successfully attack a link e, the adversary has to spend some effort which is quantified by μ_e. The players' payoffs are defined as follows: when the operator picks collection T and the attacker targets link e, the operator loses $\lambda(T, e)$ (as defined above), and the attacker gets a net reward of $\lambda(T, e) - \mu_e$. The attacker also has the option not to launch an attack, which results in zero loss for the operator and zero gain for the attacker.

We consider mixed strategy Nash equilibria, where the network operator chooses a distribution (denoted by α) over the set $\mathcal{T}$, and the attacker chooses a distribution (denoted by β) over the set E or the option of not attacking. We assume that the operator tries to minimize her *expected loss*, while the attacker tries to maximize her *expected net reward*. Formally, the operator chooses α to minimize $L(\alpha, \beta)$ defined as

$$L(\alpha, \beta) = \sum_{T \in \mathcal{T}} \sum_{e \in E} \alpha_T \beta_e \lambda(T, e) , \tag{1}$$

while the attacker chooses β to maximize $R(\alpha, \beta)$ defined as

$$R(\alpha, \beta) = L(\alpha, \beta) - \sum_{e \in E} \beta_e \mu_e \tag{2}$$

or not attacking if the maximum is negative.

Since the attacker has the option not to attack and get a payoff of zero, it is not hard to show that there does not exist an equilibrium in which the attacker receives a negative expected payoff. We let θ^* be the attacker's equilibrium payoff, which has been shown [13] to be the same in all equilibria. As a consequence, θ^* is uniquely defined. The next subsection gives a characterization of θ^* using the theory of blocking pairs of polyhedra.

2.3   Equilibrium Characterization Based on Blocking Pairs of Polyhedra

Here, we recall the notions of polyhedra and blockers, and discuss how they can be used to characterize the Nash equilibria of the game (see [13, Chap. 4] for more details).

Let Λ be the operator's payoff matrix, whose rows are $(\boldsymbol{\lambda}_T,\ T \in \mathcal{T})$, where the entries of the vector $\boldsymbol{\lambda}_T \in \mathbb{R}^{|E|}_{\geq 0}$ are given by $\lambda(T, e),\ e \in E$. We define its associated polyhedron P_Λ as the vector sum of the convex hull of the row vectors $(\boldsymbol{\lambda}_T,\ T \in \mathcal{T})$ and the nonnegative orthant. This polyhedron can be represented as

$$P_\Lambda = \left\{ \boldsymbol{x} \in \mathbb{R}^{|E|}_{\geq 0} \,\middle|\, \exists \boldsymbol{\alpha} \in \mathbb{R}^{|\mathcal{T}|}_{\geq 0} \left(\Lambda' \boldsymbol{\alpha} \leq \boldsymbol{x} \ \wedge \ \boldsymbol{\alpha}' \mathbf{1} \geq 1 \right) \right\}. \tag{3}$$

The *blocker* of P_Λ is the polyhedron defined as

$$bl(P_\Lambda) := \left\{ \boldsymbol{y} \in \mathbb{R}^{|E|}_{\geq 0} \,\middle|\, \boldsymbol{y}' \boldsymbol{x} \geq 1 \ \forall \boldsymbol{x} \in P_\Lambda \right\}. \tag{4}$$

For each vertex $\boldsymbol{\omega} = (\omega_e, e \in E)$ of the blocker, define the quantity

$$\theta(\boldsymbol{\omega}) := \frac{1}{\sum_{e \in E} \omega_e} \left(1 - \sum_{e \in E} \omega_e \mu_e \right). \tag{5}$$

A vertex of the blocker is called critical if it maximizes the quantity $\theta(\boldsymbol{\omega})$, i.e., $\theta(\boldsymbol{\omega}) = \max_{\tilde{\boldsymbol{\omega}}} \theta(\tilde{\boldsymbol{\omega}})$. Finally, let $\tilde{\theta}$ denote the maximum quantity.

In [13], it has been shown that every Nash equilibrium strategy for the attacker is a critical vertex or a convex combination of critical vertices, and that the attacker's equilibrium payoff is $\theta^* = \max(0, \tilde{\theta})$. As a consequence, if this blocker can be "efficiently" characterized, then an efficient algorithm can be derived to solve the maximization problem and, hence, the game.

2.4   Vulnerability/Robustness Metric

In the analysis of the general NBG [13, Chap. 4], it has been shown that θ^* is a property of (i.e., solely determined by) the topology of the network, the communication model, and the attack costs $\boldsymbol{\mu}$. Furthermore, this unique equilibrium payoff reflects both the network operator's expected loss due to attack as well as the attacker's willingness to attack. For a given $\boldsymbol{\mu}$, a low θ^* indicates that operating the network has low expected loss due to attack, that is, the network is robust against attacks. If, on the other hand, θ^* is high, then the expected loss is also high, and the network can be considered vulnerable. As such, θ^* has been proposed [7] as a measure of network topology vulnerability (i.e., inverse robustness) in an adversarial environment. Another property of θ^* is that, when $\boldsymbol{\mu} = \mathbf{0}$ (the case of the most powerful attacker), it can be related to well-known graph-theory notions. For instance, in the All-to-One model, θ^* was shown to be the inverse of the *persistence* of the graph of the network [10], a metric that has previously been proposed in [14] to quantify graph robustness (although in a non-game theoretic framework). In the All-to-All model with constant loss [7], θ^* can be related the *spanning tree packing number of the graph* [15]. In the All-to-All model with linear loss [9], θ^* is closely bounded by the *Cheeger constant* [16] (also called the *edge-expansion*) of the graph. In the Supply-Demand model [6], the metric is equal to the maximum average flow traversing an edge-cut, where the average is obtained by dividing the total flow by the size of the edge-cut.

As a metric for robustness, understanding the computational complexity of calculating θ^* is of primal importance. In the next section, we discuss the complexity of computing a Nash equilibrium in the unconstrained NBG model.

3   Computational Complexity of the Unconstrained Game

In this section, we show that solving a NBG is NP-hard in general. Recall that computing a Nash equilibrium in general two-player games has been shown to be PPAD-complete. Zero-sum, two-player games, on the other hand, can be cast as linear programs and, hence, can be solved in polynomial time using linear programming tools. In all these cases, the input of the computational problem is assumed to be the payoff matrix. For NBG models however, only an *implicit* description of the payoff matrix is available. In addition, the payoff matrix is potentially exponential in size, which makes NBG models even more challenging to deal with. The following theorem shows that, indeed, computing a NE for a general blocking game is NP-hard. We prove this by reducing a well-known NP-hard problem, the Knapsack Problem (KP), to the problem of computing the attacker's equilibrium payoff, which we formalize as the Equilibrium Problem (EP). The KP and the EP are formally defined as follows.

Definition 1 (Knapsack Problem [KP]). *Given N items, where item i has weight c_i and value v_i, a capacity C, and a value V, is there a subset S whose sum weight is at most C, i.e., $\sum_{i \in S} c_i \leq C$, and whose sum value is at least V, i.e., $\sum_{i \in S} v_i \geq V$?*

Definition 2 (Equilibrium Problem [EP]). *Given a set of elements E, a polynomial-time function $I_{T \in \mathcal{T}}$ for testing $T \in \mathcal{T}$, a polynomial-time function $\lambda(T, e)$, a vector of attack costs $\boldsymbol{\mu} \in \mathbb{R}^{|E|}_{\geq 0}$, and a payoff value p, is the adversary's equilibrium payoff less than or equal to p?*

The above formulation of EP allows us to easily show the computational complexity of all the problems relevant to NBGs. First, if the adversary's equilibrium payoff can be efficiently computed, then EP can also be solved efficiently. Conversely, if EP is NP-hard, then computing the adversary's equilibrium payoff is also necessarily NP-hard. Second, for similar reasons, we also have that computing the equilibrium strategies of the game is also at least as hard as EP.

The following theorem shows that EP is NP-hard.

Theorem 1. *The Knapsack Problem is polynomial-time reducible to the Equilibrium Problem.*

The proof of the theorem can be found in Appendix A.

Thus, solving a NBG is NP-hard in general. Interestingly, however, efficient algorithms have been derived to compute a NE for the models discussed in Subsection 2.1. In the following sections, we introduce a budget constraint and revisit the complexity of computing a NE of the constrained game in those models.

4   Budget Contraints

In the unconstrained NBG model, the operator is only interested in minimizing her expected loss due to attacks, without taking her operating costs into account. In practice, however, network operators also have to take economic goals and constraints into consideration when choosing their strategies. These economic decisions are affected by the topology of the network as links and, hence, feasible collections of links can have varying usage costs.

4.1   Unit Usage / Protection Cost

In [6], a (per unit) usage cost model was introduced and discussed for the particular case of the S-D communication model. Here, we extend this cost model to the general NBG. Recall that $\lambda(T, e)$ quantifies the usage (loss) associated with collection T and link e. We assume that each link e has some *unit usage cost* w_e, so that using the link costs $w_e \lambda(T, e)$ to the operator. With this definition, the *total cost of* using *a collection* T is

$$w(T) := \sum_{e \in E} \lambda(T, e) w_e \; ; \tag{6}$$

and the network operator's *expected usage cost of a mixed strategy* α is

$$w(\alpha) := \sum_{T \in \mathcal{T}} \alpha_T w(T) = \sum_{e \in E} w_e \sum_{T \in \mathcal{T}} \alpha_T \lambda(T, e) \; . \tag{7}$$

We assume that, to run the network, the operator has a fixed *budget* $b \in \mathbb{R}_{\geq 0}$ to spend. Therefore, her objective is to minimize the expected loss (see Equation (1)) by choosing an optimal strategy that satisfies her budget constraint. This budget constraint can be formulated in multiple ways. In the following sections, we introduce and study two straightforward formulations, the maximum and the expected (or average) cost budget constraints.

4.2   Maximum Cost Budget Constraint

In the first budget constraint formulation, which we refer to as the *maximum cost constraint* (MCC), we require that for a given budget b, the operator only uses collections whose total costs (see Equation (6)) are less than or equal to b. Formally, the pure strategy set of the operator is restricted to

$$\mathcal{T}^{(b)} = \{T \in \mathcal{T} \mid w(T) \leq b\} \; . \tag{8}$$

The maximum cost constraint is best-suited for budget limits that are determined by the amount of preallocated resources available. In this case, the cost of a link can be the amount of resources needed (e.g., energy consumption) to operate the link and the budget limit can be the amount of resources available (e.g., amount of power available).

4.3 Expected Cost Budget Constraint

The maximum cost constraint misses to capture certain situations. For instance, when the amount of allocated resources can be modified during operation, e.g., resources can be leased, the budget limit should apply to the average or, equivalently, the expected cost of a strategy during continuous periods of operation. Thus, in our second budget constraint formulation, which we will refer to as the *expected cost constraint* (ECC), we only require the expected (or average) cost of the operator to not exceed the budget limit.

Under the *expected cost constraint* with a budget limit b, the operator can employ a mixed strategy only if its expected cost (see Equation (7)) is less than or equal to b. Formally, the set of mixed strategies available to the operator is

$$\mathcal{A}^{(b)} = \left\{ \boldsymbol{\alpha} \in \mathbb{R}^{|\mathcal{T}|} \middle| w(\boldsymbol{\alpha}) \leq b \right\} . \tag{9}$$

Note that the above formulation generalizes the classic notion of mixed strategies in game-theory, where the set of mixed strategies is always the set of *all* distributions over the set of pure strategies. Here, a mixed strategy is chosen from a predefined subset of distributions.

4.4 Constrained Game

Having defined the set of available strategies (pure for MCC and mixed for ECC), we can now setup the constrained game in a similar way to the unconstrained game presented in Subsection 2.2. We are interested in mixed strategy Nash equilibria, where the operator picks a distribution $\boldsymbol{\alpha}$ over $\mathcal{T}^{(b)}$ (for MCC) or from the set $\mathcal{A}^{(b)}$ (for ECC), while the attacker chooses a distribution $\boldsymbol{\beta}$ over the set of links. The attacker's Nash equilibrium payoff is denoted $\theta^*(b)$ for a game with budget limit b.

Using the same interpretation as in Subsection 2.2, the attacker's NE payoff $\theta^*(b)$ can be used to quantify the vulnerability (i.e., inverse robustness) of the network when the operator's budget is b. By varying b, one can draw the Pareto frontier between the region of achievable vulnerability/budget points and the region of unachievable ones, as was done in [6] for the particular case of S-D networks with the maximum cost constraint.

Remark. In the next two sections, we discuss the complexity of solving the constrained blocking game. However, since the unconstrained NBG is in general NP-hard (see Theorem 1), we readily have that solving a NBG under a budget constraint[4] is also NP-hard in general. Therefore, we focus our discussion on

---

[4] The unconstrained game is the special case of $b \to \infty$.

the communication models introduced in Subsection 2.1, for which there exist efficient algorithms to compute the NE payoff in the unconstrained game.

5   NP-Hardness of the Maximum Cost Constraint

In this section, we show that computing the equilibrium payoff of the network blocking game with a maximum cost budget constraint is NP-hard for the models that were previously shown to be efficiently computable without a budget constraint.

Theorem 2. *Computing the NE payoff with a maximum cost budget constraint is NP-hard for the (a) S-D communication model, the (b) All-to-All communication model, and the (c) All-to-One communication model.*

Proof. We show NP-hardness by reducing a well-known NP-hard problem, the *Partition Problem (PP)* [17], to the problem of deciding whether the equilibrium payoff in a given network model with a maximum cost constraint is at most a certain value. We refer to the latter problem as the *Equilibrium Problem with Maximum Cost Constraint* (EPMAX).

Definition 3 (Partition Problem [PP]). *Given a multiset of positive integers* $\{x_1, \ldots, x_n\}$*, is there a partitioning of the multiset into two disjoint subsets A and B such that $\sum_{x \in A} x = \sum_{x \in B} x$?*

Definition 4 (Equilibrium Problem with Maximum Cost Constraint [EPMAX]). *Given a communication model, a network G, a budget limit b, and a payoff value p, is the adversary's equilibrium payoff less than or equal to p?*

For each communication model, we show how an instance of *EPMAX* (i.e., a network, a budget limit and a payoff value) can be constructed in polynomial time from an instance of *PP*. Since the proof techniques follow the same lines for all models, we only give a full proof for the S-D model. For the All-to-All model, we describe the main points of the proof in Appendix B without providing the details. For the All-to-One model, the proof can be found in [12].

To simplify the notations in our proofs, we also define the *expected loss* of an edge $e \in E$ in a given operator strategy α as

$$L(e) = \sum_{T \in \mathcal{T}} \alpha_T \lambda(T, e) . \tag{10}$$

Proof of Theorem 2 for the S-D Communication Model

Given an instance of *PP*, we build an instance of *EPMAX* as follows.
  - Let the topology of the network be the following (see Figure 1): There is one source node, denoted by s, one sink node, denoted by d, and $3n - 1$ other nodes, which are denoted by 1_a, 1_b, 1, 2_a, 2_b, 2, ..., n_a, and n_b.

Fig. 1. Illustration for the proof of Theorem 2 for the S-D model. Numbers along the edges indicate unit costs.

Node s is connected to nodes 1_a and 1_b with edges having unit costs of x_1 and 0, respectively. Nodes i_a and i_b, $i < n$, are connected to node i with edges having zero unit cost. Node i is connected to nodes $(i+1)_a$ and $(i+1)_b$ with edges having unit costs of x_{i+1} and 0, respectively. Finally, nodes n_a and n_b are connected to node d with edges having zero unit cost.
- Let the capacity of the links and the amount of goods to be moved from s to d be 1.
- Let the operator's budget be $b = \frac{1}{2}\sum_{i=1}^{n} x_i$.
- Let the equilibrium payoff value be $p = \frac{1}{2}$.

We claim that the equilibrium payoff in the above network is greater than p iff PP does not have a solution.

First, we assume that the set can be partitioned into two subsets A and B of equal sum, that is, PP has a solution. In this case, we have to show that the equilibrium payoff is at most $\frac{1}{2}$. First, notice that since the total amount of goods to be moved from s to d is 1, the set of feasible integer flows is equal to the set of s-d paths as the amount of flow on each edge is either 0 or 1. Now, we show that there exist two disjoint paths (or flows) that satisfy the budget constraint. The first path (i.e., set of links with positive flow values) consists of the edges $(i-1, i_a)$ and (i_a, i) for each $x_i \in A$ and $(i-1, i_b)$ and (i_b, i) for each $x_i \notin A$. The second path consists of the remaining edges. In other words, the first flow takes the "path above" whenever $x_i \in A$ and the "path below" whenever $x_i \notin A$, while the second flow does the contrary. It is easy to see that the cost of both flows is $\sum_{x_i \in A} x_i = \sum_{x_i \in B} x_i = \frac{1}{2}\sum_i x_i$; thus, they satisfy the maximum budget constraint. By assigning $\frac{1}{2}$ probability to each flow, we obtain an operator strategy in which the expected loss of every edge is at most $\frac{1}{2}$. If the operator employs this strategy, the payoff of every pure and, consequently, every mixed adversarial strategy is at most $\frac{1}{2}$. Therefore, the equilibrium payoff has to be at most $\frac{1}{2}$.

Second, we assume that the set cannot be partitioned into two subsets of equal sum, that is, PP does not have a solution. If the equilibrium payoff of the game were at most $\frac{1}{2}$, then there would exist an operator strategy α in which the expected loss of every edge is at most $\frac{1}{2}$. We show that no such strategy can exist.

Because of the maximum cost budget constraint, the cost of every pure strategy is less than or equal to $b = \frac{1}{2}\sum_i x_i$. Moreover, this inequality is *strict* as every pure strategy is an s-d path and, if its cost is equal to b, there must exist a subset of links $I \subsetneq \{1, 2, \ldots, n\}$ such that $\sum_{i \in I} x_i = b$. By letting $A = \{x_i \mid i \in I\}$ and $B = \{x_i \mid i \notin I\}$ we get a solution for PP, which would contradict the assumption that the set cannot be partitioned. Thus, the cost of every pure strategy is strictly less than b and, as a consequence, the expected cost of every mixed strategy is also strictly less than b; formally,

$$\sum_{e \in E} L(e) w_e < b = \frac{1}{2} \sum_{i=1}^{n} x_i = \sum_{e \in E} \frac{1}{2} w_e \ . \tag{11}$$

Now, recall that the expected loss $L(e)$ of an edge e in the S-D model is equal to the expected amount of flow on that edge. Since the total amount of goods to be moved is equal to 1 and since each pair of "above" and "below" edges (e.g., e_a and e_b) is an s-d cut, the sum of the flows on any pair of "above" and "below" edges is at least 1. Thus, for every pair of edges e_a and e_b, $L(e_a) + L(e_b) \geq 1 = \frac{1}{2} + \frac{1}{2}$. Combined with our initial assumption that the expected loss of each edge is at most $\frac{1}{2}$, we have that

$$\forall e \in E: \ L(e) = \frac{1}{2} \tag{12}$$

and

$$\sum_{e \in E} L(e) w_e = \sum_{e \in E} \frac{1}{2} w_e \ . \tag{13}$$

But this leads to a contradiction with Equation 11, showing that if PP does not have a solution, then the equilibrium payoff is greater than $\frac{1}{2}$, which concludes our proof. □

6   Efficient Algorithms for the Expected Cost Constraint

In this section, we show how the expected cost constrained game can be solved efficiently for the models introduced in Subsection 2.1. In Subsection 2.3, we gave a derivation of the attacker's Nash equilibrium payoff in the unconstrained game model using the theory of blocking pairs of polyhedra. In this section, we use a similar derivation to show how polynomial-time algorithms can be derived to solve the game with the expected cost constraint. The same detailed analytical steps presented in [13, Chap. 4] (for the unconstrained game) can be followed to show the same results for the constrained game. In this case, the definition of the polyhedron P_Λ in Equation (3) includes an additional linear inequality (given by Equation (9)) that corresponds the budget constraint. Since the expected cost $w(\alpha)$ in Equation (9) can also be formulated as $w(\alpha) = w'\Lambda'\alpha$, the constrained polyhedron can be written as

$$P_\Lambda := \left\{ x \in \mathbb{R}_{\geq 0}^{|E|} \ \middle| \ \exists \alpha \in \mathbb{R}_{\geq 0}^{|T|} \left(\Lambda'\alpha \leq x \ \wedge \ \alpha'1 \geq 1 \ \wedge \ w'\Lambda'\alpha \leq b \right) \right\}. \tag{14}$$

Notice that the definition of P_Λ above involves the matrix Λ, which is generally exponential in size. As a consequence, this definition of P_Λ cannot be directly used to efficiently solve the game.

To derive a polynomial-time solution for the ECC model, we first characterize the blocker $bl(P_\Lambda)$ of P_Λ using a set of linear equations whose cardinality is polynomial in the size of the network. We do so by showing that if a polynomial-size characterization exists for the unconstrained polyhedron, then there also exists one for the blocker of the constrained game. We then show how one can use linear programming tools to efficiently compute the equilibrium payoff based on a polynomial-size characterization of the blocker. Finally, we provide a characterization for each of the models discussed in Section 2.1.

Let the polynomial-size linear characterization of the polyhedron P_Λ be

$$P_\Lambda = \{x \mid \exists f\,(Sf \le x \wedge Cf \ge d)\} \qquad (15)$$

for the unconstrained game, where $f \in \mathbb{R}^k_{\ge 0}$ is a vector of polynomial length (i.e., k is a polynomial function of the network size), S is a mapping to the mixed strategy space, and C, d are linear constraints. Then, the expected cost constrained polyhedron is characterized by

$$P_\Lambda = \{x \mid \exists f\,(Sf \le x \wedge Cf \ge d \wedge w'Sf \le b)\} \ . \qquad (16)$$

The following theorem gives a polynomial-size characterization of the blocker in the expected cost constrained game.

Theorem 3. *The blocker of the polyhedron defined as*

$$P_\Lambda = \{x \mid \exists f\,(Sf \le x \wedge Cf \ge d \wedge w'Sf \le b)\} \qquad (17)$$

is

$$bl(P_\Lambda) = \{y \mid \exists K, g, h\,(g \le y \wedge C'h \le S'wK + S'g \wedge d'h - bK \ge 1)\} \ , \qquad (18)$$

where $K \in \mathbb{R}_{\ge 0}$, $g \in \mathbb{R}^{|E|}_{\ge 0}$, and $h \in \mathbb{R}^l_{\ge 0}$ (l is the length d).

Proof. We prove Equation (18) in two steps:

- RHS of Equation (18) $\subseteq bl(P_\Lambda)$: We have to show that, for any $\tilde{y}$ that satisfies the constraints of the RHS with some $\tilde{g}$, $\tilde{h}$ and $\tilde{K}$, it holds that $\tilde{y}'x \ge 1$ for every $x \in P_\Lambda$. We can formulate this as a linear programming problem as follows:

$$\text{Minimize } \tilde{y}'x \qquad (19)$$

  subject to

$$w'Sf \le b \qquad (20)$$
$$Sf \le x \qquad (21)$$
$$Cf \ge d \ , \qquad (22)$$

  where $f \in \mathbb{R}^k_{\ge 0}$.
  Observe that the constraints of the above LP correspond to the characterization of P_Λ; consequently, the above linear program's set of feasible solutions projected to x is P_Λ. Thus, we have to show that the value of the above linear program is at least 1. To see this, consider the dual linear program:

$$\text{Maximize } d'h - bK \qquad (23)$$

subject to

$$g \leq \tilde{y} \tag{24}$$

$$C'h \leq S'wK + S'g \,, \tag{25}$$

where $K \in \mathbb{R}_{\geq 0}$, $g \in \mathbb{R}_{\geq 0}^{|E|}$, and $h \in \mathbb{R}_{\geq 0}^{l}$.
Since $\tilde{y}$ satisfies the constraints of the RHS of Equation (18) with $\tilde{K}, \tilde{g}, \tilde{h}$, we have that $(\tilde{K}, \tilde{g}, \tilde{h})$ is a feasible solution. Furthermore, we also have that the objective function for this solution is at least 1. Thus, the value of the linear program has to be at least 1.

- $bl(P_\Lambda) \subseteq$ RHS of Equation (18): We have to show that every $\tilde{y} \in bl(P_\Lambda)$ satisfies the constraints of the RHS. To see this, first consider the linear program from the first part of the proof. Since $\tilde{y}$ blocks every $x \in P_\Lambda$, we have that the value of the linear program is at least 1. Now, consider an optimal solution $\tilde{K}, \tilde{g}, \tilde{h}$ of the dual linear program. Since the value of the linear program is at least 1, we have that $1 \leq d'\tilde{h} - b\tilde{K}$. We also have $\tilde{g} \leq \tilde{y}$ and $C'\tilde{h} \leq S'w\tilde{K} + S'\tilde{g}$ from the constraints. Thus, $\tilde{y}$ satisfies the constraints of the RHS of Equation (18) with $\tilde{K}, \tilde{g}, \tilde{h}$. $\qquad\square$

Recall that our goal is to compute $\theta^* = \max\{\tilde{\theta}, 0\}$ in polynomial time. The most straightforward solution is to formulate this as an optimization problem subject to the set of linear constraints given by the above characterization. Unfortunately, the objective function θ cannot be expressed as a linear function because of the division with $\mathbf{1}'y$. Thus, to formulate the problem as a linear program, we introduce a variable ϕ which measures $\frac{1}{\mathbf{1}'y}$ and scale the original variables. The resulting linear program is

$$\text{Maximize } \phi - \boldsymbol{\mu}'\boldsymbol{\beta} \tag{26}$$

subject to

$$\mathbf{1}'\boldsymbol{\beta} = 1 \tag{27}$$

$$g \leq \boldsymbol{\beta} \tag{28}$$

$$C'h \leq S'wK + S'g \tag{29}$$

$$d'h - bK \geq \phi \,, \tag{30}$$

where $K, \phi \in \mathbb{R}_{\geq 0}$, $\boldsymbol{\beta}, g \in \mathbb{R}_{\geq 0}^{|E|}$, and $h \in \mathbb{R}_{\geq 0}^{l}$.

All-to-All Communication Model. In [9], it was shown that the mixed strategy space of the operator in the All-to-All model be characterized using multicommodity flows. In this characterization, there exists a commodity for each node. For each commodity, the corresponding node is a sink, while all the other nodes are a sources with a uniform supply. It was shown that, if the total amount of flow transported is at least 1, the vector representing the sum flows on each edge is an element of the polyhedron, and vice versa.

This can formulated as a set of linear constraints with $|V| + |V| \cdot |E|$ variables (the uniform supply value for each commodity and the flow along each edge for each commodity) and $|V| \cdot |V| + 1$ constraints (flow conservation at each node for each commodity and the constraint on the total amount of flow transported). Then, by applying Theorem 3, we have a polynomial-size characterization of the constrained blocker:

$$bl(P_\Lambda) = \left\{ \boldsymbol{y} \in \mathbb{R}^{|E|}_{\geq 0} \;\middle|\; \exists \pi : V \times V \mapsto \mathbb{R}_{\geq 0}, K \in \mathbb{R}_{\geq 0} \Big(\right.$$

$$\forall r \in V : \sum_{v \in V} \pi_r(v) - bK \geq 1 \;\wedge$$

$$\left. \forall r \in V, e = (u, v) \in E : |\pi_r(u) - \pi_r(v)| \leq \boldsymbol{y}_e + \boldsymbol{w}_e K \Big) \right\}, \quad (31)$$

where $\pi_r(r) \equiv 0$ by definition to simplify the notation.

S-D Communication Model. Based on [6], we can characterize the polyhedron for the S-D model using network flows. Then, from Theorem 3, we have that the constrained blocker has the following polynomial-size characterization:

$$bl(P_\Lambda) = \left\{ \boldsymbol{y} \in \mathbb{R}^{|E|}_{\geq 0} \;\middle|\; \exists \pi : V \setminus \{r\} \mapsto \mathbb{R}, K \in \mathbb{R}_{\geq 0} \Big(\right.$$

$$\sum_{v \in V} \pi(v)(s(v) - d(v)) - bK \geq 1 \;\wedge$$

$$\left. \forall e = (u, v) \in E : \pi(u) - \pi(v) \leq \boldsymbol{y}_e + \boldsymbol{w}_e K \Big) \right\}. \quad (32)$$

All-to-One Communication Model. In [10], it was shown that the mixed strategy space of the operator in the All-to-One model can be characterized using special multi-source flows. By combining this result with Theorem 3, we can show that the constrained blocker has the following polynomial-size characterization:

$$bl(P_\Lambda) = \left\{ \boldsymbol{y} \in \mathbb{R}^{|E|}_{\geq 0} \;\middle|\; \exists \pi : V \setminus \{r\} \mapsto \mathbb{R}_{\geq 0}, K \in \mathbb{R}_{\geq 0} \Big(\right.$$

$$\sum_{v \in V} \pi(v) - bK \geq 1 \;\wedge$$

$$\left. \forall e = (u, v) \in E : \pi(u) - \pi(v) \leq \boldsymbol{y}_e + \boldsymbol{w}_e K \Big) \right\}, \quad (33)$$

where $\pi(r) \equiv 0$ by definition to simplify the notation.

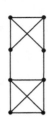

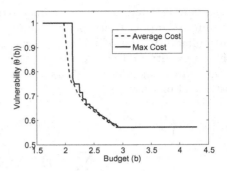

Fig. 2. All-to-All network   **Fig. 3.** Vulnerability/budget tradeoff

7   Application Example: Vulnerability/Budget Tradeoff

As it was mentioned earlier, by varying the budget limit b, one can draw the
Pareto frontier between the region of achievable vulnerability/budget points and
the region of unachievable ones. Here, we illustrate this using the All-to-All
communication model on the topology depicted in Figure 2. The link costs w_e are
randomly chosen between 0 and 0.6, which makes the average cost of a spanning
tree equal to 2.1. For each value of b, a game is played with the defender's
strategy set given by Equation (8) for the maximum cost constraint (MCC)
and by Equation (9) for the expected (or average) cost constraint (ECC). In all
games, the attacker's strategy set is the set of all links and the cost of attack is
$\mu = 0$. Figure 3 shows the vulnerability $\theta^*(b)$ as a function of the budget b for
both the MCC and the ECC. Observe that the two curves are very close to each
other, but vulnerability for the MCC is always at least as high as for the ECC.

8   Conclusions and Future Work

In this paper, we have generalized *network blocking games* by introducing budget
constraints on the operator. This generalization allows the application of network
blocking games in scenarios where the budget of the network operator is limited.
We have studied two budget constraint formulations: the *maximum cost* and the
expected (or *average*) *cost constraints*.

   Network blocking games are used to quantify the robustness of topologies in
the presence of a strategic adversary, and the equilibrium payoffs of the games
are used as such quantifications. As the greatest challenge to computing the equilibrium in practice is the exponential size of the implicitly given payoff matrix,
we have focused our work on computational complexity: we have shown that the
maximum cost formulation leads to NP-hard problems, and proposed efficient
solutions for the expected cost formulation.

   Proving that the maximum cost formulation leads to NP-hard problems was a
very important first step. Since we now know that no polynomial-time algorithm

can solve the game under the MCC (for the discussed models), an interesting future work is finding polynomial-time approximation algorithms or efficient heuristics. Another interesting future direction is the study of the cost-security tradeoff problem, where the operator has to maximize security and minimize budget at the same time.

Acknowledgement. This paper has been supported by HSN Lab, Budapest University of Technology and Economics [5], NIST-ARRA Program award 70NANB10H026, and NIST grant award 70NANB13H012, through the Univ. of Maryland, College Park.

A   Proof of Theorem 1

Proof. Given an instance (c, v, C, V) of the Knapsack Problem, we construct an instance $(E, I_{T \in \mathcal{T}}, \lambda(T, e), p)$ of the Equilibrium Problem as follows:
- Let $E = \{1, \ldots, N\}$,
- $I_{T \in \mathcal{T}} = \begin{cases} \text{true} & \text{if } \sum_{i \in T} c_i \leq C \\ \text{false} & \text{otherwise,} \end{cases}$
- $\lambda(T, e) = \frac{1}{\sum_{i \in T} v_i}$,
- $\mu = 0$,
- $p = \frac{1}{V}$.

Observe that we define $\lambda(T, e)$ such that its value does not depend on e. Consequently, the payoff of the game does not depend on the adversary's strategy, it only depends on the operator's strategy. To simplify our proof, we will let $\lambda(T)$ denote $\lambda(T, e)$ for any e.

It is easy to see that both $I_{T \in \mathcal{T}}$ and $\lambda(T)$ can be computed in polynomial time as they only require summing over a given set (and comparing the sum with a constant or computing a reciprocal). Furthermore, every step of the reduction can be carried out in time and space that is polynomial in the size of the Knapsack Problem instance.

We claim that there exists a subset $S \subseteq \{1, \ldots, N\}$ whose sum weight is at most W and whose sum value is at least V if and only if the adversary's equilibrium payoff is less than or equal to p (since $\mu = 0$).

First, assume that there exists a subset S satisfying the constraints of the Knapsack Problem. Then, consider the operator strategy $\alpha_S^* = 1$ (i.e., the strategy that uses only subset S). If the operator uses this strategy, her loss is

$$\lambda(S) = \frac{1}{\sum_{i \in S} v_i} = \frac{1}{V} = p . \tag{34}$$

Therefore, the operator's equilibrium loss and, hence, the adversary's equilibrium payoff is at most p.

---

[5] http://www.hsnlab.hu

Second, assume that there does not exist a subset satisfying the constraints of the Knapsack Problem. This implies that, for every $T \in \mathcal{T}$,

$$\lambda(T) = \frac{1}{\sum_{i \in T} v_i} > \frac{1}{V} = p \ . \tag{35}$$

Consequently, the expected loss for any operator strategy α^* is

$$\sum_{T \in \mathcal{T}} \alpha_T^* \underbrace{\lambda(T)}_{>p} > p \ . \tag{36}$$

Thus, the adversary's equilibrium payoff has to be greater than p.   □

B   Proof of Theorem 2 for the All-to-All model

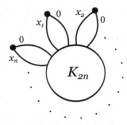

Fig. 4. Illustration for the proof of Theorem 2 for the All-to-All model.

For the All-to-All communication model, we construct an instance of *EPMAX* from an instance of *PP* as follows:

- Let the network topology be the following (see Figure 4): There is a large clique that consists of $2n$ nodes, and there are n "outer" nodes, to which we refer as node 1, node 2, ..., node n. Each node i, $i = 1, \ldots, n$, is connected to two distinct nodes of the clique with edges having unit costs of x_i and 0, such that every node in the clique is connected to exactly one outer node. Finally, edges between two nodes in the clique have zero unit cost.
- Let the operator's budget be $b = \frac{1}{2} \sum_{i=1}^{n} x_i$.
- Let the equilibrium payoff value be $p = \frac{1}{2}$.

We claim that the equilibrium payoff in the above network is greater than $\frac{1}{2}$ iff *PP* does not have a solution.

As in the previous proof, we first assume that *PP* has a solution (A, B) and use it to derive an operator strategy in which the expected loss of every edge is at most $\frac{1}{2}$. According to this strategy, the operator chooses a spanning tree as follows. First, she chooses either A or B with equal probability $(\frac{1}{2}, \frac{1}{2})$. Second, she connects each outer node i to the clique with exactly one edge: if x_i belongs to the chosen set, she uses the edge that has cost x_i; otherwise, she uses the other edge. Third, she completes the spanning tree by choosing a star subgraph of the clique uniformly at random. We show that the expected loss of every link is at most $\frac{1}{2}$: First, each outer edge e is used with probability $\frac{1}{2}$ and its removal cuts off at most 1 node; thus, $L(e) \leq \frac{1}{2}$. Second, each link e inside the clique is used with probability $\frac{1}{n}$ and its removal cuts off at most 2 nodes; thus, $L(e) \leq \frac{2}{n}$.

Next, we assume that PP does not have a solution and use the same argument as before to show that the cost of every pure strategy and, hence, the expected cost of every mixed strategy is strictly less than b, i.e.,

$$\sum_{e \in E_{\text{outer}}} w_e L(e) < b = \frac{1}{2} \sum_i x_i = \sum_{e \in E_{\text{outer}}} \frac{1}{2} w_e \, , \tag{37}$$

where E_{outer} is the set of outer links. Now, consider an arbitrary pair of edges e_a and e_b that connect an outer node to the clique. It can be shown that $L(e_a) + L(e_b) \geq 1$. If there were an operator strategy in which the expected loss of every edge is at most $\frac{1}{2}$, then it would follow that $\forall e \in E_{\text{outer}} : L(e) = \frac{1}{2}$. This would lead to a contradiction with Equation 37; thus, no such strategy can exist.   □

References

1. Holme, P., Kim, B., Yoon, C., Han, S.: Attack vulnerability of complex networks. Physical Review E 65(5), 056109 (2002)
2. Schneider, C., Moreira, A., Andrade Jr., J., Havlin, S., Herrmann, H.: Mitigation of malicious attacks on networks. Proceedings of the National Academy of Sciences 108(10), 3838–3841 (2011)
3. Grubesic, T., Matisziw, T., Murray, A., Snediker, D.: Comparative approaches for assessing network vulnerability. International Regional Science Review 31(1), 88–112 (2008)
4. Estrada, E.: Network robustness to targeted attacks. the interplay of expansibility and degree distribution. Eur. Phys. Journal B 52(4), 563–574 (2006)
5. Dall Asta, L., Barrat, A., Barthélemy, M., Vespignani, A.: Vulnerability of weighted networks. J. of Stat. Mech., 2006(4), P04006 (2006)
6. Gueye, A., Marbukh, V.: A game-theoretic framework for network security vulnerability assessment and mitigation. In: Grossklags, J., Walrand, J. (eds.) GameSec 2012. LNCS, vol. 7638, pp. 186–200. Springer, Heidelberg (2012)
7. Gueye, A., Walrand, J.C., Anantharam, V.: Design of network topology in an adversarial environment. In: Alpcan, T., Buttyán, L., Baras, J.S. (eds.) GameSec 2010. LNCS, vol. 6442, pp. 1–20. Springer, Heidelberg (2010)
8. Gueye, A., Walrand, J.C., Anantharam, V.: A network topology design game: How to choose communication links in an adversarial environment? In: Proc. of 2nd Int. ICST Conf. on Game Theory for Networks (2011)
9. Laszka, A., Szeszlér, D., Buttyán, L.: Linear loss function for the network blocking game: An efficient model for measuring network robustness and link criticality. In: Grossklags, J., Walrand, J. (eds.) GameSec 2012. LNCS, vol. 7638, pp. 152–170. Springer, Heidelberg (2012)
10. Laszka, A., Szeszlér, D., Buttyán, L.: Game-theoretic robustness of many-to-one networks. In: Krishnamurthy, V., Zhao, Q., Huang, M., Wen, Y. (eds.) Gamenets 2012. LNICST, vol. 105, pp. 88–98. Springer, Heidelberg (2012)

11. Daskalakis, C., Goldberg, P., Papadimitriou, C.: The complexity of computing a Nash equilibrium. In: Proc. of 38th Annu. ACM Symp. on Theory of Computing, pp. 71–78 (2006)
12. Laszka, A., Gueye, A.: Quantifying All-to-One network topology robustness under budget constraints. In: Proc. of Workshop on Pricing and Incentives in Networks and Systems. ACM (June 2013)
13. Gueye, A.: A Game Theoretical Approach to Communication Security. PhD thesis, EECS Department, University of California, Berkeley (March 2011)
14. Cunningham, W.: Optimal attack and reinforcement of a network. Journal of the ACM 32(3), 549–561 (1985)
15. Palmer, E.M.: On the spanning tree packing number of a graph: A survey. Discrete Mathematics 230(1), 13–21 (2001)
16. Chung, F.: Laplacians and the Cheeger inequality for directed graphs. Annals of Combinatorics 9(1), 1–19 (2005)
17. Mertens, S.: The easiest hard problem: Number partitioning. Comput. Complex. and Stat. Phys. 125(2), 125–139 (2006)

Mitigation of Targeted and Non-targeted Covert Attacks as a Timing Game

Aron Laszka[1], Benjamin Johnson[2], and Jens Grossklags[3]

[1] Department of Networked Systems and Services,
Budapest University of Technology and Economics, Hungary
[2] Department of Mathematics, University of California, Berkeley, USA
[3] College of Information Sciences and Technology,
Pennsylvania State University, USA

Abstract. We consider a strategic game in which a defender wants to maintain control over a resource that is subject to both targeted and non-targeted covert attacks. Because the attacks are covert, the defender must choose to secure the resource in real time without knowing who controls it. Each move by the defender to secure the resource has a one-time cost and these defending moves are not covert, so that a targeted attacker may time her attacks based on the defender's moves. The time between when a targeted attack starts and when it succeeds is given by an exponentially distributed random variable with a known rate. Non-targeted attackers are modeled together as a single attacker whose attacks arrive following a Poisson process. We find that in this regime, the optimal moving strategy for the defender is a periodic strategy, so that the time intervals between consecutive moves are constant.

Keywords: Game Theory, Computer Security, Games of Timing, Covert Compromise, Targeted Attacks, Non-Targeted Attacks.

1 Introduction

A growing trend in computer security is the prevalence of continuous covert attacks on networked resources. In contrast to one-time attacks with immediate benefit, such as initiating a wire transfer from a compromised bank account, a covert attack seeks to maintain control of a resource while keeping the compromise a secret. This type of attack is ubiquitous in the formation of botnets, as individual computer owners rarely know that their computer is a botnet member. Routers that are used to conduct man-in-the-middle attacks are also typically covertly compromised; and when web servers are used to compromise client's computers, the initial infection is typically covert.

In light of the prevalence of covert attacks, it behooves the user to consider what mitigation strategies can be taken to minimize the losses resulting from such attacks. Mitigation strategies include resetting passwords, changing private keys, re-installing servers, or re-instantiating virtual servers. Such strategies have notable characteristics in that they are often effective at securing the resource,

S.K. Das, C. Nita-Rotaru, and M. Kantarcioglu (Eds.): GameSec 2013, LNCS 8252, pp. 175–191, 2013.

but they reveal little about past attacks or compromises. For example, if a server is re-installed, knowledge of when the server was compromised may be lost. Similarly, resetting a password does not reveal any information about the integrity of the previous password.

A second dimension of the attack space is the extent to which an attack is targeted or customized for a particular user [4,2]. DoS attacks and incidents of cyber-espionage are examples of targeted attacks. Typical examples of non-targeted attacks include spam and phishing. The dichotomy between targeted and non-targeted attacks is explained by Cormac Herley as a consequence of economic considerations of the attacker [4]. In that framework, an outsized number of users are both susceptible to and subject to scalable attacks which compromise their computer systems, but most are never targeted simply because they cannot be distinguished from low value targets. See Table 1 for a comparison between targeted and non-targeted attacks.

Table 1. Comparison of Targeted and Non-Targeted Attacks

	Targeted	Non-Targeted
Number of attackers	low	high
Number of targets	low	high
Effort required for each attack	high	low
Success probability of each attack	high	low

Whether or not an attack is targeted is also important for the defender, because targeted and non-targeted attacks do different types of damage. For example, targeted attackers might read all of an organization's secret e-mails, causing economic damages of one type, while a non-targeted attacker might use the same compromised machine to send out spam, causing reputation loss, or machine blacklisting, or another separate type of damage. This dichotomy suggests that damages resulting from targeted and non-targeted attacks should be modeled additively.

The presence of both targeted and non-targeted covert attacks presents an interesting dilemma for a common user to choose a mitigation strategy against covert attacks. Strategies which are optimal against non-targeted covert attacks may not be the best choice against targeted attacks. At the same time, mitigation strategies against targeted attacks may not be economically cost-effective against only non-targeted attackers.

This paper fills the research gap induced by the aforementioned dichotomy, by considering the strategy spaces of users who may be subject to both targeted and non-targeted attacks. In our game, a defender must vie for a contested resource that is subject to the risk of compromise from both targeted and non-targeted covert attacks. We explore the strategy space to find good mitigation strategies against this combination.

2   Related Work

2.1   Games of Timing

Cybersecurity economics has been concerned with how to reduce the impact of the actions of financially or politically motivated adversaries who threaten computing resources and their users. Previous research in this domain has primarily focused on the choice between different canonical actions to prevent, deter or otherwise mitigate harm (e.g., [3,5,6]).

However, being successful in dynamic environments shifts the focus from selecting the most suitable option from a pool of alternatives to a decision problem of *when* to act to get an advantage over an opponent. For example, in tactical security scenarios it is important to jump to action at the right time to avoid a loss of money or even human life (see, for example, timing of interventions in international conflicts). To understand these scenarios, so-called games of timing have been studied with the tools of non-cooperative game theory since the cold war era (see, for example, [11,14]). For a detailed survey and summary of the theoretical contributions in this area, we refer the interested reader to [10].

2.2   FlipIt: Modeling Targeted Attacks

In response to recent high-profile stealthy attacks, researchers at RSA proposed the FlipIt model [13] to study such scenarios. In the original model, there are two players, a defender and an attacker, and a resource that they are both interested in maintaining control of. For each unit of time that a player is controlling the resource, she gains a fixed amount of benefit. Conversely, when a player is not in control, she gains no benefit from the resource. At any time instance, either player may "flip" the resource to gain control of it for some cost. Flipping while in control does not give the opponent control of the resource, therefore the players have to be careful not to make too many unnecessary flips to keep their costs low. This game can model, for example, the case of a password-protected account. Benefit is derived from using the account, and flipping the resource is analogous to the defender resetting the password or the attacker compromising it.

In the original FlipIt paper, dominant strategies and equilibria are studied for some simple cases [13]. Other researchers have worked on extensions [9,7]. For example, Laszka et al. extended the FlipIt game to the case of multiple resources. In addition, the usefulness of the FlipIt game has been investigated for various application scenarios [1,13].

In comparison to previous work, the FlipIt game is of interest because it combines a number of important decision-making factors [8]. First, it covers aspects of uncertainty about the game status by assuming that moves by the players are "stealthy". Second, the game is played in continuous time and asynchronous fashion. Hence, ex-post the game appears to be divided in multiple periods of uneven length. Similarly, the number of actions that can be taken by the players is quasi-unlimited (if agents have an unrestricted budget). Third, action have a cost. That is, players do not only value the time in which they have possession of the board, but they also have to balance these benefits with the cost of gaining possession of the board.

The original FlipIt game has also been studied in an experiment with human subjects [8]. In that paper, the experimenters matched human participants with computerized opponents in several fast-paced rounds of the FlipIt game. The results indicate that participant performance improves over time; but that it is dependent on age, gender, and a number of individual difference variables. The researchers also show that human participants generally perform better when they have more information about the strategy of the computerized player; i.e., they are able to make use of such game-relevant information. This experimental work was extended to also include different visual presentation modalities for the available feedback during the experiment [12].

3   Model Definition

We model the covert compromise scenario as a non-zero-sum game. The player who is the rightful owner of the resource is called the defender, while the other players are called the attackers. The game starts at time $t = 0$ with the defender in control of the resource, and it is played indefinitely as $t \to \infty$. We assume that time is continuous.

We let D, A, and N denote the defender, the targeted attacker, and the non-targeted attackers respectively. At any time instance, player i may make a move, which costs her C_i. When the defender makes a move, the resource immediately becomes uncompromised for every attacker. When the targeted attacker makes a move, she starts her attack, which takes some random amount of time. If the defender makes a move while an attack is in progress, the attack fails. We assume that the time required by the attack follows an exponential distribution. Formally, the probability that the attack has successfully finished in a amount of time is $1 - e^{-\lambda_A a}$, where λ_A is the rate parameter of the targeted attacker's attack time.

The attackers' moves are stealthy; i.e., the defender does not know when the resource got compromised or if it is compromised at all. On the other hand, the defender's moves are non-stealthy. In other words, the attackers learn immediately when the defender has made a move.

The cost rate for player i up to time t, denoted by $c_i(t)$, is the number of moves per unit of time, made by player i up to time t, multiplied by the cost per move C_i for player i.

For attacker $i \in \{A, N\}$, the benefit rate $b_i(t)$ up to time t is the fraction of time up to t that the resource has been compromised by i, multiplied by B_i. Note that if multiple attackers have compromised the resource, they all receive benefit until the defender's next move. For the defender D, the benefit rate $b_D(t)$ up to time t is defined to be $-\sum_{i \in \{A, N\}} b_i(t)$ (i.e., what has been lost to the attackers). The relation between the defender's and attackers' benefits implies that the game would be zero-sum if we only considered the players' benefits. Because our players' payoffs also consider move costs, our game is *not* zero-sum. Player i's payoff is defined as

$$\liminf_{t \to \infty} b_i(t) - c_i(t) \ . \tag{1}$$

Table 2. List of Symbols

C_D	move cost for the defender
C_A	move cost for the targeted attacker
B_A	benefit received per unit of time for the targeted attacker
B_N	benefit received per unit of time for the non-targeted attackers
λ_A	rate of the targeted attacker's attack time
λ_N	rate of the non-targeted attacks' arrival

3.1   Types of Strategies for the Defender and the Targeted Attacker

Adaptive Strategies for Attackers. Let $\mathcal{T}(n) = \{T_0, T_1, \ldots, T_n\}$ denote the move times of the defender up to her nth move (or in the case of $T_0 = 0$, the start of the game). The attacker uses an *adaptive strategy* if she waits for $W(\mathcal{T}(n))$ time until making a move after the defender's nth move (or after the start of the game), where W is a non-deterministic function. If the defender makes her $n + 1$st move before the chosen wait time is up, the attacker chooses a new wait time $W(\mathcal{T}(n + 1))$, which also considers the new information that is the defender's $n + 1$st move time. This class is a simple representation of all the rational strategies available to an attacker, since the function W depends on all the information that the attacker has, and we don't have any constraints on W.

Renewal Strategies. Player i uses a *renewal strategy* if the time intervals between consecutive moves are identically distributed independent random variables, whose distribution is given by the cumulative function F_{R_i}. Renewal strategies are well-motivated by the fact that the defender is playing blindly; thus, she has the same information available after each move. So it makes sense to use a strategy which always chooses the time until her next flip according to the same distribution Note that every renewal strategy is a special case of an adaptive strategy.

Periodic Strategies. Player i uses a *periodic strategy* if the time intervals between her consecutive moves are identical. This period is denoted by δ_i. Every periodic strategy is a special case of a renewal strategy.

3.2   Non-targeted Attacks

Suppose that there are N non-targeted attackers. In practice, N is very large, but the expected number of successful compromises is finite. As N goes to infinity, the probability that a given non-targeted attacker targets the defender approaches zero. Since the non-targeted attackers operate independently, *successful* non-targeted attacks arrive following a Poisson process. Furthermore, as the economic decisions of the non-targeted attackers depend on a very large pool of possible targets, the defender's effect on the decisions is negligible. Thus, the non-targeted attackers' strategies (that is, the attack rate) can be considered exogenously given. We let λ_N denote the expected number of arrivals that occur per unit of time; and we model all the non-targeted attackers together as a single attacker whose benefit per unit of time is B_N.

3.3   Comparison to FlipIt

Even though our game-theoretic model is in many ways similar to FlipIt, it differs in three key assumptions. First, we assume that the defender's moves are *not stealthy*. The motivation for this is that an attacker must know whether she is in control of a resource if she receives benefits from it continuously. For example, if the attacker uses the compromised password of an account to regularly spy on its e-mails, she will learn of a password reset immediately the next time she tries to log in. Second, we assume that the targeted attacker's moves are *not instantaneous*, but take some time. The motivation for this is that an attack requires some time and effort to be carried out in practice. Furthermore, the time required for a successful attack may vary, which we model using a random variable for the attack time. Third, we assume that the defender faces *multiple attackers*, not only a single one.

Moreover, to the authors' best knowledge, papers published on FlipIt so far give analytical results only on a very restricted set of strategies. In contrast, we completely describe our game's equilibria and give optimal defender strategies based on very mild assumptions, which effectively do not limit the power of players (see the introduction of Section 4).

4   Analytical Results

In this section, we give analytical results on the game. We first consider the special case of a targeted attacker only (i.e., $\lambda_N = 0$), and then the general case of both targeted and non-targeted attackers.

We start with a discussion on the players' strategies. First, recall that the defender has to play blindly, which means that she has the same information available after each one of her moves. Consequently, it makes sense for her to choose the time until her next flip according to the same distribution each time. In other words, a rational defender can use a renewal strategy.

Now, if the defender uses a renewal strategy, the time of her next move depends only on the time elapsed since her last move T_n, and the times of previous moves (including T_n) are irrelevant to the future of the game. Therefore, it is reasonable to assume that the attacker's response strategy to a renewal strategy also does not depend on $T_0, T_1, \ldots, T_n$. For the remainder of the paper, when the defender plays a renewal strategy, the attacker uses a fixed probability distribution – given by the density function f_W – over her wait times for when to begin her attack. Note that it is clear that there always exists a best response strategy for the attacker of this form against a renewal strategy.

Since the attacker always waits an amount of time that is chosen according to a fixed probability distribution after the defender's each move, the amount of time until the resource would be successfully compromised after the defender's move also follows a fixed probability distribution. Let S be the random variable measuring the time after the defender has moved until the attacker's attack would finish. The probability density function f_S of S can be computed as

$$f_S(s) = \int_{w=0}^{s} f_W(w) \int_{a=0}^{(s-w)} \lambda_A e^{-\lambda_A a} \, da \, dw \ . \tag{2}$$

We let F_S denote the cumulative distribution function of S. Since $\lambda_A e^{-\lambda_A a} > 0$ for every $a \in \mathbb{R}_{\geq 0}$, if there exists an s for which $F_S(s) > 0$, then F_S is strictly increasing on $[s, \infty)$.

4.1   Nash Equilibrium for Targeted Attacker and Renewal Defender

Defender's Best Response. We begin our analysis with finding the defender's best response strategy.

Lemma 1. *Suppose that the attacker uses an adaptive strategy with a fixed probability distribution for choosing the time to wait until starting the attack. Then,*

– *not moving is the only best response if*

$$\frac{C_D}{B_A} = l F_S(l) - \int_{s=0}^{l} F_S(s) \, ds \tag{3}$$

  has no solution for l;
– *a periodic strategy whose period is the unique solution of Equation (3) is the only best response otherwise.*

Even though we cannot express the solution of Equation (3) in closed form, it can be easily found using numerical methods, as the right hand side is continuous and increasing.[1] Note that the equations presented in the subsequent lemmas and theorems of this paper can also be solved using numerical methods.

Proof. When playing a renewal strategy, the defender randomly selects the intervals between her consecutive moves according to the distribution generating the renewal strategy. In a best response, her strategy and, hence, every interval length in the support of the strategy's distribution has to minimize the defender's

---

[1] We show that the right hand side is continuous and increasing in the proof of the lemma.

loss per unit of time. The defender's expected loss per unit of time for an interval of length l is

$$\frac{1}{l}\left(B_A \int_{s=0}^{l} f_S(s)(l-s)\ ds + C_D\right) \tag{4}$$

$$=\frac{1}{l}\left(B_A\left([F_S(s)(l-s)]_{s=0}^{l} - \int_{s=0}^{l} F_S(s)\cdot(-1)\ ds\right) + C_D\right) \tag{5}$$

$$=\frac{1}{l}\left(B_A\left(0 + \int_{s=0}^{l} F_S(s)\ ds\right) + C_D\right) \tag{6}$$

$$=\frac{1}{l}\left(B_A \int_{s=0}^{l} F_S(s)\ ds + C_D\right)\ . \tag{7}$$

To find the minimizing interval lengths (if there exists any), we take the derivative of (7) and solve it for equality with 0 as follows:

$$0 = \frac{d}{dl}\left[\frac{1}{l}\left(B_A \int_{s=0}^{l} F_S(s)\ ds + C_D\right)\right] \tag{8}$$

$$0 = -\frac{1}{l^2}\left(B_A \int_{s=0}^{l} F_S(s)\ ds + C_D\right) + \frac{1}{l}B_A F_S(l) \tag{9}$$

$$\int_{s=0}^{l} F_S(s)\ ds + \frac{C_D}{B_A} = l F_S(l) \tag{10}$$

$$\frac{C_D}{B_A} = l F_S(l) - \int_{s=0}^{l} F_S(s)\ ds\ . \tag{11}$$

Suppose that l^* is the least number for which this equation is satisfied. Then $l^* > 0$, and also $F(l^*) > 0$. This in turn implies that F_S is strictly increasing on $[l^*, \infty)$; and thus also the right hand side of the above equation is strictly increasing as a function of l on $[l^*, \infty)$. Therefore, if there is any solution to the above equation, then it is unique. Furthermore, this value of l is a minimizing value for the expected loss per unit of time as the second derivative at this minimizing l^* is greater than zero:

$$\frac{d}{dl}\left[-\frac{1}{l^2}\left(B_A \int_{s=0}^{l} F_S(s)\ ds + C_D\right) + \frac{1}{l}B_A F_S(l)\right] \tag{12}$$

$$=\frac{2}{l^3}\left(B_A \int_{s=0}^{l} F_S(s)\ ds + C_D\right) + \left(-\frac{1}{l^2}\right)B_A F_S(l)$$

$$+\left(-\frac{1}{l^2}\right)B_A F_S(l) + \frac{1}{l}B_A f_S(l) \tag{13}$$

$$=\frac{2}{l^3}\left(B_A \int_{s=0}^{l} F_S(s)\ ds + C_D\right) + \left(-\frac{2}{l^2}\right)B_A F_S(l) + \frac{1}{l}B_A f_S(l)\ . \tag{14}$$

We care about the value of this expression when the first derivative is zero. Using this constraint, we obtain

$$\frac{2}{l^3}\left(B_A \int_{s=0}^{l} F_S(s)\,ds + C_D\right) + \left(-\frac{2}{l^2}\right)B_A F_S(l) + \frac{1}{l}B_A f_S(l) \tag{15}$$

$$= -\frac{2}{l}\left(-\frac{1}{l^2}\left(B_A \int_{s=0}^{l} F_S(s)\,ds + C_D\right) + \frac{1}{l}B_A F_S(l)\right) + \frac{1}{l}B_A f_S(l) \tag{16}$$

$$= -\frac{2}{l}(0) + \frac{1}{l}B_A f_S(l) > 0 \ . \tag{17}$$

Consequently, the only best response is the periodic strategy with the minimizing l^* as the period.

On the other hand, if Equation (11) is not satisfiable for l, then the only best response for the defender is to never move. When $l \to \infty$, the defender's expected loss per unit of time approaches B_A, which is equal to her loss for never moving. When $l \to 0$, her expected loss per unit of time goes to infinity due to the ever increasing costs. Consequently, if the expected loss per unit of time does not have a minimizing l, then it is always greater than B_A. $\qquad\square$

Attacker's Best Response. We continue our analysis with finding the attacker's best response strategy.

Lemma 2. *Against a defender who uses a periodic strategy with period δ_D,*

- *never attacking is the only best response if*

$$\frac{C_A}{B_A} > \frac{e^{-\delta_D \lambda_A} - 1}{\lambda_A} + \delta_D \ ; \tag{18}$$

- *attacking immediately after the defender moved is the only best response if*

$$\frac{C_A}{B_A} < \frac{e^{-\delta_D \lambda_A} - 1}{\lambda_A} + \delta_D \ ; \tag{19}$$

- *both not attacking and attacking immediately are best responses otherwise.*

The lemma shows that the attacker should either attack immediately or not attack at all, but she should never wait to attack. Consequently, if the attacker uses her best response strategy, the defender can determine the optimal period of her strategy *solely based on the distribution of A*, which is an exponential distribution with parameter λ_A. This observation will be of key importance for characterizing the game's equilibria.

Proof. First, assume that the attacker does attack. Given that the attacker waits $w < \delta_D$ time before making her move, the expected amount of time she has the resource compromised until the defender's next move is

$$\int_{a=0}^{\delta_D - w} \lambda_A e^{-\lambda_A a}(\delta_D - w - a)da \ . \tag{20}$$

It is easy to see that the maximum of this equation is attained for $w = 0$. Therefore, if the attacker does attack, she attacks immediately. The expected amount of time she has the resource compromised until the defender's next move is

$$\int_{a=0}^{\delta_D} \lambda_A e^{-\lambda_A a} (\delta_D - a) da \qquad (21)$$

$$= \left[\left(1 - e^{-\lambda_A a} \right) (\delta_D - a) \right]_{a=0}^{\delta_D} - \int_{a=0}^{\delta_D} \left(1 - e^{-\lambda_A a} \right) (-1) da \qquad (22)$$

$$= \left(1 - e^{-\lambda_A \delta_D} \right) \underbrace{(\delta_D - \delta_D)}_{0} - \underbrace{\left(1 - e^{-\lambda_A 0} \right)(\delta_D - 0)}_{0} + \int_{a=0}^{\delta_D} 1 - e^{-\lambda_A a} \, da \qquad (23)$$

$$= \int_{a=0}^{\delta_D} 1 - e^{-\lambda_A a} \, da = \delta_D - \left[-\frac{e^{-\lambda_A a}}{\lambda_A} \right]_{a=0}^{\delta_D} = \frac{e^{-\delta_D \lambda_A} - 1}{\lambda_A} + \delta_D \, . \qquad (24)$$

Therefore, if the attacker does attack, her asymptotic benefit rate is

$$B_A \frac{\frac{e^{-\delta_D \lambda_A} - 1}{\lambda_A} + \delta_D}{\delta_D} \, , \qquad (25)$$

and her payoff is

$$B_A \frac{\frac{e^{-\delta_D \lambda_A} - 1}{\lambda_A} + \delta_D}{\delta_D} - \frac{C_A}{\delta_D} \, . \qquad (26)$$

Thus, when the above value is less than or equal to zero, never attacking is a best-response strategy; when the above value is greater than or equal to zero, always attacking immediately is a best-response strategy. When the above value is equal to zero, the attacker can decide whether to attack immediately or to not attack at all after each move of the defender. $\qquad \square$

Equilibrium. Based on the above lemmas, we can describe all the equilibria of the game (if there are any) as follows.

Theorem 1. *Suppose that the defender uses a renewal strategy and the attacker uses an adaptive strategy. Then the game's equilibria can be described as follows.*

1. *If $\frac{C_D}{B_A} = -le^{-\lambda_A l} + \frac{1-e^{-\lambda_A l}}{\lambda_A}$ does not have a solution for l, then there is a unique equilibrium in which the defender does not move and in which the attacker attacks exactly once at the beginning of the game.*
2. *If $\frac{C_D}{B_A} = -le^{-\lambda_A l} + \frac{1-e^{-\lambda_A l}}{\lambda_A}$ does have a solution δ_D for l, then*
   (a) *if $\frac{C_A}{B_A} \le \frac{e^{-\delta_D \lambda_A} - 1}{\lambda_A} + \delta_D$, then there is a unique equilibrium in which the defender plays a periodic strategy with period δ_D, and the attacker attacks immediately after the defender's each move;*
   (b) *if $\frac{C_A}{B_A} > \frac{e^{-\delta_D \lambda_A} - 1}{\lambda_A} + \delta_D$, then there is no equilibrium.*

In the first case, the attacker is at an overwhelming advantage, as the relative cost of defending the resource is prohibitively high. Consequently, the defender simply "gives up" the game since any effort to gain control of the resource is not profitable for her, and the attacker will have control of the resource all the time. In the second case, no player is at an overwhelming advantage. Both the defender and the attacker are actively trying to gain control of the resource, and both succeed from time to time. In the third case, the defender is at an overwhelming advantage. However, this does not lead to an equilibrium. If the defender moves with a sufficiently high rate, she makes moving unprofitable for the attacker. But if the attacker decides not to move, the defender is also better off not moving, as this decreases her cost. However, once the defender stops moving, it is again profitable for the attacker to move, which in turn triggers the defender to start moving.

Proof. First, we have from Lemma 1 that in any equilibrium, the defender either never moves or uses a periodic strategy. If the defender never moves, then the best strategy for the attacker is to attack immediately after the game starts. Now, if the defender moves using a periodic strategy, we have from Lemma 2 that the attacker either never attacks or attacks immediately. This leaves us with two strategies for defender and two strategies for attacker from which all equilibria must be composed.

Second, we show that there is no equilibrium in which the attacker never attacks. To see this, suppose that the attacker never attacks. Then the defender's best response is to never move, because this preserves control of the resource while minimizing the defender's cost. But if the defender never moves, then it is advantageous for the attacker to compromise the resource immediately after the start of the game. So this situation is not an equilibrium.

Next, we analyze the situation where a defender never moves. In this circumstance, the attacker attacks once and controls the resource for the duration of the game. From Lemma 1, we see that this is indeed a unique equilibrium if

$$\frac{C_D}{B_A} = lF_S(l) - \int_{s=0}^{l} F_S(s) \, ds \tag{27}$$

$$= l \left(1 - e^{-\lambda_A l}\right) - \int_{s=0}^{l} 1 - e^{-\lambda_A s} \, ds \tag{28}$$

$$= l - le^{-\lambda_A l} - \frac{e^{-\lambda_A l} - 1}{\lambda_A} - l \tag{29}$$

$$= -le^{-\lambda_A l} + \frac{1 - e^{-\lambda_A l}}{\lambda_A} \tag{30}$$

does not have a solution in $\mathbb{R}_{\geq 0}$ for l.

Finally, we consider the scenario where the defender plays a periodic strategy with period δ_D. In this case, Lemma 2 gives conditions for the best response of the attacker. Either the attacker never moves or the attacker attacks immediately. Since we know that there is no equilibrium in which an attacker never moves, we concern ourselves in the theorem only with the circumstances under which the attacker has a reason to attack immediately. From Lemma 2, the condition for this is $\frac{C_A}{B_A} \le \frac{e^{-\delta_D \lambda_A} - 1}{\lambda_A} + \delta_D$. □

4.2   Equilibrium for Both Targeted and Non-targeted Attackers

Defender's Best Response. Again, we begin our analysis by finding the defender's best response strategy.

Lemma 3. *Suppose that the non-targeted attacks arrive according to a Poisson process with rate λ_N, and the targeted attacker uses an adaptive strategy with a fixed wait time distribution given by the cumulative function F_S. Then,*

 – *not moving is the only best response if*

$$C_D = B_A \left(l F_S(l) - \int_{s=0}^{l} F_S(s) \; ds \right) + B_N \left(-l e^{-\lambda_N l} + \frac{1 - e^{-\lambda_N l}}{\lambda_N} \right) \quad (31)$$

 has no solution for l;
 – *a periodic strategy whose period is the solution to Equation (31) is the only best response otherwise.*

Proof. The outline of the proof is similar to that of Lemma 1.

The defender's expected loss per unit of time for an interval of length l is

$$\frac{1}{l} \left(B_A \int_{s=0}^{l} f_S(s)(l - s) \; ds + B_N \int_{a=0}^{l} (l - a)\lambda_N e^{-\lambda_N a} da + C_D \right) \quad (32)$$

$$= \frac{1}{l} \left(B_A \left([F_S(s)(l - s)]_{s=0}^{l} \int_{s=0}^{l} F_S(s) \; ds \right) + B_N \left(\frac{e^{-\lambda_N l} - 1}{\lambda_N} + l \right) + C_D \right) \quad (33)$$

$$= \frac{1}{l} \left(B_A \int_{s=0}^{l} F_S(s) \; ds + B_N \left(\frac{e^{-\lambda_N l} - 1}{\lambda_N} + l \right) + C_D \right) . \quad (34)$$

To find the minimizing interval lengths (if there exists any), we take the derivative of (34) and solve it for equality with 0 as follows:

$$0 = \frac{d}{dl} \left[\frac{1}{l} \left(B_A \int_{s=0}^{l} F_S(s) \; ds + B_N \left(\frac{e^{-\lambda_N l} - 1}{\lambda_N} + l \right) + C_D \right) \right] \quad (35)$$

$$0 = -\frac{1}{l^2} \left(B_A \left(\int_{s=0}^{l} F_S(s) \; ds - l F_S(l) \right) \right.$$

$$\left. + B_N \frac{e^{-\lambda_N l}(\lambda_N l - e^{\lambda_N l} + 1)}{\lambda_N} + C_D \right) \quad (36)$$

$$C_D = B_A \left(l F_S(l) - \int_{s=0}^{l} F_S(s) \; ds \right) + B_N \left(-l e^{-\lambda_N l} + \frac{1 - e^{-\lambda_N l}}{\lambda_N} \right) . \quad (37)$$

From the proof of Lemma 1, we have that the first term of the right hand side is monotonically increasing. Furthermore, the second term is strictly increasing, as its derivate is $\lambda_N l e^{-\lambda_N l} > 0$. Thus, the right hand side is strictly increasing, which implies that if there is an l^* for which the equality holds, it has to be unique. Furthermore, this l^* is a minimizing value as the second derivative is greater than zero:

$$\frac{d}{dl}\left[-\frac{1}{l^2}\left(B_A\left(\int_{s=0}^{l} F_S(s)\, ds - lF_S(l) \right) \right. \right.$$
$$\left. \left. + B_N \frac{e^{-\lambda_N l}(\lambda_N l - e^{\lambda_N l} + 1)}{\lambda_N} + C_D \right) \right] \tag{38}$$

$$= \frac{1}{l^3}\left(B_A\left(2\int_{s=0}^{l} F_S(s)\, ds - 2lF_S(l) + l^2 f_S(l) \right) \right.$$
$$\left. + B_N \frac{e^{\lambda_N l}(\lambda_N^2 l^2 + 2\lambda_N l - 2e^{\lambda_N l} + 2)}{\lambda_N} + 2C_D \right). \tag{39}$$

We care about the value of this expression when the first derivative is zero. Using this constraint, we obtain

$$\frac{1}{l^3}\left(B_A\left(2\int_{s=0}^{l} F_S(s)\, ds - 2lF_S(l) + l^2 f_S(l) \right) \right.$$
$$\left. + B_N \frac{e^{\lambda_N l}(\lambda_N^2 l^2 + 2\lambda_N l - 2e^{\lambda_N l} + 2)}{\lambda_N} + 2C_D \right) \tag{40}$$

$$= -\frac{2}{l}\left(B_A\left(\int_{s=0}^{l} F_S(s)\, ds - lF_S(l) \right) + B_N \frac{e^{-\lambda_N l}(\lambda_N l - e^{\lambda_N l} + 1)}{\lambda_N} + C_D \right)$$
$$+ \frac{1}{l}\left(B_A f_S(l) + B_N e^{-\lambda_N l}\lambda_N \right) \tag{41}$$

$$= -\frac{2}{l}(0) + \frac{1}{l}\left(B_A f_S(l) + B_N e^{-\lambda_N l}\lambda_N \right) > 0. \tag{42}$$

Consequently, the only best response is the periodic strategy with the minimizing l^* as the period.

On the other hand, if Equation (11) is not satisfiable for l, then the only best response for the defender is to never move. When $l \to \infty$, the defender's expected loss per unit of time approaches $B_A + B_N$, which is equal to her loss for never moving. When $l \to 0$, her expected loss per unit of time goes to infinity due to the ever increasing costs. Therefore, if there is no minimizing l, then the expected loss per unit of time is always greater than $B_A + B_N$. □

Equilibrium. Since the targeted attacker's payoff and, consequently, best response are not directly affected by the presence of non-targeted attackers, we can use Lemma 2 and the above lemma to describe the equilibria of the game.

Theorem 2. *Suppose that the defender uses a renewal strategy, the targeted attacker uses an adaptive strategy, and the non-targeted attacks arrive according to a Poisson process with rate λ_N. Then the game's equilibria can be described as follows.*

1. *If $C_D = B_A \left(-le^{-\lambda_A l} + \frac{1-e^{-\lambda_A l}}{\lambda_A} \right) + B_N \left(-le^{-\lambda_N l} + \frac{1-e^{-\lambda_N l}}{\lambda_N} \right)$ does not have a solution for l, then there is a unique equilibrium in which the defender does not move and in which the attacker attacks exactly once at the beginning of the game.*

2. *If $C_D = B_A \left(-le^{-\lambda_A l} + \frac{1-e^{-\lambda_A l}}{\lambda_A} \right) + B_N \left(-le^{-\lambda_N l} + \frac{1-e^{-\lambda_N l}}{\lambda_N} \right)$ does have a solution δ_D for l, then:*

   (a) *If $\frac{C_A}{B_A} \leq \frac{e^{-\delta_D \lambda_A}-1}{\lambda_A} + \delta_D$, then there is a unique equilibrium in which the defender plays a periodic strategy with period δ_D, and the targeted attacker moves immediately after the defender's each move.*

   (b) *If $\frac{C_A}{B_A} > \frac{e^{-\delta_D \lambda_A}-1}{\lambda_A} + \delta_D$, then:*

      - *if $C_D = B_N \left(-le^{-\lambda_N l} + \frac{1-e^{-\lambda_N l}}{\lambda_N} \right)$ has a solution δ'_D for l, and $\frac{C_A}{B_A} \geq \frac{e^{-\delta'_D \lambda_A}-1}{\lambda_A} + \delta'_D$, then there is a unique equilibrium in which the defender plays a periodic strategy with period δ'_D and the targeted attacker never moves;*

      - *otherwise, there is no equilibrium.*

By comparing the equation determining the defender's strategy in the theorem above to the equation in Theorem 1, we see that the parameter values B_A and C_D for which there is a solution is larger in the theorem above. Thus, the defender is more likely to move instead of giving it up when there is a threat of non-targeted attacks.

Proof. Cases 1. and 2. (a) follow from Lemma 2 and Lemma 3 using the argument as the proof of Theorem 1.

In Case 2. (b), there could be no equilibrium when the defender faced only a targeted attacker (Theorem 1), since the defender had no incentives to move if the targeted attacker did not move. However, when there are non-targeted attacker present as well, the defender moving periodically and the targeted attacker never moving can be an equilibrium. The necessary and sufficient conditions for this are that moving periodically is a best response for the defender against non-targeted attackers only (the existence of δ'_D) and that never attacking is a best-response for the targeted attacker against this period δ'_D.   □

5   Numerical Illustrations

In this section, we present numerical results on our game.

First, in Figure 1, we study the effects of varying the value of the resource, that is, the unit benefit B_A received by the targeted attacker. Figure 1a shows both players' payoffs for various values of B_A (the defender's periods for the

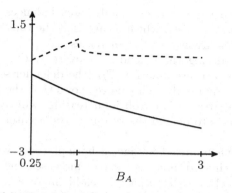

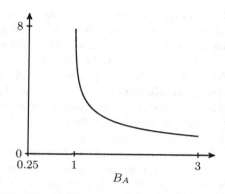

(a) The defender's and the targeted attacker's payoffs (solid and dashed lines, respectively) as a function of B_A.

(b) The defender's optimal period as a function of B_A.

Fig. 1. The effects of varying the unit benefit B_A received by the targeted attacker

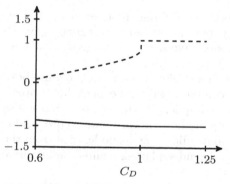

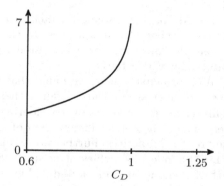

(a) The defender's and the targeted attacker's payoffs (solid and dashed lines, respectively) as a function of C_D.

(b) The defender's optimal period as a function of C_D.

Fig. 2. The effects of varying the defender's move cost C_D

same setup are shown by Figure 1b). The figure shows that the defender's payoff is strictly decreasing, which is not surprising: the more valuable the resource is, the higher the cost of security is for the defender. The attacker's payoff, on the other hand, starts growing linearly, but then suffers a sharp drop, and finally converges to a finite positive value.

For lower values ($B_A < 1$), the defender does not protect the resource, as it is not valuable enough to defend. Accordingly, Figure 1b shows no period for this region. In this case, the attacker's payoff is equal to simply the value of the resource. However, once the value of the resource reaches 1, the defender starts protecting it. At this point, the attacker's payoff drops as she no longer has the resource compromised all the time. For higher values, the defender balances between losses due to compromise and moving costs, which means that the time the resource is compromised decreases steadily as its value increases.

In Figure 2, we study the effects of varying the defender's move cost C_D. Figure 2a shows both players' payoffs for various values of C_D (the defender's periods for the same setup are shown by Figure 2b). The figure shows that the defender's payoff is decreasing, while the attacker's payoff is increasing, which is again not surprising: the more costly it is to defend the resource, the greater the attacker's advantage is.

For lower costs, no player is at an overwhelming advantage, as both players try to control the resource and succeed from time to time. As the cost increases, the defender's payoff steadily decreases, while the attacker's payoff steadily increases. For higher costs, the attacker is at an overwhelming advantage. In this case, the defender never moves, while the attacker moves once. Hence, their payoffs are -1 and 1, respectively.

6   Conclusions

Targeted and non-targeted attacks are born of different motivations and have different types of consequences. In this paper, we modeled a regime in which a defender must vie for a contested resource against both targeted and non-targeted covert attacks.

As a principal result, we found that the most effective strategy against both types of attacks (and also against their combination) is the periodic strategy. This result can be surprising considering the simplicity of this strategy, but it also serves as a theoretical justification of the periodic password and cryptographic key renewal practices. Furthermore, this contradicts the lesson learned from the FlipIt model [13], which suggests that a defender playing against an adaptive attacker should use an unpredictable strategy.

We also found that a defender is more likely to stay in play and bear the costs of periodic risk mitigation if she is threatened by non-targeted attacks. While this result seems very intuitive, it is not obvious, as we also demonstrated that a very high level of either threat type can force the defender to abandon all hope and stop moving.

Our work can be extended in multiple directions. First, even though the exponential attack time distribution can be well-motivated for a number of resources, it would be worthwhile to extend our model to general distributions with some mild assumptions only. Second, our model focuses on medium-profile targets that are susceptible to both targeted and non-targeted attacks, but it could be easily extended to a broader range by having a susceptibility probability for each type.

Acknowledgements. We gratefully acknowledge the support of the Penn State Institute for CyberScience. We also thank the reviewers for their comments on an earlier draft of the paper.

References

1. Bowers, K.D., van Dijk, M., Griffin, R., Juels, A., Oprea, A., Rivest, R.L., Triandopoulos, N.: Defending against the unknown enemy: Applying FlipIt to system security. In: Grossklags, J., Walrand, J. (eds.) GameSec 2012. LNCS, vol. 7638, pp. 248–263. Springer, Heidelberg (2012)
2. Casey, E.: Determining intent - opportunistic vs targeted attacks. Computer Fraud & Security 2003(4), 8–11 (2003)
3. Grossklags, J., Christin, N., Chuang, J.: Secure or insure? A game-theoretic analysis of information security games. In: Proceedings of the 17th International World Wide Web Conference (WWW), pp. 209–218 (2008)
4. Herley, C.: The plight of the targeted attacker in a world of scale. In: Proceedings of the 9th Workshop on the Economics of Information Security (WEIS) (2010)
5. Johnson, B., Böhme, R., Grossklags, J.: Security games with market insurance. In: Baras, J.S., Katz, J., Altman, E. (eds.) GameSec 2011. LNCS, vol. 7037, pp. 117–130. Springer, Heidelberg (2011)
6. Laszka, A., Felegyhazi, M., Buttyán, L.: A survey of interdependent security games. Technical Report CRYSYS-TR-2012-11-15, CrySyS Lab, Budapest University of Technology and Economics (November 2012)
7. Laszka, A., Horvath, G., Felegyhazi, M., Buttyan, L.: FlipThem: Modeling targeted attacks with FlipIt for multiple resources. Technical report, Budapest University of Technology and Economics (2013)
8. Nochenson, A., Grossklags, J.: A behavioral investigation of the FlipIt game. In: Proceedings of the 12th Workshop on the Economics of Information Security (WEIS) (2013)
9. Pham, V., Cid, C.: Are we compromised? Modelling security assessment games. In: Grossklags, J., Walrand, J. (eds.) GameSec 2012. LNCS, vol. 7638, pp. 234–247. Springer, Heidelberg (2012)
10. Radzik, T.: Results and problems in games of timing. Lecture Notes-Monograph Series, Statistics, Probability and Game Theory: Papers in Honor of David Blackwell 30, 269–292 (1996)
11. Radzik, T., Orlowski, K.: A mixed game of timing: Investigation of strategies. Zastosowania Matematyki 17(3), 409–430 (1982)
12. Reitter, D., Grossklags, J., Nochenson, A.: Risk-seeking in a continuous game of timing. In: Proceedings of the 13th International Conference on Cognitive Modeling (ICCM), pp. 397–403 (2013)
13. van Dijk, M., Juels, A., Oprea, A., Rivest, R.: FlipIt: The game of "stealthy takeover". Journal of Cryptology 26, 655–713 (2013)
14. Zhadan, V.: Noisy duels with arbitrary accuracy functions. Issledovanye Operacity 5, 156–177 (1976)

New Efficient Utility Upper Bounds
for the Fully Adaptive Model of Attack Trees

Ahto Buldas[1,2,3] and Aleksandr Lenin[1,3,*]

[1] Cybernetica AS
[2] Guardtime AS
[3] Tallinn University of Technology

Abstract. We present a new fully adaptive computational model for attack trees that allows attackers to repeat atomic attacks if they fail and to play on if they are caught and have to pay penalties. The new model allows safer conclusions about the security of real-life systems and is somewhat (computationally) easier to analyze. We show that in the new model optimal strategies always exist and finding the optimal strategy is (just) an NP-complete problem. We also present methods to compute adversarial utility estimation and utility upper bound approximated estimation using a bottom-up approach.

1 Introduction

Protection of information systems becomes an integral part in the deployment of technologies that operate sensitive information, the leakage of which may cause irreversible damage to affected parties. In new technology deployment, its protection and security are the concerns in the first place. It is impossible to achieve 100% level of protection. By applying various security measures one can just approach this limit. Various methods of risk assessment have been suggested, and each of them has its advantages and disadvantages. Quantitative security analysis based on attack trees has become a subject of extensive research [8, 1–6, 9].

Attack trees may be used for visualization purposes only, but also for computing *adversarial utility*, i.e. attacking the system can be modeled as an economic single-player game played by the attacker. It is assumed that the attacker behaves rationally—attacks only if the attack game is beneficial for him. Such a *rational attacker paradigm* was first introduced by Buldas *et al.* [1]. In order to estimate the adversarial utility, their model used computational rules for AND and OR nodes to compute a list of parameters for every node based on the parameters of its successor nodes. Buldas and Stepanenko [3] introduced the

---

* Part of the research leading to these results has received funding from the European Union Seventh Framework Programme (FP7/2007-2013) under grant agreement ICT-318003 (TREsPASS). This publication reflects only the authors' views and the Union is not liable for any use that may be made of the information contained herein.

S.K. Das, C. Nita-Rotaru, and M. Kantarcioglu (Eds.): GameSec 2013, LNCS 8252, pp. 192–205, 2013.

so-called *fully adaptive model* where adversaries are allowed to try atomic attacks in an arbitrary order, depending on the results of the previous trials. They
also introduced the *upper bound ideology* by pointing out that in order to verify
the security of the system, it is not necessary to compute the exact adversarial
utility but only *upper bounds*—if adversarial utility has a negative upper bound
in the fully adaptive model, it is safe to conclude that there are no beneficial
ways of attacking the system, assuming that all reasonable atomic attacks are
captured in the attack tree. A similar approach is used in civil engineering—it
is more practical to know an upper bound of the stress value that will definitely
not break a construction than the precise stress value at which the construction breaks. In [3] two ways were introduced to compute the upper bounds: (1)
simplified computational rules (for AND and OR nodes); and (2) assuming more
powerful adversaries in a way that simplifies the computational model, for example, in their *infinite repetition model* the attacker is allowed to repeat atomic
attacks (any number of times) if they fail.

Motivation of This Work: Even the fully adaptive model of Buldas and Stepanenko [3] does not completely follow their upper bound ideology. Mostly the
atomic attacks are associated with criminal behavior and hence in the attack
tree models [1, 4, 5, 9, 3] atomic attacks are associated with *penalties* that the
attacker has to pay if he is caught. In the model [3] an additional restriction
is introduced—the attacker is not able to play on after getting caught. As this
seems not to be true in all real-life cases, either the penalties in their model
contain the potential future profits of attackers (and hence be larger than they
are in real life) or the model does not give reliable upper bounds. Moreover,
it seems that such a *game over* assumption actually makes the computational
model more complex.

The Aim of This Work: We present a new fully adaptive computational model
for attack trees that allows the adversary to repeat atomic attacks if they fail
and to play on if he is caught. We show that such a model will be somewhat easier
to analyze. For example, in the case of conjunctive composition $\mathcal{X}_1 \wedge \ldots \wedge \mathcal{X}_n$ of
atomic attacks the order in which they are tried by the adversary is unimportant.

Summary of Results: We show (Sec. 3) that in the new model optimal strategies always exist and they are in the form of directed single-branched BDDs with
self-loops. We introduce methods to compute a precise estimation of the adversarial utility and an approximated estimation of the utility upper bound using
a bottom-up *utility propagation* approach. We also show (Sec. 4) that solving
the attack game in the new model is an NP-complete problem. We also present
efficient methods to compute the lower bound for expenses (Sec. 5).

2   Definitions and Related Work

In this section we will formally define some common terms and definitions within
the current model which will be used further throughout the paper.

2.1 Definitions

Definition 1 (Derived function). *If $\mathcal{F}(x_1,\ldots,x_m)$ is a Boolean function and $v \in \{0,1\}$, then by the derived Boolean function $\mathcal{F}|_{x_j=v}$ we mean the function $\mathcal{F}(x_1,\ldots,x_{j-1},v,x_{j+1},\ldots,x_m)$ derived from $\mathcal{F}$ by the assignment $x_j := v$.*

Definition 2 (Constant functions). *By 1 we mean a Boolean function that is identically true and by 0 we mean a Boolean function that is identically false.*

Definition 3 (Min-term). *By a min-term of a Boolean function $\mathcal{F}(x_1,\ldots,x_m)$ we mean a conjunction of variables $x_{i_1} \wedge x_{i_2} \wedge \ldots \wedge x_{i_k}$ such that $x_{i_1} \wedge x_{i_2} \wedge \ldots \wedge x_{i_k} \Rightarrow \mathcal{F}(x_1,\ldots,x_m)$ is a tautology.*

Definition 4 (Critical min-term). *A min-term $x_1 \wedge \ldots \wedge x_k$ of $\mathcal{F}$ is critical if none of the sub-terms $x_1,\ldots,x_{j-1},x_{j+1},\ldots,x_m$ is a min-term of $\mathcal{F}$.*

Definition 5 (Satisfiability game). *By a satisfiability game we mean a single-player game in which the player's goal is to satisfy a monotone Boolean function $\mathcal{F}(x_1,x_2,\ldots,x_k)$ by picking variables x_i one at a time and assigning $x_i = 1$. Each time the player picks the variable x_i he pays some amount of expenses $\mathcal{E}_i$, which is modeled as a random variable. With a certain probability p_i the move x_i succeeds. Function $\mathcal{F}$ representing the current game instance is transformed to its derived form $\mathcal{F}|_{x_i=1}$ and the next game iteration starts. The game ends when the condition $\mathcal{F} \equiv 1$ is satisfied and the player wins the prize $\mathcal{P} \in \mathcal{R}$, or when the condition $\mathcal{F} \equiv 0$ is satisfied, meaning the loss of the game, or when the player stops playing. With a probability $1 - p_i$ the move x_i fails. The player may end up in a different game instance represented by the derived Boolean function $\mathcal{F}|_{x_i\equiv 0}$ in the case of a game without move repetitions, and may end up in the very same instance of the game $\mathcal{F}$ in the case of a game with repetitions. Under certain conditions with a certain probability the game may end up in a forced failure state, i.e. if the player is caught and this implies that he cannot continue playing, i.e. according to the Buldas-Stepanenko model [3]. The rules of the game are model-specific and may vary from model to model. Thus we can define three common types of games:*

1. *SAT Game Without Repetitions - the type of a game where an adversary can perform a move only once.*
2. *SAT Game With Repetitions - the type of a game where an adversary can re-run failed moves again an arbitrary number of times.*
3. *Failure-Free SAT Game - the type of a game in which all success probabilities are equal to 1. It can be shown that any game with repetitions is equivalent to a failure-free game (Thm. 5).*

Definition 6 (Line of a game). *By a line of a satisfiability game we mean a sequence of assignments $\lambda = \langle x_{j_1} = v_1,\ldots,x_{j_k} = v_k \rangle$ (where $v_j \in \{0,1\}$) that represent the player's moves, and possibly some auxiliary information. We say that λ is a **winning line** if the Boolean formula $x_{i_1} \wedge \ldots \wedge x_{i_k} \Rightarrow \mathcal{F}(x_1,\ldots,x_n)$ is a tautology, where $\mathcal{F}$ is a Boolean function of the satisfiability game.*

Definition 7 (Strategy). *By a strategy S for a game G we mean a rule that for any line λ of G either suggests the next move $x_{j_{k+1}}$ or decides to give up.*

Strategies can be represented graphically as binary decision diagrams (BDDs).

Definition 8 (Line of a strategy). *A line of a strategy S for a game G is the smallest set $\mathcal{L}$ of lines of G such that (1) $\langle\rangle \in \mathcal{L}$ and (2) if $\lambda \in \mathcal{L}$, and S suggests x_j as the next move to try, then $\langle\lambda, x_j = 0\rangle \in \mathcal{L}$ and $\langle\lambda, x_j = 1\rangle \in \mathcal{L}$.*

Definition 9 (Branch). *A branch β of a strategy S for a game G is a line λ of S for which S does not suggest the next move. By $\mathcal{B}_S$ we denote the set of all branches of S.*

For example, all winning lines of S are branches.

Definition 10 (Expenses of a branch). *If $\beta = \langle x_{i_1=v_1}, \dots, x_{i_k=v_k}\rangle$ is a branch of a strategy S for G, then by expenses $\epsilon_G(S, \beta)$ of β we mean the sum $\overline{\mathcal{E}}_{i_1} + \dots + \overline{\mathcal{E}}_{i_k}$ where by $\overline{\mathcal{E}}_{i_j}$ we mean the mathematical expectation of $\mathcal{E}_{i_j}$.*

Definition 11 (Prize of a branch). *The prize $\mathcal{P}_G(S, \beta)$ of a branch β of a strategy S is $\mathcal{P}$ if β is a winning branch, and 0 otherwise.*

Definition 12 (Utility of a strategy). *By the utility of a strategy S in a game G we mean the sum: $\mathcal{U}(G, S) = \sum_{\beta \in \mathcal{B}_S} Pr(\beta) \cdot [\mathcal{P}_G(S, \beta) - \epsilon_G(S, \beta)]$. For the empty strategy $\mathcal{U}(G, \emptyset) = 0$.*

Definition 13 (Prize and Expenses of a strategy). *By the expenses $\mathcal{E}(G, S)$ of a strategy S we mean the sum $\sum_{\beta \in \mathcal{B}_S} Pr(\beta) \cdot \epsilon_G(S, \beta)$. The prize $\mathcal{P}(G, S)$ of S is $\sum_{\beta \in \mathcal{B}_S} Pr(\beta) \cdot \mathcal{P}_G(S, \beta)$.*

It is easy to see that $\mathcal{U}(G, S) = \mathcal{P}(G, S) - \mathcal{E}(G, S)$.

Definition 14 (Utility of a satisfiability game). *The utility of a SAT game G is the limit $\mathcal{U}(G) = \sup_S \mathcal{U}(G, S)$ that exists due to the bound $\mathcal{U}(G, S) \leqslant \mathcal{P}$.*

Definition 15 (Optimal strategy). *By an optimal strategy for a game G we mean a strategy S for which $\mathcal{U}(G) = \mathcal{U}(G, S)$.*

It can be shown that for satisfiability games optimal strategies always exist.

2.2 Related Work

In the *fully-adaptive model* introduced by Buldas and Stepanenko [3] the attacker does not use a specific attack suite or a specific ordering, but picks the next atomic attack arbitrarily based on the results of the previously tried atomic attacks. Their so-called *infinite repetition model* assumes that the adversary has a possibility to re-run failed attacks again immediately or later after trying some other atomic attacks. The so-called *failure-free model* [3] assumes all success probabilities are equal to 1, meaning that the player will achieve his goal anyway, but with either positive or negative utility. Due to the failure-free model concept a strategy can be represented as a set of moves $\{\mathcal{X}_1, \ldots, \mathcal{X}_k\}$ and the order in which those moves will be launched is not a concern.

The model [3] assumes that the adversary stops playing immediately upon attack detection that is an unnatural restriction placed on the adversarial actions since in reality countermeasures cannot be applied immediately and usually the adversary has some time within which he may continue attacking and may achieve his goal. Therefore it would be natural to expect that the attacker does not stop his actions if his attack is detected by the defensive security measures deployed on the analyzed system. Apart from that, Buldas-Stepanenko model is arithmetically more complex due to the force-failure states and an optimal strategy depends on the order of atomic attacks.

3   The New Model

In the new model the adversary does not stop when launched attacks are detected and continues attacking until he achieves his goal. The new model is similar to the parallel model by Jürgenson and Willemson [4–6], except that it applies the infinite repetition model concept and introduces new methods that allow us to compute the adversarial utility upper bounds. Due to the slightly simplified rules of the game the new model became more simple and manageable than Jürgenson-Willemson and Buldas-Stepanenko models, thus easier to use and analyze. The new model allowed us to elaborate efficient upper bound computation methods that run in time linear in the size of the attack tree.

Lemma 1. *For every repeatable satisfiability game $\mathcal{G}$ with $\mathcal{U}(\mathcal{G}) > 0$ there is x_j such that* $\sup_{\mathcal{S} \in \mathcal{S}_{x_j}} \mathcal{U}(\mathcal{G}, \mathcal{S}) = \mathcal{U}(\mathcal{G})$, *where $\mathcal{S}_{x_j}$ is the set of all non-empty strategies with x_j as the first move.*

Proof. As every $\mathcal{S} \neq \emptyset$ has the first move x_i, we have $\mathcal{U}(\mathcal{G}) = \sup_{\mathcal{S}} \mathcal{U}(\mathcal{G}, \mathcal{S}) = \max_i \sup_{\mathcal{S} \in \mathcal{S}_{x_i}} \mathcal{U}(\mathcal{G}, \mathcal{S})$, and hence there is x_j such that $\mathcal{U}(\mathcal{G}) = \sup_{\mathcal{S} \in \mathcal{S}_{x_j}} \mathcal{U}(\mathcal{G}, \mathcal{S})$.   □

Lemma 2. *For every repeatable satisfiability game G and for every atomic variable x_j:* $\sup_{\mathcal{S} \in \mathcal{S}_{x_j}} \mathcal{U}(\mathcal{G}, \mathcal{S}) = -\mathcal{E}_j + p_j \mathcal{U}(\mathcal{G}|_{x_j=1}) + (1 - p_j)\mathcal{U}(\mathcal{G})$.

Proof. This is because the part $\mathcal{S}'$ of $\mathcal{S}$ for playing $\mathcal{G}|_{x_j=1}$ and the part $\mathcal{S}''$ of $\mathcal{S}$ for playing $\mathcal{G}$ after an unsuccessful trial of x_j can be chosen independently.   □

Theorem 1. *Repeatable satisfiability games have optimal strategies.*

Proof. If $\mathcal{U}(\mathcal{G}) = 0$, then $\mathcal{S} = \emptyset$ is optimal. For the case $\mathcal{U}(\mathcal{G}) > 0$ we use induction on the number m of atomic variables. If $m = 0$, there are no moves and $\emptyset$ is the only possible strategy and is optimal by definition. In case $m > 0$ and supposing that every repeatable satisfiability game with $m - 1$ atomic variables has an optimal strategy, by *Lemma 1*, there is x_j such that $\mathcal{U}(\mathcal{G}) = \sup\limits_{\mathcal{S} \in \mathcal{S}_{x_j}} \mathcal{U}(\mathcal{G}, \mathcal{S})$.

Let $\mathcal{S}_0$ be the strategy that repeats x_j until x_j succeeds and then behaves like an optimal strategy for $\mathcal{G}|_{x_j=1}$ (a game with $m - 1$ atomic variables). Utility of $\mathcal{S}_0$ is $\mathcal{U}(\mathcal{G}, \mathcal{S}_0) = -\frac{\mathcal{E}_j}{p_j} + \mathcal{U}(\mathcal{G}|_{x_j=1})$. On the other hand, by *Lemma 2* we have $\mathcal{U}(\mathcal{G}) = \sup\limits_{\mathcal{S} \in \mathcal{S}_{x_j}} \mathcal{U}(\mathcal{G}, \mathcal{S}) = -\mathcal{E}_j + p_j \mathcal{U}(\mathcal{G}|_{x_j=1}) + (1 - p_j)\mathcal{U}(\mathcal{G})$, that implies

$$\mathcal{U}(\mathcal{G}) = -\frac{\mathcal{E}_j}{p_j} + \mathcal{U}(\mathcal{G}|_{x_j=1}) = \mathcal{U}(\mathcal{G}, \mathcal{S}_0) \text{ and hence } \mathcal{S}_0 \text{ is optimal.} \qquad \square$$

Corollary 1. *In every repeatable satisfiability game there exist optimal strategies in the form of directed single-branched BDDs with self-loops.*

Proof. Let $\mathcal{S}$ be an optimal strategy for $\mathcal{G}$ and $\mathcal{X}_{j_1}$ be the first move suggested by $\mathcal{S}$. In case of a failure, $\mathcal{X}_{j_1}$ remains the best move and hence $\mathcal{S}$ has a self-loop at $\mathcal{X}_{j_1}$. In case of success, the Boolean function of the game reduces to $\mathcal{F}|_{\mathcal{X}_{j_1}=1}$. Let $\mathcal{X}_{j_2}$ be the next move suggested by $\mathcal{S}$. Similarly, we conclude that there is a self-loop at $\mathcal{X}_{j_2}$ in case of a failure, and so on. This leads to the BDD in Fig. 1. $\quad \square$

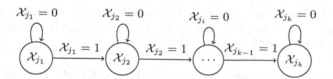

Fig. 1. A strategy in the form of a directed single-branched BDD with self-loops

Theorem 2. *If $\mathcal{S}$ is a strategy in the form of a self-looped BDD (Fig. 1), then*

$$\mathcal{E}(\mathcal{G}, \mathcal{S}) = \frac{\overline{\mathcal{E}}_{j_1}}{p_{j_1}} + \ldots + \frac{\overline{\mathcal{E}}_{j_k}}{p_{j_k}} .$$

Proof. The probability that a move succeeds at the n-th try is $p(1 - p)^{n-1}$ and the average expenses are $\overline{\mathcal{E}}(1 - p)^{n-1}$ and hence the total success probability is $p \cdot \sum\limits_{n=1}^{\infty} (1 - p)^{n-1} = 1$ and the average expenses are $\overline{\mathcal{E}} \cdot \sum\limits_{n=1}^{\infty} (1 - p)^{n-1} = \frac{\overline{\mathcal{E}}}{p}.$ $\quad \square$

It is obvious that in the new model $\mathcal{P}(\mathcal{G}, \mathcal{S}) \in \{0, \mathcal{P}\}$.

Definition 16 (Winning strategy). *A strategy $\mathcal{S}$ is winning if $\mathcal{P}(\mathcal{G}, \mathcal{S}) = \mathcal{P}$.*

Definition 17 (Expenses of a game). *By the average expenses of a game $\mathcal{G}$ we mean $\mathcal{E}(\mathcal{G}) = \inf\limits_{\mathcal{S}} \mathcal{E}(\mathcal{G}, \mathcal{S})$, where $\mathcal{S}$ varies over all winning strategies.*

Theorem 3. *For any satisfiability game* $\mathcal{G}$: $\mathcal{U}(\mathcal{G}) = \max\{0, \mathcal{P} - \mathcal{E}(\mathcal{G})\}$.

Proof. Let $\mathcal{S}$ be an optimal strategy for $\mathcal{G}$. If $\mathcal{S}$ is not winning, $\mathcal{U}(\mathcal{G}, \mathcal{S}) = -\mathcal{E}(\mathcal{G}, \mathcal{S}) \leqslant 0$ and hence $\emptyset$ is optimal and hence $\mathcal{U}(\mathcal{G}) = 0$. If $\mathcal{S}$ is winning then $\mathcal{U}(\mathcal{G}) = \sup_{\mathcal{S}} \mathcal{U}(\mathcal{G}, \mathcal{S}) = \mathcal{P} - \inf_{\mathcal{S}} \mathcal{E}(\mathcal{G}, \mathcal{S}) = \mathcal{P} - \mathcal{E}(\mathcal{G})$.   □

Due to the features of the new model, in order to compute the utility it is sufficient to compute the expenses, that allows us use the expenses propagation technique introduced below. Moreover, we will show that solving a satisfiability game in the new model is equivalent to solving a weighted monotone satisfiability problem.

3.1   Precise Utility Computation

The algorithm described in [3] is good because it is independent of the Boolean circuit structure and only depends on the Boolean function of the game. It is described formally as the following recursive relation:

$$\mathcal{U}(\mathcal{G}) = \max\left\{ 0, \ -\frac{\overline{\mathcal{E}}_i}{p_i} + \mathcal{U}\left(\mathcal{G}|_{x_i=1}\right), \ \mathcal{U}\left(\mathcal{G}|_{x_i=0}\right) \right\} , \qquad (1)$$

with initial conditions $\mathcal{U}(\mathbf{1}) = \mathcal{P}$ and $\mathcal{U}(\mathbf{0}) = 0$, and where x_i is any variable that $\mathcal{G}$ contains. The algorithm allows us to compute the precise adversarial utility value in time exponential in the size of the game. The computational complexity of the algorithm is $\mathcal{O}(2^n)$.

3.2   Utility Upper Bound Estimation Using Utility Propagation

The model of [3] uses attack tree representation of the Boolean formula for utility propagation using the following inequalities:

$$\mathcal{U}(\mathcal{G}_1 \wedge \ldots \wedge \mathcal{G}_k) \leqslant \min\{\mathcal{U}(\mathcal{G}_1), \ldots, \mathcal{U}(\mathcal{G}_k)\} ,$$
$$\mathcal{U}(\mathcal{G}_1 \vee \ldots \vee \mathcal{G}_k) \leqslant \mathcal{U}(\mathcal{G}_1) + \ldots + \mathcal{U}(\mathcal{G}_k) .$$

Firstly, we can show that even in the model of [3] we can actually use more precise folmula (2). Secondly, in the new model we can use expenses propagation approach that turns out to be even more precise.

Theorem 4. *In the model of [3] and in the new model:*

$$\mathcal{U}(\mathcal{G}_1 \vee \ldots \vee \mathcal{G}_k) = \max\{\mathcal{U}(\mathcal{G}_1), \ldots, \mathcal{U}(\mathcal{G}_k)\} . \qquad (2)$$

The proof is based on the fact that optimal strategies can be represented by critical min-terms of $\mathcal{F}$.

Proof. Let $\mathcal{S}$ be the optimal strategy for the game $\mathcal{G} = \mathcal{G}_1 \vee \ldots \vee \mathcal{G}_k$. According to *Thm. 8* in [3], $\mathcal{S}$ can be represented as a critical min-term $\mathcal{X}_{j_1} \wedge \ldots \wedge \mathcal{X}_{j_m}$ of the Boolean function of $\mathcal{G} = \mathcal{G}_1 \vee \ldots \vee \mathcal{G}_k$, this means that there exists such j that $\mathcal{X}_{j_1} \wedge \ldots \wedge \mathcal{X}_{j_m}$ is a min-term of $\mathcal{G}_j$. Hence $\mathcal{X}_{j_1} \wedge \ldots \wedge \mathcal{X}_{j_m} \Rightarrow \mathcal{G}_j$ is a tautology. This means that

$$\mathcal{U}(\mathcal{G}_1 \vee \ldots \vee \mathcal{G}_k) = \mathcal{U}(\mathcal{G}_1 \vee \ldots \vee \mathcal{G}_k, \mathcal{S}) \leqslant \mathcal{U}(\mathcal{G}_j) \leqslant \max\{\mathcal{U}(\mathcal{G}_1), \ldots, \mathcal{U}(\mathcal{G}_k)\} \ .$$

On the other hand, for any j let $\mathcal{S}_j$ be the optimal strategy of $\mathcal{G}_j$. As $\mathcal{G}_j \Rightarrow \mathcal{G}_1 \vee \ldots \vee \mathcal{G}_k$ is a tautology, $\mathcal{U}(\mathcal{G}_j) = \mathcal{U}(\mathcal{G}_j, \mathcal{S}_j) \leqslant \mathcal{U}(\mathcal{G}_1 \vee \ldots \vee \mathcal{G}_k)$. As j was arbitrary, this implies $\max\{\mathcal{U}(\mathcal{G}_1), \ldots, \mathcal{U}(\mathcal{G}_k)\} \leqslant \mathcal{U}(\mathcal{G}_1 \vee \ldots \vee \mathcal{G}_k)$. Combining these two inequalities we reach equation (2).   □

Algorithm 3.1 utilizes the conjunctive and disjunctive bottom-up adversarial utility propagation rules in every game instance starting from the atomic moves and ending up in the root instance of the game. The algorithm allows us to compute the adversarial utility upper bound in time linear in the size of the game, thus complexity is $\mathcal{O}(n)$.

ALGORITHM 3.1. Iterated utility propagation in conjunctive/disjunctive game instances

---

Input: Satisfiability game instance $\mathcal{G}$
Output: Utility upper bound (real number)
1   Procedure `ComputeUtilityUpperBound` $(\mathcal{G})$
2   **if** *m is an instance of a conjunctive game* **then**
3 | /* $\mathcal{G}_1, \ldots, \mathcal{G}_k$ are the sub-games of game m */
4 | **return** min $\{\mathcal{U}(\mathcal{G}_1), \ldots, \mathcal{U}(\mathcal{G}_k)\}$
5   **else if** *m is an instance of a disjunctive game* **then**
6 | /* $\mathcal{G}_1, \ldots, \mathcal{G}_k$ are the sub-games of game m */
7 | **return** max $\{\mathcal{U}(\mathcal{G}_1), \ldots, \mathcal{U}(\mathcal{G}_k)\}$
8   **else** m is leaf
9 | **return** $\mathcal{U}(m)$

---

4   Computational Complexity of the New Model

Definition 18 (Weighted Monotone Satisfiability /WMSAT/). *Given a threshold value $\mathcal{P}$ and a monotone Boolean function $\mathcal{F}(x_1, \ldots, x_m)$ with corresponding weights $w(x_i) = w_i$ decide whether there is a satisfying assignment $\mathcal{A}$ with a total weight $w(\mathcal{A}) < \mathcal{P}$.*

Theorem 5. *In the new model, the problem of deciding whether $\mathcal{U}(\mathcal{G}) > 0$ is equivalent to the weighted monotone satisfiability problem with the same Boolean function as in $\mathcal{G}$ and with weights of the input variables x_i defined by $w_i = \frac{\overline{\varepsilon_i}}{p_i}$ with threshold value $\mathcal{P}$.*

Proof. If $\mathcal{S}$ is an optimal strategy in the *infinite repetition model* and is a single-branched BDD with self-loops with nodes $\mathcal{X}_{i_1}, \ldots, \mathcal{X}_{i_k}$, then the assignment $\mathcal{A} = \langle x_{i_1} = \ldots = x_{i_k} = 1 \rangle$ satisfies the Boolean function of the game and its total weight is $w(\mathcal{A}) = \frac{\overline{\mathcal{E}}_{i_1}}{p_{i_1}} + \ldots + \frac{\overline{\mathcal{E}}_{i_k}}{p_{i_k}}$. If $\mathcal{U}(\mathcal{G}) > 0$, then

$$0 < \mathcal{U}(\mathcal{G}) = \mathcal{U}(\mathcal{G}, \mathcal{S}) = \mathcal{P} - \frac{\overline{\mathcal{E}}_{i_1}}{p_{i_1}} - \ldots - \frac{\overline{\mathcal{E}}_{i_k}}{p_{i_k}} = \mathcal{P} - w(\mathcal{A}) \ .$$

Thus, $w(\mathcal{A}) < \mathcal{P}$. If there is an assignment $\mathcal{A} = \langle x_{i_1} = \ldots = x_{i_k} = 1 \rangle$ in the WMSAT model with a total weight $w(\mathcal{A}) = \frac{\overline{\mathcal{E}}_{i_1}}{p_{i_1}} + \ldots + \frac{\overline{\mathcal{E}}_{i_k}}{p_{i_k}} < \mathcal{P}$, then the strategy depicted in Fig. 1 has the utility

$$\mathcal{U}(\mathcal{G}, \mathcal{S}) = \mathcal{P} - \frac{\overline{\mathcal{E}}_{i_1}}{p_{i_1}} - \ldots - \frac{\overline{\mathcal{E}}_{i_k}}{p_{i_k}} > 0 \ ,$$

and hence $\mathcal{U}(\mathcal{G}) \geqslant \mathcal{U}(\mathcal{G}, \mathcal{S}) > 0$. $\square$

The parameter $\frac{\overline{\mathcal{E}}_i}{p_i}$, the cost-success ratio, is similar to the time-success ratio parameter used in cryptography [7]. This parameter can be estimated more precisely and is measurable in monetary units, as opposed to the respective probability and expenses parameters in the existing models.

Theorem 6. *The Weighted Monotone Satisfiability Problem is NP-complete.*

Proof. We will show that the Vertex Cover problem can be polynomially reduced to the WMSAT problem. Let $\mathcal{G}$ be the graph with a vertex set $\{v_1, \ldots, v_m\}$. We define a Boolean function $\mathcal{F}(x_1, \ldots, x_m)$ as follows. For each edge (v_i, v_j) of $\mathcal{G}$ we define the clause $\mathcal{C}_{i_j} = x_i \vee x_j$. The Boolean function $\mathcal{F}(x_1, \ldots, x_m)$ is defined as the conjunction of all $\mathcal{C}_{i_j}$ such that (v_i, v_j) is an edge of $\mathcal{G}$. Let the weight w_i of each x_i be equal to 1.

It is obvious that $\mathcal{G}$ has a vertex cover $\mathcal{S}$ of size $|\mathcal{S}| < \mathcal{P}$ iff the monotone Boolean function $\mathcal{F}(x_1, \ldots, x_m)$ has a satisfying assignment with a total weight less than $\mathcal{P}$. $\square$

5   Efficient Computation of Expenses Lower Bounds

This section presents some examples of computing the adversarial expenses lower bound using expenses propagation.

5.1   Expenses Propagation

Let $\mathcal{G} = \mathcal{G}_1 \vee \ldots \vee \mathcal{G}_k$ be a disjunctive game. For the disjunctive game to succeed at least one of its sub-games needs to be tried and successfully completed. We need to choose one single sub-game which is the cheapest one. Therefore, the utility upper bound of the game $\mathcal{G}$ may be computed using the method that for each sub-game computes utility estimation $\mathcal{U}(\mathcal{G}_i)$:

1. Find the cheapest sub-game (the sub-game $\mathcal{G}_i$ having minimal $\mathcal{E}\left(\mathcal{G}_i\right)$ value).
2. For this sub-game compute the utility upper bound as: $\mathcal{U}\left(\mathcal{G}\right) = \mathcal{P} - \mathcal{E}\left(\mathcal{G}_i\right)$.

It can be shown that in the new model

$$\mathcal{E}\left(\mathcal{G}_1 \vee \ldots \vee \mathcal{G}_k\right) = \min\left\{\mathcal{E}\left(\mathcal{G}_1\right), \ldots, \mathcal{E}\left(\mathcal{G}_k\right)\right\} ,$$
$$\max\{\mathcal{E}\left(\mathcal{G}_1\right), \ldots, \mathcal{E}\left(\mathcal{G}_k\right)\} \leqslant \mathcal{E}\left(\mathcal{G}_1 \wedge \ldots \wedge \mathcal{G}_k\right) \leqslant \mathcal{E}\left(\mathcal{G}_1\right) + \ldots + \mathcal{E}\left(\mathcal{G}_k\right) .$$

The last inequality turns into an equation if the games $\mathcal{G}_1, \ldots, \mathcal{G}_k$ have no common moves.

5.2   Expenses Reduction

In the following section we will discuss the problem associated with the games that have common moves and suggest a solution to it.

Let $\mathcal{G}_1$ and $\mathcal{G}_2$ be sub-games of game $\mathcal{G}$. Those sub-games may in turn contain the conjunctive as well as disjunctive sub-games alternately with no evidence if those sub-games contain no common moves. We assume that some of the sub-games may contain common moves and that the optimal strategy might utilize them. Thus some of the atomic attacks may be referenced more than once and multiply their corresponding investments into the expenses parameter that these nodes propagate.

In order to get the correct utility for the intermediate sub-games $\mathcal{G}_1, \mathcal{G}_2$ and, eventually, $\mathcal{G}$ we artificially reduce the expenses $\mathcal{E}\left(\mathcal{X}_i\right)$ for the common moves and produce the modified move parameter $\tilde{\mathcal{E}}\left(\mathcal{X}_i\right)$ that will here and further be referenced as *reduced expenses*. It is reasonable to reduce expenses by the amount of occurrences of the same move in a sub-game. In graph representation we reduce the expenses by the amount of references (incoming edges) to the atomic move. Let us denote the number of occurrences of the atomic move $\mathcal{X}_i$ as e_{x_i}. Thus $\tilde{\mathcal{E}}\left(\mathcal{X}_i\right) = \frac{\mathcal{E}(\mathcal{X}_i)}{e_{x_i}}$.

Although by reducing the expenses of the common moves we make them easier to play, the idea behind this is that if the system can be proven to be secure even if some of the atomic attacks are artificially made easier than they really are, this implies that the attacks against the real system are infeasible.

We present an *Algorithm 5.1* for expenses lower bound computation using expenses propagation which runs in time linear in the size of the game, thus complexity is $\mathcal{O}\left(n\right)$. The utility upper bound can be computed as $\mathcal{U}(\mathcal{G}) = \mathcal{P} - \mathcal{E}(\mathcal{G})$, where $\mathcal{E}(\mathcal{G})$ is the expenses lower bound of the game. The local optimum decisions that are made in the disjunctive games are not subject to the problem discovered by Jürgenson and Willemson in [6], as the optimal strategy, according to *Thm. 8* in [3], is a critical min-term of the Boolean function representing the game instance, and a critical min-term is not redundant. Thus local optimum decisions are global optimum decisions in the new model and allow us to use the expenses propagation approach and the recursive algorithm 5.1.

Algorithm 5.1. Iterated *Expenses* propagation

Input: The game $\mathcal{G}$
Output: The expenses of the game $\mathcal{E}(\mathcal{G})$ (a real number)

```
1   Procedure ComputeExpenses (G)
2   if m is a conjunctive game instance then
3   |   expenses := 0
4   |   forall the sub-games i of m do
5   |   |   expenses += ComputeExpenses (i)
6   else if m is a disjunctive game instance then
7   |   cheapest := FindCheapestSubGame (m)
8   |   return cheapest
9   else m is an atomic move
10  |   return E (m)
```

6   Interpretation of Results

The new model allows us to compute the adversarial utility upper bound. In case it is positive the analyzed system lacks security at some point and profitable attack vectors, that can result in a positive outcome for an attacker, are likely to exist. If the utility upper bound is 0, we may conclude that the system is potentially secure against rational gain-oriented attackers. The presented model still relies on the ability of analysts to construct an attack tree precisely enough to capture all feasible attack vectors and reflect the real system being modeled. Security has to be a continuous cyclic process where the list of threats and vulnerabilities is being continuously revised.

7   Open Questions and Future Research

The research on the presented model is still unfinished. Future research will focus on the unsolved problems and open questions.

As mentioned earlier, attack tree models including the new one, depend on the metrics assigned to the atomic attacks. Unfortunately, efficient frameworks for metrics estimation do not exist yet. Should one be developed, it would be a valuable addition not only to this model, but to all the models that utilize attack trees.

Secondly, current attack tree models using the game-theoretic approach have one single node—the root node that is assigned the prize parameter—the revenue for an attacker. However, in reality some intermediate nodes may have their own value for an attacker and in the model may be assigned with their own prize. These nodes can represent the secondary goals of an attacker and affect the strategy in certain cases.

Finally, it would be useful to extend the model capabilities to take possible defensive measures into account and to extend the notion of *attack trees* to the notion of *attack-defense trees*. Those defensive measures, if applied, can affect the parameters the respective nodes propagate.

Although quantified security analysis is an area that has been thoroughly studied, we cannot say that the results meet the requirements of real life. Further research in this area is required to ensure the reliability and trustworthiness of the developed models.

References

1. Buldas, A., Laud, P., Priisalu, J., Saarepera, M., Willemson, J.: Rational choice of security measures via multi-parameter attack trees. In: López, J. (ed.) CRITIS 2006. LNCS, vol. 4347, pp. 235–248. Springer, Heidelberg (2006)
2. Buldas, A., Mägi, T.: Practical security analysis of e-voting systems. In: Miyaji, A., Kikuchi, H., Rannenberg, K. (eds.) IWSEC 2007. LNCS, vol. 4752, pp. 320–335. Springer, Heidelberg (2007)
3. Buldas, A., Stepanenko, R.: Upper bounds for adversaries' utility in attack trees. In: Grossklags, J., Walrand, J. (eds.) GameSec 2012. LNCS, vol. 7638, pp. 98–117. Springer, Heidelberg (2012)
4. Jürgenson, A., Willemson, J.: Computing exact outcomes of multi-parameter attack trees. In: Meersman, R., Tari, Z. (eds.) OTM 2008, Part II. LNCS, vol. 5332, pp. 1036–1051. Springer, Heidelberg (2008)
5. Jürgenson, A., Willemson, J.: Serial model for attack tree computations. In: Lee, D., Hong, S. (eds.) ICISC 2009. LNCS, vol. 5984, pp. 118–128. Springer, Heidelberg (2010)
6. Jürgenson, A., Willemson, J.: Efficient Semantics of Parallel and Serial Models of Attack Trees, TUT (2010)
7. Luby, M.: Pseudorandomness and Cryptographic Applications. In: Hanson, D.R., Tarjan, R.E. (eds.) Princeton Computer Science Notes. Princeton University Press (1996)
8. Mauw, S., Oostdijk, M.: Foundations of attack trees. In: Won, D.H., Kim, S. (eds.) ICISC 2005. LNCS, vol. 3935, pp. 186–198. Springer, Heidelberg (2006)
9. Niitsoo, M.: Optimal adversary behavior for the serial model of financial attack trees. In: Echizen, I., Kunihiro, N., Sasaki, R. (eds.) IWSEC 2010. LNCS, vol. 6434, pp. 354–370. Springer, Heidelberg (2010)

A   Computational Example

Example 1. In order to demonstrate the application of the proposed model, an example is presented. Consider the attack tree shown in Fig. 2. Attack tree leaves parameters are shown in Table 1 and the computed expenses and utilities in Table 2.

The *Expenses* parameter present in the model is represented as a function of attack preparation costs $\mathcal{C}$, attack detection probability r and the penalty Π such that $\mathcal{E}(\mathcal{X}_i) = \mathcal{C}_{\mathcal{X}_i} + r_{\mathcal{X}_i}\Pi_{\mathcal{X}_i}$. Firstly, each of the atomic attacks $\mathcal{X}_i$ parameters *success probability* $p_{\mathcal{X}_i}$ and *expenses* $\mathcal{E}(\mathcal{X}_i)$ are transformed to the form applicable for the failure-free Model: $p_{\mathcal{X}_i} \to 1$; $\mathcal{E}(\mathcal{X}_i) \to \mathcal{E}_{inf}(\mathcal{X}_i) = \frac{\mathcal{E}(\mathcal{X}_i)}{p}$.

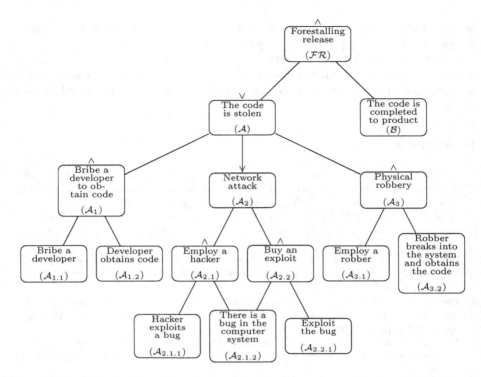

Fig. 2. A sample attack tree for a software developing company. The conjunctive game instances are depicted with red background and with ∧ label above the node, the disjunctive game instances are depicted with blue background and with ∨ label above the node and atomic moves are depicted with green background.

Table 1. Estimated and calculated values of atomic attacks

Threat	Description	p	r	$\mathcal{C}$	Π	Expenses	FFM Expenses	RED Expenses	Utility
$\mathcal{B}$	Stolen code is used in products	0.9	0.9	10^6	10^6	190000	2111111.1	2111111.1	-1010111.1
$\mathcal{A}_{1.1}$	Bribe a developer	0.1	0.2	10^6	10^3	1000200	10002000	10002000	-8901000
$\mathcal{A}_{1.2}$	Developer obtains code	0.9	0.005	0	10^5	500	555.5	555.5	1100444.4
$\mathcal{A}_{2.1.1}$	Hacker exploits a bug	0.5	0.5	10^3	1	1000.5	2001	2001	1098999
$\mathcal{A}_{2.1.2}$	An exploitable bug exists	0.006	0.005	0	0	0	0	0	1101000
$\mathcal{A}_{2.2.1}$	Exploit the bug	0.5	0.1	0	1	0.1	0.2	0.2	1100999.8
$\mathcal{A}_{3.1}$	Employ a robber	0.9	0.001	10^5	10^4	100010	111122.222	111122.222	989877.778
$\mathcal{A}_{3.2}$	Robber obtains the code	0.5	0.9	10^3	10^5	91000	182000	182000	919000

Afterwords the *expenses reduction technique* is applied producing the reduced expenses $\tilde{\mathcal{E}}(\mathcal{X}_i)$ of each of the atomic attacks. In this particular case the expenses parameters of the nodes remain the same, except for the $\mathcal{A}_{2.1.2}$ node which is referenced twice and thus its reference count $e = 2$, thus $\mathcal{E}_{inf}(\mathcal{A}_{2.1.2}) = \frac{\mathcal{E}(\mathcal{A}_{2.1.2})}{2} =$

$\frac{0}{2} = 0$. Secondly, the adversarial utility is calculated using two methodologies, the expenses propagation approach as well as the utility propagation approach.

Calculated parameters of intermediate nodes properties are introduced below. The root node $\mathcal{FR}$ prize $\mathcal{P} = 1,101,000$.

Table 2. Attack Tree Expenses and Utility

Node	Description	Type	Expenses	Utility[1]	Utility[2]
$\mathcal{R}$	Forestalling release	AND	2111111.311	-1010111.311	-1010111.111
$\mathcal{A}$	Steal the code	OR	0.2	-	1100999.8
$\mathcal{A}_1$	Insider attack	AND	10002555.556	-	-8901000
$\mathcal{A}_2$	Network attack	OR	0.2	-	1100999.8
$\mathcal{A}_{2.1}$	Employ a hacker	AND	2001	-	1098999
$\mathcal{A}_{2.2}$	Buy an exploit	AND	0.2	-	1100999.8
$\mathcal{A}_3$	Physical robbery	AND	293122.222	-	919000

[1] Expenses propagation
[2] Utility propagation

Optimizing Active Cyber Defense

Wenlian Lu[1,2], Shouhuai Xu[3], and Xinlei Yi[1]

[1] School of Mathematical Sciences, Fudan University
Shanghai, P.R. China, 200433
{wenlian,11210180008}@fudan.edu.cn
[2] Department of Computer Science, University of Warwick
Coventry CV4 7AL, UK
[3] Department of Computer Science, University of Texas at San Antonio
San Antonio, Texas 78249, USA
shxu@cs.utsa.edu

Abstract. Active cyber defense is one important defensive method for combating cyber attacks. Unlike traditional defensive methods such as firewall-based filtering and anti-malware tools, active cyber defense is based on spreading "white" or "benign" worms to combat against the attackers' malwares (i.e., malicious worms) that also spread over the network. In this paper, we initiate the study of *optimal* active cyber defense in the setting of strategic attackers and/or strategic defenders. Specifically, we investigate infinite-time horizon optimal control and fast optimal control for strategic defenders (who want to minimize their cost) against non-strategic attackers (who do not consider the issue of cost). We also investigate the Nash equilibria for strategic defenders and attackers. We discuss the cyber security meanings/implications of the theoretic results. Our study brings interesting open problems for future research.

Keywords: cyber security model, active cyber defense, optimization, epidemic model.

1   Introduction

The importance of cyber security is well recognized now. However, our understanding of cyber security is still at its infant stage. In general, the attackers are constantly escalating their attack power and sophistication, while the defenders largely lag behind. To be specific, we mention the following *asymmetry* between cyber attack and cyber defense: The effect of malware-like attacks is *automatically amplified* by the network connectivity, while the defense effect is not. This phenomenon had been implied by many previous results (e.g., [28,9,6,26,34]), but was not explicitly pointed out until very recently [35]. The asymmetry is fundamentally caused by that the defense is *reactive*, including intrusion detection systems, firewalls and anti-malware tools. The asymmetry can be eliminated by the idea of *active cyber defense* [35], where the defender also aims to take advantage of the network connectivity. The concept of active cyber defense is not completely new because researchers have proposed for years the idea of using

S.K. Das, C. Nita-Rotaru, and M. Kantarcioglu (Eds.): GameSec 2013, LNCS 8252, pp. 206–225, 2013.

the defender's "white" or "benign" worms to combat against the attackers' malwares [5,1,29,23,16,18,13,30]. In a sense, active cyber defense already happened in practice; for example, the `Welchia` worm attempted to "kill" the Blaster worm in compromised computers [23,20]. It appears that full-fledged active cyber defense is perhaps inevitable in the near future according to some recent reports [18,24,31]. It is therefore more imperative than ever to systematically characterize the effectiveness of active cyber defense. This motivates the present study.

1.1   Our Contributions

This paper is inspired by the recent mathematical model of active cyber defense dynamics [35], which characterizes the effect of various model parameters (including the underlying complex network structures) in the setting where *neither the attacker nor the defender is strategic* (i.e., both the attacker and the defender do not consider the issue of cost). Here we study a new perspective of active cyber defense, namely the strategic interaction between the attacker and the defender. On one hand, our study moves a step beyond [35] because we incorporate control-theoretic and game-theoretic models to accommodate strategic interactions. On the other hand, our study assumes away the underlying complex network structures that are explicitly investigated in [35]. This means that our study is essentially based on the *homogeneous* (or *well-mixed*) assumption that each compromised computer can attack the same portion of computers. Tackling the problem of strategic attack-defense interactions with explicit complex network structures is left for future research. Therefore, we deem the present paper as a significant first step toward ultimately understanding the effectiveness of strategic active cyber defense. Specifically, we make the following contributions.

First, we investigate two flavors of optimal control for strategic defenders against non-strategic attackers: *infinite-time horizon* optimal control and *fast* optimal control. In the setting of infinite-time horizon optimal control for the defender, we characterize the conditions under which the defender should adjust its active cyber defense power in a certain quantitative fashion. For example, we identify a condition under which the defender should give up using active cyber defense alone, and instead should resort to other defense methods as well (e.g., proactive defense). In the setting of fast optimal control, where the defender wants to occupy a certain portion of the network as soon as possible and at the minimal cost, there is a significant difference between the case that the active defense cost is linear and the case that the active defense cost is quadratic.

Second, we identify the Nash equilibrium strategies when both the defender and the attacker are strategic. The findings are interesting. For example, when the defender (or attacker) is reluctant to use/expose its advanced active cyber defense tools (or zero-day exploits), it will give up escalating its active defense (or attack) power; otherwise, there are three scenarios: (i) If the defender (or attacker) initially occupies only a certain small portion of the network, it will give up escalating its active defense (or attack). (ii) If the defender (or attacker) initially occupies a certain significant portion of the network, it will escalate its active defense (or attack) as much as possible. (iii) If the defender (or attacker) initially occupies a certain large portion of the network, it will not escalate its active defense (or attack) — a sort of diminishing returns.

The rest of the paper is structured as follows. Section 2 briefly reviews the related prior work. Section 3 describes the basic active cyber defense model under the homogeneous assumption. Section 4 investigates optimal control for strategic defenders against non-strategic attackers. Section 5 studies Nash equilibria for strategic defenders and attackers. Section 6 concludes the paper with some open problems. Lengthy proofs are deferred to the Appendix. The main notations used in the paper are listed below:

α_B, α_R	defender $\mathbf{B}$'s defense power α_B and attacker $\mathbf{R}$'s attack power α_R
$i_B(t), i_R(t)$	portions of the nodes occupied respectively by the defender and the attacker at time t, where $i_B(t) + i_R(t) = 1$
$\pi_B, \pi_B(t)$	π_B is control variable and $\pi_B(t)$ is control function
$\hat{\pi}_B$	solution in the infinite-time horizon optimal control case
π_B^*, π_B^{**}	solutions in the case of fast optimal control with linear and quadratic cost functions, respectively
z	discount rate
k_B	normalization ratio between the defender's detection cost and recovery cost
λ	normalization ratio between the unit of time and the defender's active defense cost
k_R	normalization ratio between the attacker's maintenance cost and penetration cost

2   Related Work

Our investigation is built on recent studies in *mathematical computer malware models*. These models originated in the *mathematical biological epidemic models* introduced in the 1920's [19,12], which were first adapted to study the spreading of computer virus in the 1990's [10,11]. All these models made the homogeneous assumption that each individual (e.g., computer) in the population has equally infection effect on the other individuals in the population, and the assumption that the infected individuals recover because of reactive defense (e.g., anti-malware tools). In the past decade, there were many studies that aim to eliminate the aforementioned homogeneous assumption, by explicitly incorporating the heterogeneous network structures [28,9,6,26,34,32]. The mathematical tools used for these studies are *Dynamical Systems* in nature. These studies demonstrated that the attack effect of malware spreading against reactive defense is automatically amplified by the largest eigenvalue of the adjacency matrix, which represents the underlying complex network structure. This is the attack-defense asymmetry phenomenon mentioned above.

The attack-defense asymmetry phenomenon motivated the study of mathematical models of *active cyber defense* [35], which is a relatively new sub-field in cyber security [18,24,31] as previous explorations were mainly geared toward

legal and policy issues [5,1,29,23,16,18,13,30]. One real-life incident of the fla-
vor of active cyber defense is that the `Welchia` worm attempted to "kick out"
another kind of worms (e.g., the Blaster worm) [23,20]. In the first mathematical characterization of active cyber defense [35], neither the attacker nor the
defender is strategic (i.e., they do not consider the issue of cost), albeit the
model accommodates the underlying complex network structure. In the present
paper, we move a step toward ultimately understanding *optimal* active cyber
defense, where the attacker and/or the defender are/is strategic (i.e., they want
to minimize their cost). Finally, we note that automatic patching [27] is not active cyber defense because automatic patching aims to *prevent* attacks, whereas
active cyber defense aims to identify and possibly clean up infected computers.

There have been many studies (e.g., [33,21,8,4,14,22,15,25]) on applying Control Theory and Game Theory to understand various issues related to computer
malware spreading. Our study is somewhat inspired by the botnet-defense model
investigated in [4]. All the studies mentioned above only considered *reactive* defense; whereas we investigate how to optimize *active* cyber defense. For general
information about the applications of Control Theory and Game Theory to cyber
security, we refer to [2,17] and the references therein.

3   The Basic Active Cyber Defense Model

Consider a population of nodes, which can abstract computers in a cyber system.
At any point in time, a node is either occupied by defender **B** (i.e., the node is
secure), or occupied by attacker **R** (i.e., the node is compromised). Denote by
$i_B(t)$ the portion of nodes that are occupied by the defender at time t, and by
$i_R(t)$ the portion of nodes that are occupied by the attacker at time t, where
$i_B(t)+i_R(t) = 1$ for any $t \geq 0$. In the interaction between cyber attack and active
cyber defense, the defender and the attacker can "grab" nodes from each other
in the same fashion. Let α_B abstract defender **B**'s power in grabbing attacker-
occupied nodes using active cyber defense, and α_R abstract attacker **R**'s power in
compromising defender-occupied nodes using malware-like cyber attacks. Under
the homogeneous assumption that (i) each secure node has the same power in
"grabbing" the attacker-occupied nodes and (ii) each compromised node has
the same power in compromising the defender-occupied nodes, we obtain the
following Dynamical System model:

$$
\begin{cases}
\frac{di_B(t)}{dt} = \alpha_B i_B(t) i_R(t) - \alpha_R i_R(t) i_B(t) \\
\frac{di_R(t)}{dt} = \alpha_R i_R(t) i_B(t) - \alpha_B i_B(t) i_R(t),
\end{cases}
$$

where $i_B(t) + i_R(t) = 1$, $i_B(t) \geq 0$, and $i_R(t) \geq 0$ for all $t \geq 0$. Due to the
symmetry, we only need to consider

$$
\frac{di_B(t)}{dt} = \alpha_B i_B(t)(1 - i_B(t)) - \alpha_R i_B(t)(1 - i_B(t)). \tag{1}
$$

If neither the attacker nor the defender is strategic (i.e., they do not consider
the issue of cost), the dynamics of system (1) can be characterized as follows.

- If the attacker is more powerful than the defender, namely $\alpha_R > \alpha_B$, the attacker will occupy the entire network in the fashion of the Logistic equation (i.e., when i_R is small, i_R increases slowly; when i_R is around a threshold value, i_R increases exponentially; when i_R is large, i_R increases slowly).
- If the defender is more powerful than the attacker, namely $\alpha_B > \alpha_R$, the defender will occupy the network in the same fashion as in the above case.
- If the attacker and the defender are equally powerful, namely $\alpha_R = \alpha_B$, the system state is in equilibrium. In other words, $i_B(t) = i_B(0)$ and $i_R(t) = i_R(0) = 1 - i_B(0)$ for any $t > 0$.

The above model accommodates non-strategic attackers and non-strategic defenders, and is the starting point for our study of optimal active cyber defense.

4   Optimal Control for Strategic Defender against Non-strategic Attacker

4.1   Infinite-Time Horizon Optimal Control

In this setting, the non-strategic attacker **R** maintains a *fixed* degree of attack power α_R, while the defender **B** is strategic. That is, the strategic defender aims to minimize its cost (specified below) by adjusting its defense power α_B via

$$\alpha_B = b + \pi_B(a - b),$$

while obeying the dynamics of (1), where $\pi_B \in [0, 1]$ is the control variable and $\alpha_B \in [b, a]$ is the defender's defense power with $1 \geq a > b \geq 0$. The cost to the defender consists of two parts.

- The *recovery cost* for recovering the compromised nodes to secure states (e.g., re-installing the operating systems and updating the backup data files, interference with the computers' routine functions). We represent this cost by $f_B(i_B(t))$ for some real-valued function $f_B(\cdot)$. We assume $f_B'(\cdot) < 0$ because the more nodes the defender occupies, the lower the cost for the defender to recover the compromised nodes.
- The *detection cost* for detecting (or recognizing) compromised nodes via active cyber defense, which partly depends on the attack's evasiveness. We represent this cost by $k_B \cdot \pi_B(\cdot)$, where k_B is the normalization ratio between the detection cost and the recovery cost, and $\pi_B(\cdot)$ is the control function that specifies the adjustable degree of active cyber defense power. This is plausible because using more powerful active defense mechanisms (e.g., more sophisticated/advanced "white" worms) causes a higher cost but allows the defender to fight against the attacks more effectively.

The above definition of cost accommodates at least the following family of active cyber defense: The defender uses "white" worms to detect the compromised nodes, then possibly manually recovers the compromised nodes. This is perhaps the most probable scenario because for example, the attacker's malware may have corrupted or deleted some data files in the compromised computers. Note that the detection cost highlights the difference between (i) active-cyber-defense based detection, where the defender's detection tools (i.e., "white" worms) do not reside on the compromised computers, and (ii) reactive-cyber-defense based detection such as the current generation of anti-virus software, where the detection tools do not spread over the network.

Assuming that the attacker maintains a fixed degree of attack power α_R, the defender's *optimization goal* is to minimize the total cost with a constant discount rate z over an infinite-time horizon, namely

$$\inf_{0 \leq \pi_B(\cdot) \leq 1} \left\{ J_B(\pi_B(\cdot)) = \int_0^\infty e^{-zt}(f_B(i_B(t)) + k_B \cdot \pi_B(t))dt \right\}, \qquad (2)$$

where $f_B'(\cdot) < 0$, $\pi_B(\cdot) \in [0, 1]$, and the attacker's fixed degree of attack power α_R is treated as a constant. Now the optimization problem reduces to identifying the optimal defense strategy $\hat{\pi}_B$. To solve the minimization problem, we use Pontryagin's Minimum Principle to find the Hamiltonian associated to (2):

$$\begin{aligned} &H_B(i_B, \pi_B, p) \\ &= f_B(i_B) + k_B\pi_B + p[\alpha_B i_B(1 - i_B) - \alpha_R i_B(1 - i_B)] \\ &= (k_B + pi_B(1 - i_B)(a - b))\pi_B + f_B(i_B) + pbi_B(1 - i_B) - p\alpha_R i_B(1 - i_B),(3) \end{aligned}$$

where p is the adjoint equation

$$\begin{cases} \dot{p} &= -\frac{\partial H_B}{\partial i_B} + zp = -f_B'(i_B) + p[z - (\alpha_B - \alpha_R)(1 - 2i_B)] \\ p_1(\infty) &= 0. \end{cases} \qquad (4)$$

The *optimal strategy* $\hat{\pi}_B$ is obtained by minimizing the Hamiltonian $H_B(i_B, \pi_B, p)$. Since $H_B(i_B, \pi_B, p)$ is linear in π_B, the optimal control strategy $\hat{\pi}_B$ takes the following bang-bang control form:

$$\hat{\pi}_B = \begin{cases} 1 & if \ \frac{\partial H_B}{\partial \pi_B} < 0 \\ u_B \ (0 < u_B < 1, \text{ to be determined}) & if \ \frac{\partial H_B}{\partial \pi_B} = 0 \\ 0 & if \ \frac{\partial H_B}{\partial \pi_B} > 0 \end{cases} \qquad (5)$$

where $\frac{\partial H_B}{\partial \pi_B} = k_B + pi_B(1 - i_B)(a - b)$. In the singular form $\frac{\partial H_B}{\partial \pi_B} = 0$ and for a period of time, we have

$$p = \frac{-k_B}{i_B(1 - i_B)(a - b)}. \qquad (6)$$

Further differentiating $\frac{\partial H_B}{\partial \pi_B}$ with respect to t, we have

$$\frac{d}{dt}\left(\frac{\partial H_B}{\partial \pi_B}\right) = \dot{p}i_B(1 - i_B)(a - b) + p(1 - 2i_B)\dot{i}_B(a - b)$$

$$= i_B(1 - i_B)(a - b)\left\{-f'_B(i_B) + p[z - (\alpha_B - \alpha_R)(1 - 2i_B)]\right\}$$

$$+p(1 - 2i_B)(a - b)\left\{\alpha_B i_B(1 - i_B) - \alpha_R i_B(1 - i_B)\right\}$$

$$= -i_B(1 - i_B)(a - b)f'_B(i_B) - k_B z$$

Define $F_B(i_B) = -i_B(1 - i_B)(a - b)f'_B(i_B) - k_B z$. Then we need to study the roots of $F_B(\cdot) = 0$.

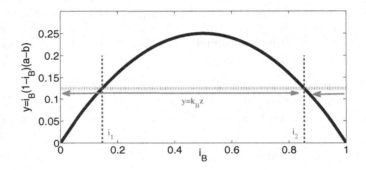

Fig. 1. Illustration of the roots of $F_B(i_B) = 0$ with $f_B(i_B) = 1 - i_B$, $a - b = 1$ and $k_B z = 1/8$, where the x-axis represents i_B and the y-axis represents $y(i_B) = i_B(1 - i_B)(a - b)$. The arrows indicate the directions the outcome under optimal control will head for.

Before presenting the results, we discuss the ideas behind them. In this paper, we focus on case $f_B(i_B) = 1 - i_B$, which can be easily extended to any linear recovery-cost function. If $k_B z < \frac{1}{4}(a - b)$, then $F_B(i_B) = 0$ has two roots:

$$i_1 = \frac{1 - \sqrt{1 - 4\frac{k_B z}{a - b}}}{2} \quad \text{and} \quad i_2 = \frac{1 + \sqrt{1 - 4\frac{k_B z}{a - b}}}{2}$$

with $0 < i_1 < i_2 < 1$. As illustrated in Figure 1, this implies

$$\begin{cases} F_B(i_B) < 0 & if \ i_B < i_1 \\ F_B(i_B) > 0 & if \ i_1 < i_B < i_2 \\ F_B(i_B) < 0 & if \ i_B > i_2. \end{cases}$$

Then, the optimal strategy $\hat{\pi}_B$ of the singular form can be obtained by solving $\dot{i}_B \mid_{i_B = i_1 \ or \ i_B = i_2} = 0$.

Theorem 1. *Suppose the non-strategic attacker maintains a fixed degree of attack power* α_R, $f_B(i_B) = 1 - i_B$ *and* $k_B z < \frac{1}{4}(a - b)$. *Let* $i_1 < i_2$ *be the roots of* $F_B(i_B) = 0$. *Let* $u_B = \frac{\alpha_R - b}{a - b}$. *The optimal control strategy for defender* **B** *is:*

$$
\hat{\pi}_B = \begin{cases} 0 & if \ i_B < i_1 \\ u_B & if \ i_B = i_1 \\ 1 & if \ i_1 < i_B < i_2 \ . \\ u_B & if \ i_B = i_2 \\ 0 & if \ i_B > i_2 \end{cases} \tag{7}
$$

Proof of Theorem 1 is deferred to Appendix A. In practice, i_1 and i_2 can be obtained numerically. Theorem 1 (also as illustrated in Figure 1) shows that the outcome of the infinite-time horizon optimal control, namely $\lim_{t \to \infty} i_B(t)$, depends on the initial system state $i_B(0)$ as follows:

- If $1 > i_B(0) > i_2$, the defender should use the least powerful/costly active defense mechanisms (i.e., $\alpha_B = b$) because $\hat{\pi}_B = 0$. Moreover, the outcome of the optimal defense is that the defender will occupy i_2 portion of the network, namely $\lim_{t \to \infty} i_B(t) = i_2$. This suggests a sort of *diminishing returns* in active cyber defense: It is more cost-effective to pursue "good enough" security (i.e., $\lim_{t \to \infty} i_B(t) = i_2 < 1$) than to pursue "perfect" security (i.e., $\lim_{t \to \infty} i_B(t) = 1$) even if it is possible.
- If $0 = i_B(0) < i_1$, the defender should use the least powerful/costly active defense mechanisms (i.e., $\alpha_B = b$) because $\hat{\pi}_B = 0$. Moreover, the outcome of the optimal defense is that the defender should give up (using active cyber defense as the only defense methods), as the attacker will occupy the entire network, namely $\lim_{t \to \infty} i_B(t) = 0$. In other words, the defender should resort to other defense methods as well (e.g., proactive defense).
- If $i_B(0) \in (i_1, i_2)$, the defender should use the most powerful/costly active defense mechanisms (i.e., $\alpha_B = a$) because $\hat{\pi}_B = 1$. Moreover, the outcome of the optimal defense is that the defender will occupy i_2 portion of the network, namely $\lim_{t \to \infty} i_B(t) = i_2$. This also suggests a sort of *diminishing returns* mentioned above.
- If $i_B(0) = i_1$ or $i_B(0) = i_2$, the defender should adjust its deployment of active cyber defense mechanisms according to $u_B = \frac{\alpha_R - b}{a - b}$, which means $\alpha_B = \alpha_R$. Moreover, the outcome of the optimal defense is that $i_B(t) = i_B(0)$ for all $t > 0$.

Now we consider the degenerated scenarios of $k_B z \geq 1/4(a - b)$. The proof is similar to, but much simpler than, the proof of Theorem 1, and thus omitted.

Theorem 2. *Suppose the non-strategic attacker maintains a fixed degree of attack power* α_R *and* $f_B(i_B) = 1 - i_B$.

- *If* $k_B z = 1/4(a - b)$, *then* $F_B(i_B) = 0$ *has only one root,* $i_1 = i_2 = \frac{1}{2}$. *The optimal control strategy is*

$$
\hat{\pi}_B = \begin{cases} 0 & if \ i_B < i_1 \\ u_B = \frac{\alpha_R - b}{a - b} & if \ i_B = i_1 \\ 0 & if \ i_B > i_1. \end{cases} \tag{8}
$$

- If $k_B z > 1/4(a - b)$, then $F_B(i_B) = 0$ has no root. The optimal control strategy is $\hat{\pi}_B = 0$.

The *cyber security implications* of Theorem 2 are the following. In the case $k_B z = \frac{1}{4}(a - b)$, the outcome under the optimal control depends on the initial system state as follows:

- If $1 > i_B(0) > i_1$, the defender should use the least powerful/costly active cyber defense mechanisms because $\hat{\pi}_B = 0$. The outcome is that the defender will occupy i_1 portion of the network, namely $\lim_{t \to \infty} i_B(t) = i_1$.
- If $0 = i_B(0) < i_1$, the defender should use the least powerful/costly active cyber defense mechanisms because $\hat{\pi}_B = 0$. The outcome is that the defender will give up using active cyber defense alone, as the attacker will occupy the entire the network, namely $\lim_{t \to \infty} i_B(t) = 0$. In other words, the defender should resort to other defense methods as well (e.g., proactive defense).
- If $i_B(0) = i_1$, the defender will adjust its degree of active cyber defense power according to $\hat{\pi}_B = u_B = \frac{\alpha_R - b}{a - b}$, which means $\alpha_B = \alpha_R$. The outcome is that $i_B(t) = i_B(0)$ for all $t > 0$.

In the case $k_B z > 1/4(a - b)$, the defender should use the least powerful/costly active cyber defense mechanisms because $\hat{\pi}_B = 0$. The outcome is that $\lim_{t \to \infty} i_B$ $(t) = 0$, meaning that the defender should give up using active cyber defense alone and resort to other defense methods as well (e.g., proactive defense).

By considering Theorems 1 and 2 together, we draw some deeper insights. Specifically, for a given z, different k_B's suggest different optimal active defense strategies. More specifically, if $k_B > \frac{1}{4z}(a - b)$, meaning that the cost of optimal control is dominating, then defender **B** should use the least powerful/costly active cyber defense mechanisms because $\hat{\pi}_B(t) = 0$ for all t and the outcome is $\lim_{t \to \infty} i_B = 0$. In other words, the defender should give up using active cyber defense alone, and resort to other kinds of defense methods as well (e.g., proactive defense). If $k_B < \frac{1}{4z}(a - b)$, meaning that the cost of control is not dominating, the defender should enforce optimal control according to the initial state $i_B(0)$. In particular, if $k_B = 0$, meaning that the special case that the cost of control is not counted, defender **B** should use the most powerful/costly active defense mechanisms as $\hat{\pi}_B(t) = 1$ for all t, and the outcome is that $\lim_{t \to \infty} i_B = 1$, namely that the defender will occupy the entire network.

4.2 Fast Optimal Control for Strategic Defenders against Non-strategic Attackers

Now we consider *fast* optimal control for strategic defenders against non-strategic attackers, as motivated by the following question: Suppose the attacker maintains a fixed degree of attack power α_R and the defender initially occupies $i_B(0) = i_0 < i_e$ portions of the nodes, how can the defender use optimal control to occupy the desired i_e portions of the nodes *as soon as possible*? More precisely, the optimization is to minimize *the sum of active defense cost and time* (after appropriate normalization), which can be described by the following functional:

$$J_F(\pi_B(\cdot)) = T + \lambda \int_0^T h(\pi_B(t))dt$$

where $h(\cdot)$ is the cost function with respect to the control function $\pi_B(\cdot)$. We consider two scenarios of cost functions: linear and quadratic. In both scenarios, we need to identify defender **B**'s optimal strategy with respect to the dynamics of (1) and a given objective $i_e > i_0$ for some *hitting time* T that is to be identified.

Scenario I: Fast optimal control with linear cost functions. In this scenario, we have $h(\pi_B) = \pi_B$. The optimization task is to minimize the active defense cost plus the time T:

$$\inf_{0 \le \pi_B(\cdot) \le 1} \left\{ J_F(\pi_B(\cdot)) = T + \lambda \int_0^T \pi_B(t)dt \right\} \tag{9}$$

$$\text{subject to } \begin{cases} \frac{di_B(t)}{dt} = \alpha_B i_B(t)(1 - i_B(t)) - \alpha_R i_B(t)(1 - i_B(t)) \\ i_B(0) = i_0 \\ i_B(T) = i_e \end{cases}$$

where $\lambda > 0$ is the normalization ratio between the unit of time and the active defense cost $\int_0^T \pi_B(t)dt$, and $i_0 < i_e$. That is, λ, i_0 and i_e are given, but T is free. Note that the active defense cost $\int_0^T \pi_B(t)dt$ includes both detection and recovery cost, where $\pi_B(t)$ is the control function.

Theorem 3. *The solution to the fast optimal control problem (9) is*

$$(\pi_B^*, T^*) = (1, T_1), \tag{10}$$

where $T_1 = \frac{1}{a - \alpha_R} \ln \left(\frac{i_e}{1 - i_e} \frac{1 - i_0}{i_0} \right)$.

Proof of Theorem 3 is deferred to Appendix B. The *cyber security implication* of Theorem 3 is the following. In order to achieve fast optimal control, the defender should use the most powerful/costly active cyber defense mechanisms, namely $\pi_B(t) = 1$ for $t < T^*$, until the system state becomes $i_B(T^*) = i_e$ at time T^*. After time T^*, if the defender continues enforcing $\pi_B(t) = 1$ for $t > T^*$, then $\lim_{t \to \infty} i_B(t) = 1$, meaning that the defender will occupy the entire network.

Scenario II: Fast optimal control with quadratic cost functions. In this scenario, we have $h(\pi_B) = \pi_B^2$. The optimization task is to minimize the following sum of active defense cost and time, which differs from the linear cost (9) in that the cost function π_B is replaced with cost function π_B^2:

$$\inf_{0 \le \pi_B(\cdot) \le 1} \left\{ J_F(\pi_B(\cdot)) = T + \lambda \int_0^T \pi_B^2(t)dt \right\} \tag{11}$$

$$\text{subject to } \begin{cases} \frac{di_B(t)}{dt} = \alpha_B i_B(t)(1 - i_B(t)) - \alpha_R i_B(t)(1 - i_B(t)) \\ i_B(0) = i_0 \\ i_B(T) = i_e \end{cases}$$

where $\lambda > 0$ is the ratio between the unit of time and the active defense cost $\int_0^T \pi_B^2(t)dt$ (including both recovery cost and detection cost), and $i_0 < i_e$. That is, λ, i_0 and i_e are given, but T is free.

Theorem 4. *The solution to the fast optimal control problem (11) is*

$$(\pi_B^{**}, T^{**}) = \begin{cases} (u^*, T_2), & \text{if } \lambda \geq \frac{a-b}{a+b-2\alpha_R} \text{ and } a - b > 2(\alpha_R - b), \\ (1, T_3), & \text{otherwise} \end{cases} \tag{12}$$

where

$$u^* = \frac{\alpha_R - b}{a - b} + \sqrt{\left(\frac{b - \alpha_R}{a - b}\right)^2 + \frac{1}{\lambda}},$$

$$T_2 = \frac{1}{b + (a - b)u^* - \alpha_R} \ln\left(\frac{i_e}{1 - i_e} \frac{1 - i_0}{i_0}\right),$$

$$T_3 = \frac{1}{a - \alpha_R} \ln\left(\frac{i_e}{1 - i_e} \frac{1 - i_0}{i_0}\right).$$

Proof of Theorem 4 is deferred to Appendix C. It *cyber security implication* is: Unlike in the setting of linear cost function (Theorem 3), the defender should not necessarily enforce the most powerful/costly active cyber defense mechanisms as π_B^{**} is not always equal to 1. If the defender continues enforcing $\pi_B(t) = 1$ for $t > T^{**}$ after the system reaches state $i_B(T^{**}) = i_e$ at time T^{**}, the defender will occupy the entire network, namely $\lim_{t\to\infty} i_B(t) = 1$.

5   Nash Equilibria for Strategic Attacker and Defender

Now we ask the question: What if the attacker is also strategic? Analogous to the way of modeling strategic defenders, we assume $\alpha_R \in [b, a]$. (It is straightforward to extend the current setting $\alpha_B, \alpha_R \in [b, a]$ to the setting $\alpha_B \in [b_B, a_B]$ and $\alpha_R \in [b_R, a_R]$.) A strategic attacker can adjust its attack power

$$\alpha_R = b + \pi_R(a - b),$$

via control variable $\pi_R(\cdot) \in [0, 1]$. That is, the attacker can launch more sophisticated attacks (i.e., greater π_R leading to greater α_R), which however incurs higher cost (e.g., the investment for obtaining more powerful attack tools).

Since both the defender and the attacker are strategic, we naturally consider a game-theoretic model. Specifically, the defender **B**'s optimization task is

$$\phi_B(i_B) = \inf_{0 \leq \pi_B(\cdot) \leq 1} \left\{ J_B(\pi_B(\cdot), \pi_R(\cdot)) = \int_0^\infty e^{-zt}(f_B(i_B(t)) + k_B \cdot \pi_B(i_B(t)))dt \right\},$$

and the attacker **R**'s optimization task is

$$\phi_R(i_B) = \inf_{0 \leq \pi_R(\cdot) \leq 1} \left\{ J_R(\pi_B(\cdot), \pi_R(\cdot)) = \int_0^\infty e^{-zt}(f_R(i_B(t)) + k_R \cdot \pi_R(i_B(t)))dt \right\},$$

where $\pi_B(\cdot), \pi_R(\cdot) \in [0,1]$, $f'_B(\cdot) < 0$ (as in the infinite-time horizon optimal control case investigated above), $f'_R(\cdot) > 0$ because $f_R(i_B(t))$ represents the *maintenance* cost to the attacker, k_R is the normalization ratio between the attacker's maintenance cost and *penetration* cost (which depends on the capability of the attack tools), and $k_R \cdot \pi_R(\cdot)$ is the *penetration* cost. Note that $f'_R(\cdot) > 0$ is relevant because the attacker may need to conduct some costly (or risky) activities after "grabbing" a node from the defender (e.g., downloading attack payloads from some remote server, while this downloading operation may increase the chance that the compromised node is detected by active defense). Since $f'_R(\cdot) > 0$ implies $df_R/di_R < 0$, the attacker's optimization task for π_R is in parallel to the optimization for π_B. The Hamiltonians associated to defender **B**'s and attacker **R**'s optimization problems are:

$$H_B(i_B, \pi_B(i_B), \pi_R(i_B), p_1)$$
$$= f_B(i_B) + k_B \pi_B + p_1[\alpha_B i_B(1 - i_B) - \alpha_R i_B(1 - i_B)]$$
$$= (k_B + p_1 i_B(1 - i_B)(a - b))\pi_B + f_B(i_B) + p_1 b i_B(1 - i_B) - p_1 \alpha_R i_B(1 - i_B);$$

$$H_R(i_B, \pi_B(i_B), \pi_R(i_B), p_2)$$
$$= f_R(i_B) + k_R \pi_R + p_2[\alpha_B i_B(1 - i_B) - \alpha_R i_B(1 - i_B)]$$
$$= (k_R - p_2 i_B(1 - i_B)(a - b))\pi_R + f_R(i_B) + p_2 \alpha_B i_B(1 - i_B) - p_2 b i_B(1 - i_B).$$

The adjoint equation is

$$\begin{cases} \dot{p}_1 & = -\frac{\partial H_B}{\partial i_B} + z p_1 = -f'_B(i_B) + p_1[z - (\alpha_B - \alpha_R)(1 - 2i_B)] \\ p_1(\infty) = 0 \\ \dot{p}_2 & = -\frac{\partial H_R}{\partial i_B} + z p_2 = -f'_R(i_B) + p_2[z - (\alpha_B - \alpha_R)(1 - 2i_B)] \\ p_2(\infty) = 0. \end{cases}$$

Theorem 5. *Suppose $f_B(i_B) = 1 - i_B$, $f_R(i_B) = i_B$. Then, the Nash equilibria under various scenarios are listed in Table 1, where $F_B(i_B) = -i_B(1 - i_B)(a - b)f'_B(i_B) - k_B z$ and $F_R(i_B) = i_B(1 - i_B)(a - b)f'_R(i_B) - k_R z$.*

Proof of Theorem 5 is similar to the proof of Theorem 1 and omitted due to space limitation. Its cyber security implication is: The outcome of playing the Nash equilibrium strategies also depends on the initial system state and the relationship between k_B and k_R. As illustrated in Figure 2, if $k_B < k_R$ with $k_R z < \frac{1}{4}(a - b)$, meaning that the attacker is more concerned with its control cost (e.g., reluctant to use/expose its advanced attack tools such as zero-day exploits) than the defender, then $F_B(i_B) = 0$ has two roots i_1, i_2 and $F_R(i_B) = 0$ has two roots i_3, i_4. Then, we have $i_1 < i_3 < i_4 < i_2$ (the only possibility under the given conditions). Therefore, the outcomes under the Nash equilibrium strategies are summarized as follows:

- If $i_B(0) < i_1$, then $i_B(t) = i_B(0)$ and $i_R(t) = i_R(0)$ for all $t > 0$ because $\hat{\pi}_B = \hat{\pi}_R = 0$ are the Nash equilibrium strategies.
- If $i_3 > i_B(0) > i_1$, then $\hat{\pi}_B = 1$ and $\hat{\pi}_R = 0$ until $i_B = i_3$, which implies that $i_B(t)$ strictly increases until $i_B = i_3$. When $i_B(t) = i_3$ at some point in time $t = t_1$, $\hat{\pi}_B = \hat{\pi}_R = 1$ implies $i_B(t) = i_3$ for $t > t_1$.

Table 1. Nash equilibrium strategies for defender and attacker in various cases

k_B	k_R	Roots of $F_B(i_B)=0$	Roots of $F_R(i_B)=0$	Nash equilibria
$k_B z < \frac{1}{4}(a-b)$	$k_R z < \frac{1}{4}(a-b)$	$0 < i_1 < i_2 < 1$	$0 < i_3 < i_4 < 1$	$\hat{\pi}_B = \begin{cases} 0 & if\ i_B(0) \leq i_1 \\ 1 & if\ i_1 < i_B(0) < i_2 \\ 0 & if\ i_B(0) \geq i_2 \end{cases}$ $\hat{\pi}_R = \begin{cases} 0 & if\ i_B(0) < i_3 \\ 1 & if\ i_3 \leq i_B(0) \leq i_4 \\ 0 & if\ i_B(0) > i_4 \end{cases}$
$k_B z < \frac{1}{4}(a-b)$	$k_R z = \frac{1}{4}(a-b)$	$0 < i_1 < i_2 < 1$	$i_3 = i_4 = \frac{1}{2}$	$\hat{\pi}_B = \begin{cases} 0 & if\ i_B(0) \leq i_1 \\ 1 & if\ i_1 < i_B(0) < i_2 \\ 0 & if\ i_B(0) \geq i_2 \end{cases}$ $\hat{\pi}_R = \begin{cases} 0 & if\ i_B(0) < i_3 \\ 1 & if\ i_B(0) = i_3 \\ 0 & if\ i_B(0) > i_3 \end{cases}$
$k_B z < \frac{1}{4}(a-b)$	$k_R z > \frac{1}{4}(a-b)$	$0 < i_1 < i_2 < 1$	No real-valued roots	$\hat{\pi}_B = \begin{cases} 0 & if\ i_B(0) \leq i_1 \\ 1 & if\ i_1 < i_B(0) < i_2 \\ 0 & if\ i_B(0) \geq i_2 \end{cases}$ $\hat{\pi}_R = 0$
$k_B z = \frac{1}{4}(a-b)$	$k_R z < \frac{1}{4}(a-b)$	$0 < i_1 = i_2 = \frac{1}{2}$	$0 < i_3 < i_4 < 1$	$\hat{\pi}_B = \begin{cases} 0 & if\ i_B(0) < i_1 \\ 1 & if\ i_B(0) = i_1 \\ 0 & if\ i_B(0) > i_2 \end{cases}$ $\hat{\pi}_R = \begin{cases} 0 & if\ i_B(0) \leq i_3 \\ 1 & if\ i_3 < i_B(0) < i_4 \\ 0 & if\ i_B(0) \geq i_4 \end{cases}$
$k_B z = \frac{1}{4}(a-b)$	$k_R z = \frac{1}{4}(a-b)$	$0 < i_1 = i_2 = \frac{1}{2}$	$i_3 = i_4 = \frac{1}{2}$	$\hat{\pi}_B = \begin{cases} 0 & if\ i_B(0) < i_1 \\ \pi_R & if\ i_B(0) = i_1 \\ 0 & if\ i_B(0) > i_2 \end{cases}$ $\hat{\pi}_R = \begin{cases} 0 & if\ i_B(0) < i_3 \\ \pi_B & if\ i_B(0) = i_3 \\ 0 & if\ i_B(0) > i_3 \end{cases}$
$k_B z = \frac{1}{4}(a-b)$	$k_R z > \frac{1}{4}(a-b)$	$0 < i_1 = i_2 = \frac{1}{2}$	No real-valued roots	$\hat{\pi}_B = 0,\ \hat{\pi}_R = 0$
$k_B z > \frac{1}{4}(a-b)$	$k_R z < \frac{1}{4}(a-b)$	No real-valued roots	$0 < i_3 < i_4 < 1$	$\hat{\pi}_B = 0$ $\hat{\pi}_R = \begin{cases} 0 & if\ i_B(0) \leq i_3 \\ 1 & if\ i_3 < i_B(0) < i_4 \\ 0 & if\ i_B(0) \geq i_4 \end{cases}$
$k_B z > \frac{1}{4}(a-b)$	$k_R z = \frac{1}{4}(a-b)$	No real-valued roots	$i_3 = i_4 = \frac{1}{2}$	$\hat{\pi}_B = 0,\ \hat{\pi}_R = 0$
$k_B z > \frac{1}{4}(a-b)$	$k_R z > \frac{1}{4}(a-b)$	No real-valued roots	No real-valued roots	$\hat{\pi}_B = 0,\ \hat{\pi}_R = 0$

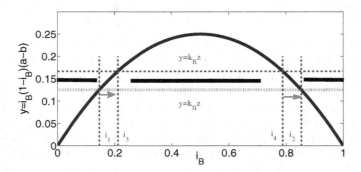

Fig. 2. Illustration of the roots of $F_B(i_B) = 0$ with $f_B(i_B) = 1 - i_B$, and the roots of $F_R(i_B) = 0$ with $f_R(i_B) = i_B$, where $a - b = 1$, $k_B z = 1/8$ and $k_R z = 1/6$. The x-axis represents i_B and the y-axis represents $y(i_B) = i_B(1 - i_B)(a - b)$. Arrows indicate the directions the outcome under the Nash equilibrium heads for. Black-colored bars indicate that the trajectory under the Nash equilibrium stays static.

- If $i_4 > i_B(0) > i_3$, then $i_B(t) = i_B(0)$ and $i_R(t) = i_R(0)$ for all $t > 0$ because $\hat{\pi}_B = \hat{\pi}_R = 1$.
- If $i_2 > i_B(0) > i_4$, then $\hat{\pi}_B = 1$ and $\hat{\pi}_R = 0$ until $i_B = i_2$, which implies that $i_B(t)$ strictly increases until $i_B = i_2$. When $i_B(t) = i_2$ at some point in time $t = t_2$, $\hat{\pi}_B = \hat{\pi}_R = 1$ implies $i_B(t) = i_3$ for $t > t_2$.
- If $i_B(0) > i_2$, then $i_B(t) = i_B(0)$ and $i_R(t) = i_R(0)$ for all $t > 0$ because $\hat{\pi}_B = \hat{\pi}_R = 1$.

If $k_R > \frac{1}{4}(a - b) > k_B$, meaning that the attacker is extremely concerned with its control cost (e.g., not willing to easily use/expose its advanced attack tools such as zero-day exploits) but the defender is not, then it always holds that $\hat{\pi}_R = 0$ because $F_R(i_B) = 0$ has no root but $F_B(i_B) = 0$ has two roots $i_1 < i_2$. From Table 1, we see that $\lim_{t\to\infty} i_B(t) = 1$ always holds, namely that the attacker gives up using its advanced attack tools.

If both $k_B > \frac{1}{4}(a-b)$ and $k_R > \frac{1}{4}(a-b)$, meaning that both the defender and the attacker are extremely concerned with their control costs (i.e., neither the defender wants to easily use/expose its advanced active defense tools, nor the attacker wants to use/expose its advanced attack tools such as zero-day exploits), then it always holds that $\hat{\pi}_B = \hat{\pi}_R = 0$ because $F_B(i_B) = 0$ and $F_R(i_B) = 0$ have no real-valued roots. As a result, $i_B(t) = i_B(0)$ for any $t > 0$.

The scenarios that one or both $F_B(i_B) = 0$ and $F_R(i_B) = 0$ have one root can be regarded as degenerated cases of the above. Moreover, the cases of $k_B > k_R$ (i.e., the defender is more concerned about its control cost, such as not willing to easily use/expose its advanced active defense tools), the outcomes under the Nash equilibria can be derived analogously.

6 Conclusion

We have investigated how to optimize active cyber defense, by presenting optimal control solutions for strategic defenders against non-strategic attackers, and identifying Nash equilibrium strategies for strategic defenders and attackers. We have discussed the cyber security implications of the theoretic results.

This paper brings interesting problems for future research. First, it is interesting to extend the models to accommodate nonlinear $f_B(\cdot)$ and $f_R(\cdot)$. Second, the models are geared toward active cyber defense. A comprehensive defense solution, as hinted in our analysis, should require the *optimal* integration of reactive, active, and proactive cyber defense. Therefore, we need to extend the models to accommodate reactive defense and proactive cyber defense. Moreover, it is interesting to investigate how to extend the models to accommodate moving target defense, which has not be systematically evaluated yet [7]. Third, how to extend the models to accommodate the underlying network structures?

Acknowledgement. Wenlian Lu was jointly supported by the Marie Curie International Incoming Fellowship from the European Commission (no. FP7-PEOPLE-2011-IIF-302421), the National Natural Sciences Foundation of China (no. 61273309), the Shanghai Guidance of Science and Technology (SGST) (no. 09DZ2272900) and the Laboratory of Mathematics for Nonlinear Science, Fudan University. Shouhuai Xu was supported in part by ARO Grant #W911NF-12-1-0286 and AFOSR Grant FA9550-09-1-0165. Any opinions, findings, and conclusions or recommendations expressed in this material are those of the author(s) and do not necessarily reflect the views of any of the funding agencies.

References

1. Aitel, D.: Nematodes – beneficial worms (September 2005), http://www.immunityinc.com/downloads/nematodes.pdf
2. Alpcan, T., Başar, T.: Network Security: A Decision and Game Theoretic Approach. Cambridge University Press (2011)
3. Bardi, M., Capuzzo-Dolcetta, I.: Optimal control and viscosity solutions of Hamilton-Jacobi-Bellman equations. Birkhauser (2008)
4. Bensoussan, A., Kantarcioglu, M., Hoe, S.R.: A game-theoretical approach for finding optimal strategies in a botnet defense model. In: Alpcan, T., Buttyán, L., Baras, J.S. (eds.) GameSec 2010. LNCS, vol. 6442, pp. 135–148. Springer, Heidelberg (2010)
5. Castaneda, F., Sezer, E., Xu, J.: Worm vs. worm: preliminary study of an active counter-attack mechanism. In: Proc. ACM WORM 2004, pp. 83–93 (2004)
6. Chakrabarti, D., Wang, Y., Wang, C., Leskovec, J., Faloutsos, C.: Epidemic thresholds in real networks. ACM Trans. Inf. Syst. Secur. 10(4), 1–26 (2008)
7. Collins, M.: A cost-based mechanism for evaluating the effectiveness of moving target defenses. In: Grossklags, J., Walrand, J. (eds.) GameSec 2012. LNCS, vol. 7638, pp. 221–233. Springer, Heidelberg (2012)
8. Fultz, N., Grossklags, J.: Blue versus Red: Towards a Model of Distributed Security Attacks. In: Dingledine, R., Golle, P. (eds.) FC 2009. LNCS, vol. 5628, pp. 167–183. Springer, Heidelberg (2009)
9. Ganesh, A., Massoulie, L., Towsley, D.: The effect of network topology on the spread of epidemics. In: Proc. of IEEE Infocom 2005 (2005)
10. Kephart, J., White, S.: Directed-graph epidemiological models of computer viruses. In: Proc. IEEE Symposium on Security and Privacy, pp. 343–361 (1991)
11. Kephart, J., White, S.: Measuring and modeling computer virus prevalence. In: Proc. IEEE Symposium on Security and Privacy, pp. 2–15 (1993)
12. Kermack, W., McKendrick, A.: A contribution to the mathematical theory of epidemics. Proc. of Roy. Soc. Lond. A 115, 700–721 (1927)
13. Kesan, J., Hayes, C.: Mitigative counterstriking: Self-defense and deterrence in cyberspace. Harvard Journal of Law and Technology (forthcoming), SSRN: http://ssrn.com/abstract=1805163
14. Khouzani, M., Sarkar, S., Altman, E.: A dynamic game solution to malware attack. In: Proc. IEEE INFOCOM, pp. 2138–2146 (2011)
15. Khouzani, M., Sarkar, S., Altman, E.: Saddle-point strategies in malware attack. IEEE Journal on Selected Areas in Communications 30(1), 31–43 (2012)

16. Lin, H.: Lifting the veil on cyber offense. IEEE Security & Privacy 7(4), 15–21 (2009)
17. Manshaei, M., Zhu, Q., Alpcan, T., Basar, T., Hubaux, J.: Game theory meets network security and privacy. In: ACM Computing Survey (to appear)
18. Matthews, W.: U.s. said to need stronger, active cyber defenses (October 1, 2010), http://www.defensenews.com/story.php?i=4824730
19. McKendrick, A.: Applications of mathematics to medical problems. Proc. of Edin. Math. Soceity 14, 98–130 (1926)
20. Naraine, R.: 'friendly' welchia worm wreaking havoc (August 19, 2003), http://www.internetnews.com/ent-news/article.php/3065761/ Friendly-Welchia-Worm-Wreaking-Havoc.htm
21. Omic, J., Orda, A., Van Mieghem, P.: Protecting against network infections: A game theoretic perspective. In: Infocom 2009, pp. 1485–1493 (2009)
22. Píbil, R., Lisý, V., Kiekintveld, C., Bošanský, B., Pěchouček, M.: Game theoretic model of strategic honeypot selection in computer networks. In: Grossklags, J., Walrand, J. (eds.) GameSec 2012. LNCS, vol. 7638, pp. 201–220. Springer, Heidelberg (2012)
23. Schneier, B.: Benevolent worms (February 19, 2008), http://www.schneier.com/blog/archives/2008/02/benevolent_worm_1.html
24. Shaughnessy, L.: The internet: Frontline of the next war? (November 7, 2011), http://www.cnn.com/2011/11/07/us/darpa/
25. Theodorakopoulos, G., Boudec, J.-Y.L., Baras, J.S.: Selfish response to epidemic propagation. IEEE Trans. Aut. Contr. 58(2), 363–376 (2013)
26. Van Mieghem, P., Omic, J., Kooij, R.: Virus spread in networks. IEEE/ACM Trans. Netw. 17(1), 1–14 (2009)
27. Vojnovic, M., Ganesh, A.: On the race of worms, alerts, and patches. IEEE/ACM Trans. Netw. 16, 1066–1079 (2008)
28. Wang, Y., Chakrabarti, D., Wang, C., Faloutsos, C.: Epidemic spreading in real networks: An eigenvalue viewpoint. In: Proc. IEEE SRDS 2003, pp. 25–34 (2003)
29. Weaver, N., Ellis, D.: White worms don't work. login: The Usenix Magazine 31(6), 33–38 (2006)
30. Homeland Security News Wire. Active cyber-defense strategy best deterrent against cyber-attacks (June 28, 2011), http://www.homelandsecuritynewswire.com/ active-cyber-defense-strategy-best-deterrent-against-cyber-attacks
31. Wolf, J.: Update 2-u.s. says will boost its cyber arsenal (November 7, 2011), http://www.reuters.com/article/2011/11/07/ cyber-usa-offensive-idUSN1E7A61YQ20111107
32. Xu, S., Lu, W., Xu, L.: Push- and pull-based epidemic spreading in arbitrary networks: Thresholds and deeper insights. ACM Transactions on Autonomous and Adaptive Systems (ACM TAAS) 7(3), 32:1–32:26 (2012)
33. Xu, S., Lu, W., Xu, L., Zhan, Z.: Adaptive epidemic dynamics in networks: Thresholds and control. ACM Transactions on Autonomous and Adaptive Systems (ACM TAAS) (to appear)
34. Xu, S., Lu, W., Zhan, Z.: A stochastic model of multivirus dynamics. IEEE Trans. Dependable Sec. Comput. 9(1), 30–45 (2012)
35. Xu, S., Lu, W., Li, H.: A stochastic model of active cyber defense dynamics. Internet Mathematics (to appear)

A   Proof of Theorem 1

Proof. By the Dynamic Programming (DP) argument [3], we know that defender **B**'s value function of the optimal solution can be defined as:

$$\phi(i_B) = \inf_{0 \le \pi_B(\cdot) \le 1} \left\{ J_B(\pi_B(\cdot)) = \int_0^\infty e^{-zt}(f_B(i_B(t)) + k_B \cdot \pi_B(t))dt \right\}. \quad (13)$$

This leads to the following Bellman equation:

$$z\phi(i_B) = \inf_{0 \le \pi_B(\cdot) \le 1} \left\{ f_B(i_B) + k_B\pi_B(t) + \phi'(i_B)[\alpha_B i_B(1 - i_B) - \alpha_R i_B(1 - i_B)] \right\}$$

$$= \inf_{0 \le \pi_B(\cdot) \le 1} H_B(i_B, \pi_B(t), \phi'(i_B))$$

$$= \inf_{0 \le \pi_B(\cdot) \le 1} H_B(i_B, \pi_B(t), p), \; where \; p = \phi'(i_B). \quad (14)$$

From (5), we know that the optimal strategy $\hat{\pi}_B$ takes the form:

$$\hat{\pi}_B = \mathbf{1}_{k_B + p i_B(1-i_B)(a-b) < 0} + u_B \mathbf{1}_{k_B + p i_B(1-i_B)(a-b) = 0}, \quad (15)$$

where **1** is the indicator function. The infimum of Hamiltonian (3) is:

$$\inf_{0 \le \pi_B(\cdot) \le 1} H_B(i_B, \pi_B, p)$$
$$= f_B(i_B) + [k_B + p i_B(1 - i_B)(a - b)]\mathbf{1}_{k_B + p i_B(1-i_B)(a-b)<0} + p(b - \alpha_R)i_B(1 - i_B).$$

Hence, we have

$$z\phi(i_B)$$
$$= f_B(i_B) + [k_B + p i_B(1 - i_B)(a - b)]\mathbf{1}_{k_B + p i_B(1-i_B)(a-b)<0} + p(b - \alpha_R)i_B(1 - i_B)$$
$$= f_B(i_B) + [k_B + \phi'(i_B)i_B(1 - i_B)(a - b)]\mathbf{1}_{k_B + \phi'(i_B)i_B(1-i_B)(a-b)<0} +$$
$$\phi'(i_B)(b - \alpha_R)i_B(1 - i_B). \quad (16)$$

Let $y(i_B) = k_B + \phi'(i_B)i_B(1 - i_B)(a - b)$. In what follows, we are to verify that (7) satisfies (15) with $\phi(i_B)$ defined by (16), which means that (7) minimizes the Hamiltonian in the term of (13). This completes the proof.

In order to verify that (7) satisfies (15) with $\phi(i_B)$ defined by (16), we differentiate (16) with respect to i_B to obtain

$$[(b - \alpha_R) + (a - b)\mathbf{1}_{y<0}]i_B(1 - i_B)y' - zy - F_B(i_B) = 0, \quad (17)$$

which can be rewritten as

$$y' - \frac{z}{[(b - \alpha_R) + (a - b)\mathbf{1}_{y<0}]i_B(1 - i_B)}y - \frac{F_B(i_B)}{[(b - \alpha_R) + (a - b)\mathbf{1}_{y<0}]i_B(1 - i_B)} = 0. \quad (18)$$

If $k_B + \phi'(i_B)i_B(1 - i_B)(a - b) < 0$ namely $y(i_B) < 0$, (18) should be

$$\frac{d}{dx}\left[y(x)e^{-\int_0^x \frac{z}{(a-\alpha_R)\xi(1-\xi)}d\xi} \right] - \frac{F_B(x)}{(a - \alpha_R)x(1 - x)}e^{-\int_0^x \frac{z}{(a-\alpha_R)\xi(1-\xi)}d\xi} = 0. \quad (19)$$

If $k_B + \phi'(i_B)i_B(1 - i_B)(a - b) > 0$ namely $y(i_B) > 0$, (18) should be:

$$\frac{d}{dx}\left[y(x)e^{\int_x^1 \frac{z}{(b-\alpha_R)\xi(1-\xi)}d\xi}\right] - \frac{F_B(x)}{(b - \alpha_R)x(1 - x)}e^{\int_x^1 \frac{z}{(b-\alpha_R)\xi(1-\xi)}d\xi} = 0. \quad (20)$$

Therefore, we only need to prove that the optimal defense strategy (7) satisfies (17), namely (19) or (20). The proof is split into cases, depending on x residing in interval $(i_2, 1)$ or $(0, i_1)$ or (i_1, i_2), or $x = i_1$, or $x = i_2$.

Case 1: $i_2 < x < 1$. By (20), we have

$$y(x)e^{\int_x^1 \frac{z}{(b-\alpha_R)\xi(1-\xi)}d\xi} - \int_{i_2}^x \frac{F_B(\zeta)}{(b - \alpha_R)\zeta(1 - \zeta)}e^{\int_\zeta^1 \frac{z}{(b-\alpha_R)\xi(1-\xi)}d\xi}d\zeta = 0.$$

Hence, we have

$$y(x) = \int_{i_2}^x \frac{F_B(\zeta)}{(b - \alpha_R)\zeta(1 - \zeta)}e^{\int_\zeta^x \frac{z}{(b-\alpha_R)\xi(1-\xi)}d\xi}d\zeta. \quad (21)$$

Since $i_B > i_2$, we have $y(x) > 0$. Therefore, we have $\hat{\pi}_B = 0$ for $i_B \in (i_2, 1)$.

Case 2: $0 < x < i_1$. By (20), we have

$$y(x)e^{-\int_0^x \frac{z}{(b-\alpha_R)\xi(1-\xi)}d\xi} - \int_{i_1}^x \frac{F_B(\zeta)}{(b - \alpha_R)\zeta(1 - \zeta)}e^{-\int_0^\zeta \frac{z}{(b-\alpha_R)\xi(1-\xi)}d\xi}d\zeta = 0.$$

Hence,

$$y(x) = \int_0^x \frac{F_B(\zeta)}{(b - \alpha_R)\zeta(1 - \zeta)}e^{\int_\zeta^x \frac{z}{(b-\alpha_R)\xi(1-\xi)}d\xi}d\zeta + k_B e^{\int_0^x \frac{z}{(b-\alpha_R)\xi(1-\xi)}d\xi}$$

$$= k_B - \int_0^x \frac{a - b}{b - \alpha_R}f_B'(\zeta)e^{\int_\zeta^x \frac{z}{(b-\alpha_R)\xi(1-\xi)}d\xi}d\zeta \quad (22)$$

Since $i_B < i_1$, we have $y(x) > 0$. Therefore we have $\hat{\pi}_B = 0$ for $i_B \in (0, i_1)$.

Case 3: $i_1 < x < i_2$. By (19), we have

$$y(x)e^{\int_x^1 \frac{z}{(a-\alpha_R)\xi(1-\xi)}d\xi} - \int_{i_2}^x \frac{F_B(\zeta)}{(a - \alpha_R)\zeta(1 - \zeta)}e^{\int_\zeta^1 \frac{z}{(a-\alpha_R)\xi(1-\xi)}d\xi}d\zeta = 0.$$

Hence

$$y(x) = \int_{i_2}^x \frac{F_B(\zeta)}{(a - \alpha_R)\zeta(1 - \zeta)}e^{\int_\zeta^x \frac{z}{(a-\alpha_R)\xi(1-\xi)}d\xi}d\zeta. \quad (23)$$

Since $i_B \in (i_1, i_2)$, we have $y(x) < 0$. This implies $\hat{\pi}_B = 1$.

Cases 4 & 5: $x = i_1$ or $x = i_2$. By (21,22,23), we have $y(x) = 0$. If $x = i_1$ or $x = i_2$, we can derive $\phi'(i_B) = \frac{-k_B}{i^*(1-i^*)(a-b)}$ from the definition of $y(\cdot)$. According to (16), we have: $z\phi(i^*) = f_B(i^*) + k_B\frac{\alpha_R - b}{a - b}$. Differentiating with respect to i^*, we have $-i^*(1 - i^*)(a - b)f_B'(i^*) - k_B z = F_B(i^*) = 0$. Consider the singular form $i_B(t) = i^*$ for a period of time. We obtain that $\dot{i}_B|_{i_B = i^*} = 0$ and thus $\hat{\pi}_B = u_B = \frac{\alpha_R - b}{a - b}$, where $i^* = i_1$ or $i^* = i_2$. $\square$

B   Proof of Theorem 3

Proof. To solve the minimization problem, we formulate the current value Hamiltonian associated with (9):

$$H_F(i_B, \pi_B, q) = \lambda \pi_B + q(\alpha_B - \alpha_R)i_B(1 - i_B)$$
$$= [\lambda + q(a - b)i_B(1 - i_B)]\pi_B + q(b - \alpha_R)i_B(1 - i_B).$$

The adjoint equation is $\dot{q} = -\frac{\partial H_F}{\partial i_B} = -q(\alpha_B - \alpha_R)(1 - 2i_B)$, with the boundary condition

$$H_F(i_B^*(T^*), \pi_B^*(T^*), q(T^*)) + 1 = 0, \tag{24}$$

where T^* denotes the optimal hitting time that $i_B(T^*) = i_e$, $\pi_B^*(\cdot)$ denotes the optimal feedback control, and $i_B^*(\cdot)$ denotes the corresponding trajectory.

The optimal control π_B^* is obtained by minimizing Hamiltonian $H_F(i_B, \pi_B, q)$. Since $H_F(i_B, \pi_B, q)$ is linear in π_B, the optimal control π_B^* takes the following bang-bang form:

$$\pi_B^* = \begin{cases} 1 & if\ \frac{\partial H_F}{\partial \pi_B} < 0 \\ u_B^*\ (0 < u_B^* < 1,\ \text{to be determined}) & if\ \frac{\partial H_F}{\partial \pi_F} = 0 \\ 0 & if\ \frac{\partial H_F}{\partial \pi_B} > 0 \end{cases}$$

where $\frac{\partial H_F}{\partial \pi_B} = \lambda + q(a - b)i_B(1 - i_B)$. From (24), there are two possibilities: (i). If $\frac{\partial H_F}{\partial \pi_B} \geq 0$, then $0 = \frac{b - \alpha_R}{a - b}(\frac{\partial H_F}{\partial \pi_B} - \lambda) + 1$, which implies $\frac{\partial H_F}{\partial \pi_B} = \frac{a - b}{\alpha_R - b} + \lambda$ is a positive constant. (ii). If $\frac{\partial H_F}{\partial \pi_B} < 0$, then $0 = \frac{\partial H_F}{\partial \pi_B} + \frac{b - \alpha_R}{a - b}(\frac{\partial H_F}{\partial \pi_B} - \lambda) + 1$, which implies $\frac{\partial H_F}{\partial \pi_B} = \frac{a - b}{a - \alpha_R}\left[\frac{b - \alpha_R}{a - b}\lambda - 1\right]$ is a negative constant. It can be seen that only under the above (ii), the constraint $i_B(T) = i_e$ can be obtained for some T. Then, the solution to the optimal fast control should be $\pi_B(t) = 1$ for all time. So $(\pi_B^*, T^*) = (1, T_1)$, where T_1 satisfies

$$i_B(T_1) = \frac{\frac{i_0}{1 - i_0}e^{(a - \alpha_R)T_1}}{1 + \frac{i_0}{1 - i_0}e^{(a - \alpha_R)T_1}} = i_e,$$

that is, $T_1 = \frac{1}{a - \alpha_R}\ln\left(\frac{i_e}{1 - i_e}\frac{1 - i_0}{i_0}\right)$. This completes the proof.   □

C   Proof of Theorem 4

Proof. To solve the optimization problem, we formulate the current value Hamiltonian associated with (11):

$$H_F(i_B, \pi_B, q) = \lambda \pi_B^2 + q(\alpha_B - \alpha_R)i_B(1 - i_B)$$
$$= \lambda \pi_B^2 + q(a - b)i_B(1 - i_B)\pi_B + q(b - \alpha_R)i_B(1 - i_B).$$

The adjoint equation is $\dot{q} = -\frac{\partial H_F}{\partial i_B} = -q(\alpha_B - \alpha_R)(1 - 2i_B)$, and the boundary condition is

$$H_F(i_B(T^{**}), \pi_B^{**}(T^{**}), q(T^{**})) + 1 = 0, \qquad (25)$$

where T^{**} denotes the optimal final time, $\pi_B^{**}(\cdot)$ denotes the optimal feedback control, $i_B(\cdot)$ denotes the corresponding trajectory, and $i_B(T^{**}) = i_e$. Let $D = q(a - b)i_B(T^{**})(1 - i_B(T^{**}))$. From (25) we have

$$H_F(i_B(T^{**}), \pi_B^{**}(T^{**}), q(T^{**})) + 1 = \lambda(\pi_B^{**})^2 + D\pi_B^{**} + \frac{b - \alpha_R}{a - b}D + 1 = 0. \quad (26)$$

The optimal control, π_B^{**}, is obtained by minimizing the Hamiltonian $H_F(i_B, \pi_B, q)$. Because the Hamiltonian $H_F(i_B, \pi_B, q)$ is quadratic in π_B, the optimal control, π_B^{**}, takes the following form:

$$\pi_B^{**} = \begin{cases} 1 & if \ -\frac{D}{2\lambda} > 1 \\ -\frac{D}{2\lambda} \ (0 < u_B^* < 1, \text{ to be determined}) & if \ 0 \leq -\frac{D}{2\lambda} \leq 1 \\ 0 & if \ -\frac{D}{2\lambda} < 0. \end{cases}$$

From (26), we know there are three possibilities. (i). If $-\frac{D}{2\lambda} < 0$, then $0 = \frac{b - \alpha_R}{a - b}D + 1$, namely that $D = \frac{a - b}{\alpha_R - b}$ is a positive constant. (ii) If $-\frac{D}{2\lambda} > 1$, then $0 = \frac{b - \alpha_R}{a - b}D + 1$, namely that $D = -\frac{a - b}{a - \alpha_R}(\lambda + 1)$ is also a constant. Note that $D < -2\lambda$ if and only if $a - b \leq 2(\alpha_R - b)$, or if and only if $\lambda < \frac{a - b}{a + b - 2\alpha_R}$ and $a - b > 2(\alpha_R - b)$. (iii). If $0 \leq -\frac{D}{2\lambda} \leq 1$, then

$$0 = \lambda\left(-\frac{D}{2\lambda}\right)^2 - 2\lambda\left(-\frac{D}{2\lambda}\right)^2 - \frac{b - \alpha_R}{a - b}2\lambda\left(-\frac{D}{2\lambda}\right) + 1,$$

namely that

$$D = 2\frac{b - \alpha_R}{a - b}\lambda - \sqrt{4\left(\frac{b - \alpha_R}{a - b}\lambda\right)^2 + 4\lambda}$$

is a constant. Note that $-\frac{D}{2\lambda} \in (0, 1)$ if and only if $\lambda \geq \frac{a - b}{a + b - 2\alpha_R}$ and $a - b > 2(\alpha_R - b)$.

In term of minimizing the Hamiltonian H_F under the above case (i), we have $\pi_B^{**} = 0$ for all time, which is impossible to obtain $i_B(T) = i_e$; under the above case (ii), we have $\pi_B^{**} = 1$ for all time; under the above case (iii), we have $\pi_B^{**} = \frac{D}{-2\lambda}$ for all time. To sum up, we have

$$(\pi_B^{**}, T^{**}) = \begin{cases} (u^*, T_2) \text{ if } \lambda \geq \frac{a - b}{a + b - 2\alpha_R} \text{ and } a - b > 2(\alpha_R - b) \\ (1, T_3) \quad \text{otherwise} \end{cases} \qquad (27)$$

where $u^* = \frac{D}{-2\lambda} = \frac{\alpha_R - b}{a - b} + \sqrt{\left(\frac{b - \alpha_R}{a - b}\right)^2 + \frac{1}{\lambda}}$, and T_2 and T_3 satisfy

$$i_B(T_2) = \frac{\frac{i_0}{1 - i_0}e^{(b + (a - b)u^* - \alpha_R)T_2}}{1 + \frac{i_0}{1 - i_0}e^{(b + (a - b)u^* - \alpha_R)T_2}} = i_e, \quad i_B(T_3) = \frac{\frac{i_0}{1 - i_0}e^{(a - \alpha_R)T_3}}{1 + \frac{i_0}{1 - i_0}e^{(a - \alpha_R)T_3}} = i_e,$$

respectively. This completes the proof. $\qquad \Box$

Equilibrium Concepts
for Rational Multiparty Computation

John Ross Wallrabenstein and Chris Clifton

Dept. of Computer Science, Purdue University, USA
{jwallrab,clifton}@cs.purdue.edu

Abstract. In this work, we build upon previous results to strengthen the equilibrium concept for *rational multiparty computation*. We consider only rational players, acting to maximize their utility functions. We consider extensive form dynamic games of imperfect information, using a computational variant of perfect Bayesian equilibrium as the solution concept. We argue that the perfect Bayesian equilibrium is a more appropriate solution concept for multiparty computation, as in cryptographic protocols information is often imperfect by design. Further, the perfect Bayesian equilibrium concept is able to address dynamic games, where players move sequentially rather than simultaneously. By considering players that move sequentially, we are able to remove the assumption of a broadcast channel. Finally, we give novel definitions of privacy, correctness and fairness solely in terms of game theoretic constructs.

Keywords: Rational Multiparty Computation, Rational Secret Sharing, Perfect Bayesian Equilibrium, Non-Cooperative Computation.

1   Introduction

A recent focus of the cryptographic literature has been to formulate a framework for analyzing the security of protocols from a game theoretic perspective. The notion of rational multi-party computation considers only a single class of players: those that are rational, seeking to maximize their utility functions. A survey of the intersection of cryptography and game theory is given by Katz [1].

Most previous work towards a general game theoretic framework for reasoning about security in rational multiparty computation has been limited to those functions that are *non-cooperatively computable* (NCC), as defined by Shoham et al. [2]. In addition to being restricted to NCC, most existing work uses computational variants of Nash, Correlated or Bayesian equilibrium [3–6] as the solution concept for games. The exception is work by Gradwohl et al. [7] where the authors consider a relaxed version of computational sequential rationality that removes non-credible threats, called *threat-free Nash equilibrium*. Halpern [8] discusses many of the problems inherent to the equilibrium concepts that have been proposed for rational multiparty computation, such as the necessity of providing a model for the computational costs imposed on players and the fact that cryptographic protocols typically are not simultaneous interactions.

S.K. Das, C. Nita-Rotaru, and M. Kantarcioglu (Eds.): GameSec 2013, LNCS 8252, pp. 226–245, 2013.

We argue that the notion of perfect Bayesian equilibrium (PBE), a solution concept for extensive form dynamic games of imperfect information, is preferable for modeling cryptographic protocols. As players commonly cannot observe the moves made by others in cryptographic protocols, PBE offers a natural method for modeling this uncertainty. Further, it formally models observable actions and auxiliary information available to players that affects their strategy selection.

The goal of a rational multiparty computation framework is to relax the requirements of the malicious and semi-honest models in secure multiparty computation. The malicious model must protect against *all* deviations from the protocol specification, including actions that do not give an adversary an advantage. Protocols secure in the semi-honest model achieve greater efficiency, but suffer from the strong assumption that parties will not deviate from the protocol even if they benefit from doing so. As we describe in Section 5, our framework requires only that parties follow the protocol if such action constitutes *rational behavior*. We argue that the assumption of rationality is more plausible than the blind obedience required in the semi-honest model, and the resulting protocols will be more efficient than their malicious model counterparts at the expense of preventing arbitrary (i.e., *non-rational*) actions. Perhaps most critically, even protocols secure under the malicious model do not prevent a party from lying about their input. Rational behavior provides a means to incorporate this into the discussion, ensuring that results reflect the true data.

Our goal in this work is to give a framework for rational multiparty computation that models the asymmetry of information available to players in cryptographic protocols, as well as how players update their strategies dynamically throughout the interaction. We will first review existing work applying game theory to cryptographic protocols. Section 3 discusses limitations with prior work, using the rational secret sharing problem to illustrate that existing approaches do not fully model information asymmetry or how players update their beliefs in response to new information. We argue that modeling imperfect information, beliefs about the game state, and non-credible threats are desirable qualities of a candidate equilibrium concept for rational multiparty computation. Section 4 illustrates the deficiencies in equilibrium concepts employed in existing work, and motivates our choice of perfect Bayesian equilibrium as the solution concept for our framework. We finally present our new game theoretic framework for analyzing cryptographic protocol in Section 5, which models both the asymmetry of information available to players, as well as allowing players' strategies to be updated dynamically throughout the protocol. We give novel definitions capturing the cryptographic concepts of privacy, correctness and fairness in terms of game theoretic constructs and prove the necessary and sufficient conditions under which they hold.

2 Related Work

The impetus for this work is largely due to a recent survey by Katz describing ongoing research into potential links between cryptographic and game theoretic notions [1]. Halpern and Teague study rational multiparty computation under the assumptions of *correctness* and *exclusivity* [5]. They show the impossibility of secret sharing and general multiparty computation for any *deterministic* mechanism under these assumptions. However, they give randomized algorithms that terminate in expected constant time for both problems, and show that they satisfy their framework. Kol and Naor expand on the work of Halpern and Teague to give protocols that are not susceptible to *backward induction*, even in the presence of exponentially many iterations [9]. These solutions assume the existence of a broadcast channel, and they give solutions for both the non-simultaneous and simultaneous cases. The authors choose the notion of a *computational Nash equilibrium*, and leave extensions to subgame perfection open. We argue that even the extension to subgame perfection is inadequate, as it assumes players are aware of the moves made by others. Limiting the information players gain through a cryptographic protocol is of critical importance, so the equilibrium concept should not require that players' moves are always observed. Instead, we consider extensive form dynamic games of imperfect information, where players' information sets are not guaranteed to be singleton nodes (e.g., not all moves are observed) and players move sequentially rather than simultaneously, removing reliance on a broadcast channel. In this setting, we can model both the inherent asymmetry of information available to players, as well as model dynamic strategies that change throughout the game, rather than being fixed prior to the interaction. Nojoumian et al. [10] introduced *socio-rational secret sharing*, where rational and malicious players engage in the same protocol more than once. A public trust network is assumed, which stores a player's believed honesty based on past protocol interactions. We go beyond this model by modeling all players as rational, rather than creating a separate class of malicious players. Further, we do not assume the existence of a public trust network, nor do we assume that players necessarily value future interaction.

The most complete game theoretic framework to date was given by Halpern et al. [11]. They consider how agents play games when computation has an associated cost and affects agents' utility functions directly. The authors formalize the notion of a computational Nash equilibrium, and demonstrate that mixed computational Nash equilibria are guaranteed to exist for the set of computational games where randomization is free. However, the framework considers only Bayesian games of *perfect* information. Bayesian Nash equilibrium can result in implausible equilibria, as it does not exclude non-credible threats. In the setting of cryptography, threatening to break the underlying cryptosystem would constitute a non-credible threat for a player bound to probabilistic polynomial time (PPT), despite the action's optimality for an unbounded player. We build on their framework to provide a computational model of extensive form dynamic games of *imperfect* information, and consider an equilibrium concept designed to handle non-credible threats.

The most complete framework from a cryptographic perspective that integrates game theoretic concepts was given by Groce et al. [12], which builds on the framework by Asharov et al. [3]. Asharov et al. demonstrate how standard cryptographic notions of security can be framed in a game theoretic view when considering malicious fail-stop adversaries. The authors demonstrate that privacy, correctness and fairness can be met using a game theoretic simulation-based framework. However, the framework only considers computational Nash equilibrium in extensive-form games of perfect information. We argue that a computational variant of PBE is preferable for constructing cryptographic protocols in a game theoretic framework, where players may not know the actions of other players when their computational abilities are bounded. The authors limit a player's strategy set to $\{\sigma^{\text{continue}}, \sigma^{\text{abort}}\}$, where at each node to follow σ^{continue} requires following the protocol specified by the mechanism designer precisely. From this the authors argue that non-credible threats in *fail-stop* games are meaningless, as a party that aborts cannot be punished. The work of Groce et al. [12] demonstrates that fairness can be achieved for a much broader class of utility functions than those specified by Asharov et al. [3]. Further, Groce et al. consider the *byzantine* case, where deviations are not limited to the fail-stop model. However, the equilibrium concept considered by Groce et al., namely Bayesian strict Nash equilibrium, does not *explicitly* model players' beliefs about the game state. Rather, this concept captures only uncertainty about the *types* of the other players. However, the players' beliefs about the current game state are modeled exogenously in Groce et al.'s setting. In cryptographic settings, a player's uncertainty about the current state is of critical importance, and we demonstrate the shortcomings of other equilibria concepts in Section 3. Our framework builds directly on Asharov et al.'s work, and as in Groce et al.'s setting, we allow for *arbitrary* deviation from the protocol beyond simple aborts.

3 Motivation

We motivate our approach by demonstrating cryptographic interactions where players' information is imperfect, and their beliefs must be formally modeled. Specifically, at the end of this section we show that a simple change to the rational secret sharing protocol of Groce et al. [12] results in a protocol where a rational player would cheat, but existing work incorrectly predicts the player behaves honestly. We first introduce the necessary background and notation from game theory[1].

Cryptographic protocols proceed in a series of rounds, where at each round some subset of the parties select and play an action. Game theory models such interactions as extensive form dynamic games, where players move sequentially through a series of rounds, rather than normal form strategic games that model a single simultaneous interaction.

Imperfect Information. The information available to a player in a crypto-graphic protocol is of critical importance. The notion of computational security

---

[1] For a proper introduction to the subject, Katz [1] describes the current effort to combine game theoretic and cryptographic concepts, while Osborne et al. [13] and Fudenberg et al. [14] give a complete introduction to game theory.

relies on the fact that players can be modeled as asymptotically bounded algorithms, and are only able to gain certain information with negligible probability. Consider for instance the *ciphertext indistinguishability* (IND-CPA) game [15]. In this game, an adversary $\mathcal{A}$ bound to probabilistic polynomial time (PPT) has two plaintext messages $\{m_0, m_1 : |m_0| = |m_1|\}$, and the challenger $\mathcal{C}$ has an asymmetric key pair $\{E_{\mathcal{C}}, D_{\mathcal{C}}\}$ from a public key cryptosystem. $\mathcal{C}$ publicizes $E_{\mathcal{C}}$, and $\mathcal{A}$ performs up to polynomially many encryptions before sending $\{m_0, m_1\}$ to $\mathcal{C}$. $\mathcal{C}$ selects a bit $b \in \{0, 1\}$ uniformly at random, and returns $E_{\mathcal{C}}(m_b)$ to $\mathcal{A}$. After performing at most polynomially many operations, $\mathcal{A}$ outputs a *guess* $b' \in \{0, 1\}$, and succeeds when $b' = b$. The cryptosystem is said to be IND-CPA secure if, for all PPT adversaries $\mathcal{A}$

$$|Pr[\mathcal{A}(E_{\mathcal{C}}(m_b)) = 1] - Pr[\mathcal{A}(E_{\mathcal{C}}(m_{1-b})) = 1]| \leq \epsilon(\lambda)$$

where $\epsilon(\cdot)$ is a negligible function and λ is the security parameter. Clearly this property reflects the inability of a computationally bounded adversary to distinguish between two cases. From a game theoretic perspective, we argue that this lack of knowledge is properly modeled as an extensive form dynamic game of *imperfect*[2] information. When some player p_0 does not observe a previous action by another player p_1, we say that the game has imperfect information and p_0's *information set* is *non-singleton*. That is, p_0 only knows that p_1 has moved, and does not know which action was played.

In the IND-CPA ciphertext indistinguishability game, $\mathcal{A}$ has imperfect information as it does not observe $\mathcal{C}$'s action $b \mapsto \{0, 1\}$. Thus, $\mathcal{C}$'s information set contains both the left ($b \mapsto 0$) and right ($b \mapsto 1$) nodes of the game tree Γ under the assumption that $\mathcal{C}$ is bound to PPT. Current rational multiparty computation frameworks consider solution concepts that require perfect information, and do not formally model players' information and beliefs. For instance, if $\mathcal{A}$ had some auxiliary information (e.g., $\mathcal{C}$'s random seed), it may be able to predict $\mathcal{C}$'s choice for b with probability non-negligibly greater than $\frac{1}{2}$. Thus, any solution concept must explicitly model the fact that moves in cryptographic interactions are frequently unobserved, and also that players may have auxiliary information or beliefs that influence their strategy selection.

Updating Beliefs. Players typically update their *beliefs* throughout cryptographic protocols based on observed events. Consider the case of interactive zero-knowledge proof systems [17]. This game is an interaction between a prover $\mathcal{P}$ in possession of a secret, and a verifier $\mathcal{V}$ that is to learn *only* whether or not $\mathcal{P}$ does, in fact, know the secret. In each round, a prover not in possession of the secret succeeds with probability $0 < p < 1$. Thus, $\mathcal{V}$ must interact with $\mathcal{P}$ through k rounds until $1 - p^k$ is acceptably close to 1. If at any round $\mathcal{P}$ fails the test, then $\mathcal{V}$ knows with certainty that $\mathcal{P}$ does not possess the secret. However, the likelihood that $\mathcal{P}$ *does* know the secret approaches 1 as $k \to \infty$. Thus, $\mathcal{V}$ is consistently updating a belief about $\mathcal{P}$ throughout the protocol.

---

[2] The well-known Harsanyi transformation [16] allows any game of incomplete information to be transformed into a game of complete and imperfect information by introducing an initial move by Nature that assigns a *type* to each player.

Dynamic Games. In game theory, games may be either *strategic* or *dynamic*. In the former, actions are played simultaneously, while in the latter actions may be played sequentially. This is equivalent to deciding between whether or not to assume the existence of a *broadcast channel*. As broadcast channels are a relaxation of real world interactions, removing this assumption is desirable as it allows players to act in a specified order. This introduces non-trivial issues into protocols that may be very basic in the semi-honest model, such as the recovery protocol for secret sharing. This protocol was modeled as an extensive form dynamic game by Groce et al. [12], who give a solution when players must move sequentially in a known order.

Non-credible Threats. Recently, Halpern et al. [11] showed that a Nash equilibrium is guaranteed to exist for all finite machine games under the assumption that randomization is free. However, their framework considers only Bayesian Nash equilibrium: an equilibrium concept susceptible to implausible equilibria through non-credible threats. A threat is not considered credible if it is "off the equilibrium path" for a player. That is, action a is not credible if player i receives a greater *expected* utility by playing action $a' \neq a$. We consider a *computational* non-credible threat to be any action a where that there exists another action a' that yields negligibly less utility and is computable subject to the player's complexity bound $\mathscr{C}$. Our definition assumes that a player will choose the optimal strategy whenever their complexity $\mathscr{C}$ allows such action to be performed.

Rational Secret Sharing. The necessity of modeling imperfect information, and the difficulty imposed when broadcast channels cannot be assumed, is easily illustrated using the most common example of rational cryptographic protocols: *rational secret sharing* [18–21, 12, 22]. The goal of threshold secret sharing is to split a secret among n parties such that any k shares are sufficient to recover the secret value, using a scheme such as the polynomial interpolation approach proposed by Shamir [23]. Rational secret sharing, introduced by Halpern and Teague [5], is particularly concerned with the process of *recovering* the secret from the shares. As noted by Halpern et al. [5], rational players' utility functions are assumed to value *exclusivity*, where preference is given to learning the output of the function while preventing other players from doing so. Under this assumption, no party has any incentive to distribute their share to the other parties. Rather, the equilibrium is to wait for other players to distribute their shares, as this is the only action that increases a player's utility function. The authors demonstrate that this implies no deterministic protocol exists where rational parties are willing to disseminate their shares to other players. However, their randomized protocol relies on the fact that parties are unaware whether the current state is terminal (allowing the secret to be recovered), or merely a "test" state (where the secret cannot be recovered, but players who do not distribute shares are caught as cheaters). This fundamental lack of information constitutes an extensive form game of *imperfect* information, for which the Nash equilibrium (and computational variants thereof) are insufficient equilibria concepts.

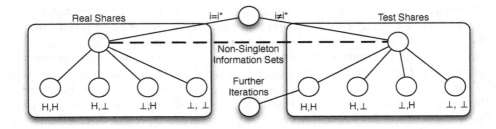

Fig. 1. Imperfect Information Sets in the Rational Secret Sharing Game

Figure 1 illustrates the two-party rational secret sharing game Γ, which proceeds in a series of rounds. At round i, player p_0's share x_0^i may be a legitimate share, such that combined with p_1's share the secret may be recovered. However, p_0's share may also be illegitimate, such that the shares combine to a pre-determined test value that is not the original secret. Players are not aware whether the given round i is the terminal round i^* where the secret may be recovered, or a test round $i \neq i^*$ where no information may be learned from the shares. Assume that a player's strategy set σ is limited to $\sigma \in \{H, \perp\}$, where H denotes the honest strategy of revealing the player's share, and $\perp$ denotes the action of not revealing the share. By choosing i^* from a geometric distribution, as in Groce et al. [12], cheating players that choose strategy $\sigma =\perp$ when $i \neq i^*$ are caught and the game may be terminated. Thus, players now have an incentive to distribute their share, as playing $\perp$ only yields μ^+ when $i = i^*$.

The difference between the Bayesian strict Nash equilibrium (BNE), used in the rational secret sharing setting of Groce et al. [12], and the perfect Bayesian equilibrium (PBE) concept we consider in our setting, requires clarification. If all moves were simultaneous, BNE and PBE would yield the same equilibria. However, in extensive form games of imperfect information, a player may not be able to observe all moves by other players. This results in non-singleton information sets, which BNE is unable to model, as it only considers uncertainty about players' types. Consequently, this uncertainty about the game state should be explicitly modeled into their expected utility. The PBE concept is able to "cut through" the non-singleton information sets present in the rational secret sharing game, as it considers players' beliefs about the type of other players *as well as beliefs about the current game state*. Thus, PBE avoids implausible equilibria that result from the presence of non-singleton information sets. The critical issue with the Groce et al. [12] approach is that the expectation for utility *exogenously* considers the probability that $i \overset{?}{=} i^*$, rather than making this belief explicit in the equilibrium concept. Thus, they restrict a player to fix their strategy at the start of the game for consistency with BNE, even as a mediator introduces auxiliary information.

Consider a game where p_0 is given auxiliary information about whether $i \overset{?}{=} i^*$ after the game has started. Suppose the share generator reveals to p_0 that the current round i is, in fact, the terminal round i^*. This information crucially affects p_0's expected utility function under PBE, as p_0's beliefs about i^* have changed from the start of the game. This information should be explicitly factored into the calculation of expected utility, but the definition of Bayesian Nash equilibria ignores this, focusing on uncertainty only about the player's types. Thus, even in the case where p_0 *knows* the correct value of i^* at some round k, the BNE for the above game predicts that the player will play honestly and reveal their share. However, PBE allows p_0 to update their belief about i^* as the game progresses, and requires that all subsequent play be optimal with respect to their beliefs. Thus, PBE predicts that p_0 should not reveal their share, and instead collect p_1's share to recover the secret. Given p_0's beliefs about the game state, this clearly maximizes p_0's expected utility. The equilibrium predicted by BNE, namely for p_0 to distribute their share, is implausible given the auxiliary information provided to p_0 and the fact that p_0 values exclusivity. This implausible equilibrium is avoided when the PBE concept is used.

4   Equilibrium Concepts

We now review the equilibrium concepts and game settings considered by existing work. Our goal is to demonstrate the shortcomings of equilibrium concepts considered by existing work, and to motivate our choice of perfect Bayesian equilibrium as the solution concept for our framework.

4.1   Normal Form Games

We begin by introducing *normal form*, or strategic, games. Normal form representation of games is ideal for modeling simultaneous interaction, rather than sequential moves, as strategies are fixed prior to playing the game. We review the formal definition from Osborne [13]:

Definition 1. *A **normal form game** Γ consists of:*

1. *A finite set N of players.*
2. *A nonempty set A_i of actions available for each player $i \in N$.*
3. *A preference relation $\precsim_i$ on $A = \times_{j \in N} A_j$ for each player i.*

Frequently, the preference relation $\precsim_i$ is represented by a *utility function* $\mu_i : A \to \mathbb{R}$, such that $\mu_i(a) \geq \mu_i(b)$ when $b \precsim_i a$. The normal form game is then denoted by $\Gamma = \langle N, (A_i), (\mu_i) \rangle$.

Normal form games are well-suited to modeling one-shot protocols where players move simultaneously[3]. In a computational setting, this requires the existence of a broadcast channel. However, it is desirable to remove the assumption of simultaneous moves (and, thus, the assumption of a broadcast channel) so that

---

[3] Technically, the notion of simultaneity only requires that players *commit* to their strategies before moving. However, this is still an assumption we would like to remove.

players may move sequentially. More importantly, modeling interactions in normal form fixes players' strategies prior to starting the game. Ideally, a player's strategy should be dependent on their beliefs about the current game state. We return to this goal when we consider extensive-form dynamic games.

Nash Equilibrium. The computational Nash equilibrium is the most widely used solution concept for rational cryptography [3, 4, 24–26]. The intuition is to account for strategies that, although optimal, occur with only negligible probability. In a cryptographic setting, an optimal strategy may be to break the underlying cryptosystem. However, for players bound to PPT, this strategy succeeds with only negligible probability. Consequently, Nash equilibrium has been refined into a *computational* variant that states players only switch strategies if the gain is not negligible with respect to the security parameter λ. The original definition of a computational Nash equilibrium was given by Dodis et al. [4]:

Definition 2. *A **computational Nash equilibrium** of a two-party extensive-form game Γ is an independent strategy profile (σ_1^*, σ_2^*), such that*

1. *both σ_1^*, σ_2^* are PPT computable.*
2. *for any other PPT computable strategies σ_1', σ_2', we have $\mu_1(\sigma_1', \sigma_2^*) \leq \mu_1(\sigma_1^*, \sigma_2^*) + negl(\lambda)$ and $\mu_2(\sigma_1^*, \sigma_2') \leq \mu_2(\sigma_1^*, \sigma_2^*) + negl(\lambda)$.*

Nash equilibria are well-suited to normal form games, where players move simultaneously and have full knowledge of the game state and payoffs. However, in the computational setting we must consider extensive form dynamic games of imperfect information, where players move sequentially and may be unaware of the game state or the payoffs of other players.

Correlated Equilibrium. A strong case for the use of *correlated* equilibrium can be made from the fact that a mediator is able to "recommend" a set of actions to the players. Thus, the action set follows a *joint* probability distribution, where each player learns the conditional distribution over the actions of other players. By recommending actions to players, greater utility may be achieved when players follow the mediator's advice. Further, correlated equilibria are computationally less expensive (in *strategic games*) to compute than general Nash equilibrium. That is, computing Nash equilibria is NP-Hard, while computing correlated equilibria can be done in polynomial time by solving a linear program [27]. Correlated equilibria were considered in a computational setting by Urbano et al. [28], and specifically in the context of rational cryptography by Dodis et al. [4] and later Gradwohl et al. [7]. Our objection to correlated equilibria is that they are defined only for strategic form games, rather than the more expressive extensive form games. The extension of correlated equilibria to extensive form games was considered by von Stengel et al. [29], but they demonstrated finding the optimal equilibria is NP-Hard.

Bayesian Nash Equilibrium. Bayesian Nash equilibria (BNE) consider uncertainty with respect to a player's *type*, chosen by the fictitious player *nature*.

Thus, the optimal strategy for a player is conditioned on the probability of the other players' types. Bayesian Nash equilibria result in implausible equilibria in extensive dynamic games as non-credible threats are not accounted for. The rational secret sharing problem was considered by Groce et al. [12] without assuming broadcast channels, using BNE as the solution concept. As BNE requires players fix their strategies before the game, they are unable to update their strategies based on information observed throughout the game. Bayesian Nash equilibria are sufficient for strategic games, but lack the notion of *sequential rationality* necessary for application in extensive form games. We introduce the refinement of Bayesian Nash equilibria, namely perfect Bayesian equilibria, in Section 4.2.

4.2   Extensive Form Games

We now leave the setting of normal form games, and consider extensive form dynamic games where players move sequentially. Extensive form games are defined by Osborne et al. [13] as follows:

Definition 3. *An **extensive form game** Γ consists of:*

1. *A finite set N of players.*
2. *A (finite) set of sequences $\mathcal{H}$. The empty sequence $\emptyset$ is a member of $\mathcal{H}$. We let k denote the current decision node. If $(a^k)_{k=1,...,K} \in \mathcal{H}$ and $L < K$ then $(a^k)_{k=1,...,L} \in \mathcal{H}$. If an infinite sequence $(a^k)_{k=1}^{\infty}$ satisfies $(a^k)_{k=1,...,L} \in \mathcal{H}$ for every positive integer L then $(a^k)_{k=1}^{\infty} \in \mathcal{H}$. A history $(a^k)_{k=1,...,K} \in \mathcal{H}$ is a terminal history if it is infinite or if there is no a^{K+1} such that $(a^k)_{k=1,...,K+1} \in \mathcal{H}$. The set of actions available after the nonterminal history h is denoted $A(h) = \{a : (h, a) \in \mathcal{H}\}$ and the set of terminal histories is denoted $\mathcal{Z}$. We let $\mathcal{H}^k$ denote the history through round k.*
3. *A player function P that assigns to each nonterminal history (each member of $\mathcal{H}/\mathcal{Z}$) a member of $N \cup \{nature\}$. When $P(h) = nature$, then nature determines the action taken after history h.*
4. *For each player $i \in N$ a partition $\mathcal{I}_i$ of $\{h \in \mathcal{H} : P(h) = i\}$ with the property that $A(h) = A(h')$ whenever h and h' are in the same member of the partition. For $I_i \in \mathcal{I}_i$ we denote by $A(I_i)$ the set $A(h)$ and by $P(I_i)$ the player $P(h)$ for any $h \in I_i$. Thus, $\mathcal{I}_i$ is the information partition of player i, while the set $I_i \in \mathcal{I}_i$ is an information set of player i.*
5. *For each player $i \in N$ a preference relation $\precsim_i$ on lotteries[4] over $\mathcal{Z}$ that can be represented as the expected value of a payoff function defined on $\mathcal{Z}$.*

Throughout, we replace the preference relation $\precsim_i$ by a *utility function* $\mu_i : A \to \mathbb{R}$, such that $\mu_i(a) \geq \mu_i(b)$ when $b \precsim_i a$.

---

[4] Even if all actions are deterministic, moves by *nature* can induce a probability distribution over the set of terminal histories.

Perfect Bayesian Equilibrium. Formal definitions of perfect Bayesian equilibria (PBE) are usually not generalizable to general extensive form games, and contain the vague requirement that beliefs be updated according to Bayes' rule "whenever possible". Bonanno [30] gives a definition of PBE that is applicable for general extensive form games, but we will use the definition by Diaz et al. [31], as they go further by extending to general extensive form games as well as clarifying the ambiguous "whenever possible" updating requirement.

We first require that, for player $i \in N$, their *assessment* (σ_i, β_i) consisting of a strategy σ_i and a *belief* β_i about the game state, be sequentially rational:

Definition 4. *An assessment (σ_i, β_i) is (computationally) **sequentially rational** if, for every player $i \in N$ and every information set $I_i \in \mathcal{I}_i$, there holds:*

$$\mu_i(\sigma_i, \beta_i | I_i) + \epsilon(\lambda) \geq \mu_i((\sigma_{-i}, \sigma_i'), \beta_i | I_i) \tag{1}$$

for every strategy σ_i', a probability distribution over actions, of player i, where (σ_{-i}, σ_i') is a strategy profiles that all players stick to the strategy σ except that player i turns to the strategy σ_i', and $\mu_i((\sigma_{-i}, \sigma_i'), \beta_i | I_i)$ denotes player i's utility induced by this strategy profile and the belief system β_i, a probability distribution over game states, conditional on I_i being reached. The term $\epsilon(\lambda)$ denotes a negligible utility gain with respect to the security parameter λ, and σ_i is an efficiently computable strategy for player i with complexity $\mathscr{C}$.

Next, we give the definition of a *weak perfect Bayesian equilibrium*, which we build on to construct the final equilibrium concept that applies to general extensive form games:

Definition 5. *Let Γ be an extensive form game. An assessment (σ, β) is a **weak perfect Bayesian equilibrium** if it is sequentially rational and, on the path of σ, β is derived from σ from Bayes' rule.*

With this, we reach the definition of a $\mathscr{C}$-simple perfect Bayesian equilibrium:

Definition 6. *Let Γ be an extensive form game. An assessment (σ, β) is a $\mathscr{C}$-**simple perfect Bayesian equilibrium** if, for each regular information set I_i^k, the restriction of (σ, β) to $\Gamma_{I_i^k}(\sigma, \beta)$ is sequentially rational and β is obtained by conditional updating from σ (i.e., the restriction of (σ, β) to $\Gamma_{I_i^k}(\sigma, \beta)$ is a weak perfect Bayesian equilibrium), where σ is efficiently computable by an interactive Turing machine (ITM) with complexity $\mathscr{C}$.*

5   Framework

In order to show the application of game theoretic models to cryptography, a proper security model must be introduced. Thus, we consider appropriate game theoretic definitions of privacy, correctness and fairness.

Our framework is an extension of Asharov et al.'s [3] model of security under *fail-stop* games. The original work considered two players with action sets limited to $\{\sigma^{\mathrm{abort}}, \sigma^{\mathrm{continue}}\}$, where σ^{abort} implied that the ITM output a special signal $\perp$ observed by all players and stopped playing the game, and σ^{continue} is the strategy of following the game specification *without deviation*. Thus, the only deviating strategy is to abort the protocol, which is similar to the standard semi-honest security model. We extend this model to assume that σ^{continue} is precisely the vector of strategies of *not aborting*, regardless of whether or not the chosen action is the honest choice. Similarly, $\sigma^{\text{deviate}} = \{\sigma^U / \{\sigma^{\text{honest}}, \sigma^{\text{abort}}\}\}$ is the *set* of all possible strategies that are *dishonest*, taking σ^U to be the universe of strategies. That is, σ^{deviate} corresponds to choosing a strategy σ that deviates from the prescribed protocol. Without loss of generality, we assume that $\sigma^{\text{continue}} = \{\sigma^{\text{honest}}, \sigma^{\text{deviate}}\}$, where σ^{honest} is equivalent to following the prescribed protocol. As multiparty computation players are assumed to be mutually distrustful in the cryptographic literature, we assume they are *risk-averse* in the game theoretic sense. Thus, when an honest player cannot distinguish between the probability of $\mathcal{A}$ selecting $\sigma_{\mathcal{A}}^{\text{deviate}}$ or $\sigma_{\mathcal{A}}^{\text{honest}}$, the honest party assumes that $\sigma_{\mathcal{A}}^{\text{deviate}}$ was selected. We consider only the two-party case, as the extension to multiple parties requires modeling player collusion. Throughout, we let μ^+ represents positive utility gain, μ^- represent negative utility, and μ^0 represents neutral utility. We now give novel definitions of privacy, correctness and fairness in purely game theoretic terms, considering a more expressive model where players may deviate *arbitrarily* from the protocol beyond simply aborting.

5.1   Privacy

We follow Asharov et al.'s [3] intuition and require that parties' utility functions reflect the loss of privacy with negative utility. This requires no assumptions about other players' utility functions with respect to the *gain* of information; *the burden is player specific* and known, as we assume players are aware of their own utility functions. Thus, players may choose to require that any subset of privacy, correctness and fairness are satisfied by the protocol.

We first introduce a new notion of indistinguishability defined in terms of a $\mathscr{C}$-bounded distinguisher $\mathcal{D}$'s ability to differentiate between information sets. We first introduce notation for an ITM's local history:

Definition 7. *Let* $\pi = (M_0, M_1)$ *be a two-party protocol between a pair of ITMs* (M_0, M_1). *Then we write*

$$\mathcal{H}_{\pi,i}^k(x_0, x_1, \lambda) = (x_i, M_R, m_1^i, \ldots, m_k^i)$$

to denote the local history of M_i *at round* k, *with input* x_i, *random tape* M_R, *security parameter* λ *and* m_j^i *represents the* j^{th} *message.*

We consider the set of infinitely many input tuples $(x_0, x_1^0, x_1^1, \lambda)$ where we have that $|x_0| = |x_1^0| = |x_1^1| = \lambda$, and party p_0's input is fixed at x_0 while p_1's input is in the set $\{x_1^0, x_1^1\}$.

Definition 8. *We say that a finite extensive form computational game Γ^λ has* **indistinguishable initial information sets in the presence of $\mathscr{C}$-bounded adversaries** *if:*

$$|Pr[(\mathcal{H}^0_{\pi,\mathcal{D}}(x_0, x_1^0, \lambda) \in I_0) = 1] - Pr[(\mathcal{H}^0_{\pi,\mathcal{D}}(x_0, x_1^1, \lambda) \in I_0) = 1]| \leq \epsilon(\lambda)$$

for some negligible function $\epsilon(\cdot)$.

That is, no $\mathscr{C}$-bounded distinguisher $\mathcal{D}$ can distinguish the *type* (i.e., private input) of party p_1 with probability non-negligibly greater than $\frac{1}{2}$. With this notion formally defined, we now give a definition for players' utility functions with respect to privacy:

Definition 9. *Let π be a two-party protocol and f be a two-party function. Then, for every x_0^0, x_0^1, x_1 such that $f(x_0^0, x_1) = f(x_0^1, x_1)$, and for every $\mathscr{C}$-bounded distinguisher $\mathcal{D}$, the* **utility function for privacy** *μ^p for party p_i, on input $x_0 \in \{x_0^0, x_0^1\}$, is defined by*

- $\mu_0^p(\mathcal{H}^\varnothing_{\pi,i}) = 0$ *when p_0 aborts immediately*
- $\mu_0^p(\mathcal{H}^k_{\pi,0}(x_0^b, x_1, \lambda)) \mapsto \begin{cases} \mu^- : \text{guess}_\pi((\mathcal{H}^k_{\pi,\mathcal{D}}(x_0^b, x_1, \lambda) \in I_\mathcal{D})) = b', x_0^b = x_0^{b'} \\ \mu^+ \qquad\qquad\qquad\qquad\qquad\qquad\qquad\qquad\quad : \textit{otherwise} \end{cases}$

where $\text{guess} : \mathcal{H} \to T_i$ *is a function from histories to player types.*

Initially π is run, then $\mathcal{D}$ is given as input the local state of π w.r.t. p_i and two auxiliary values (x_0^0, x_0^1). $\mathcal{D}$ outputs a guess $b' \in \{0, 1\}$, where $\mathcal{D}$ succeeds whenever $x_0^b = x_0^{b'}$.

For all rational players with utility functions $\mu \in \mu^p$, we have that $\mu(\sigma^{\text{continue}}) > \mu(\sigma^{\text{abort}})$ *iff*:

$$Pr[\text{guess}_\pi((\mathcal{H}^k_{\pi,\mathcal{D}}(x_0^b, x_1, \lambda) \in I_\mathcal{D})) = 1] = \frac{1}{2} + \epsilon(\lambda)$$

That is, a rational party with a utility function preferring privacy ($\mu \in \mu^p$) only continues participating in the protocol (i.e., by selecting a strategy in σ^{continue}) if for all $\mathscr{C}$-bounded adversaries, the probability of success is at most negligibly greater than $\frac{1}{2}$. We let σ^{deviate} imply that

$$Pr[\text{guess}_\pi((\mathcal{H}^k_{\pi,\mathcal{D}}(x_0^b, x_1, \lambda) \in I_\mathcal{D})) = 1] > \frac{1}{2} + \epsilon(\lambda) \tag{2}$$

That is, by playing σ^{deviate} the adversary has an advantage at breaking the privacy of the protocol with probability non-negligibly greater than $\frac{1}{2}$. Any other strategy $\sigma \notin \sigma^{\text{deviate}}$ will not affect *privacy* under this assumption, although *it may affect correctness or fairness*. We restrict our attention to privacy at the moment.

Definition 10. *Let f and π be as above. Then, π is $\mathscr{C}$-**Game-Theoretic Private** for party p_i if $\mu_i(\sigma_0^{\text{honest}}, \sigma_1^{\text{honest}})$ is a $\mathscr{C}$-PBE with respect to $\mu^p_{i,i\in\{0,1\}}$, $\beta_{i,i\in\{0,1\}}$ and all $\mathscr{C}$-bounded distinguishers $\mathcal{D}$.*

We now prove a theorem defining how protocol π may satisfy Definition 10:

Theorem 1. *Let f be a deterministic two-party function, and let π be a two-party protocol that computes f correctly. Then, π is $\mathscr{C}$-**Game-Theoretic Private** w.r.t. p_0 (resp. p_1) iff π has indistinguishable initial information sets in the presence of $\mathscr{C}$-bounded adversaries.*

Proof (Theorem 1). We first demonstrate that if π is $\mathscr{C}$-Game-Theoretic Private w.r.t. p_0, then π has indistinguishable initial information sets w.r.t. p_0 in the presence of $\mathscr{C}$-bounded adversaries.

If π is $\mathscr{C}$-Game-Theoretic Private w.r.t. p_0, then by definition we have that:

$$\mu_0(\sigma_0^{\text{honest}}|\beta_0, \mathcal{H}_0) + \epsilon(\lambda) \geq \mu_0(\sigma_0', \sigma_0^{\neg\text{honest}}|\beta_0, \mathcal{H}_0)$$

That is, if π is $\mathscr{C}$-Game-Theoretic Private, then players receive more utility by playing strategy σ^{honest} than any other strategy $\sigma^{\neg\text{honest}} = \{\sigma^U/\sigma^{\text{honest}}\}$. Assume by contradiction that π does not have indistinguishable initial information sets w.r.t. p_0. Without loss of generality, we assume $\mathcal{A}$ corrupts p_1. Then a $\mathscr{C}$-bounded adversary $\mathcal{A}$ is able to choose a strategy in the set $\sigma_1^{\text{deviate}}$, where $\mathcal{A}$ invokes a $\mathscr{C}$-bounded distinguisher $\mathcal{D}$ which succeeds in differentiating p_0's information set with probability

$$Pr[\text{guess}_\pi((\mathcal{H}_{\pi,\mathcal{D}}^k(x_0^b, x_1, \lambda) \in I_\mathcal{D})) = 1] > \frac{1}{2} + \epsilon(\lambda)$$

as given by Equation 2, which is a non-negligible advantage. Thus, we have that:

$$\begin{aligned}
\mu_0(\sigma_0^{\text{honest}}, \sigma_1^{\text{deviate}}) &= Pr[\text{guess}_\pi((\mathcal{H}_{\pi,\mathcal{D}}^k(x_0^b, x_1, \lambda) \in I_\mathcal{D})) = 1] \cdot \mu^- \\
&+ Pr[\text{guess}_\pi((\mathcal{H}_{\pi,\mathcal{D}}^k(x_0^b, x_1, \lambda) \in I_\mathcal{D})) = 0] \cdot \mu^+ \\
&< \mu^0 < \mu_0(\sigma_0^{\text{abort}}, \sigma_1^{\text{deviate}}) = \mu^0
\end{aligned}$$

thus contradicting the assumption that σ_0^{honest} is a $\mathscr{C}$-PBE w.r.t. p_0, and that π is $\mathscr{C}$-Game-Theoretic Private by Definition 10, as σ^{abort} yields more utility for p_0 than σ^{honest}.

Next, we show that if π has indistinguishable initial information sets w.r.t. p_0, then π is $\mathscr{C}$-Game-Theoretic Private. By definition, if π has indistinguishable initial information sets w.r.t. p_0, then there *does not* exist a strategy in the set $\sigma_\mathcal{A}^{\text{deviate}}$ such that, for any $\mathscr{C}$-bounded distinguisher $\mathcal{D}$ invoked by $\mathcal{A}$

$$Pr[\text{guess}_\pi((\mathcal{H}_{\pi,\mathcal{D}}^k(x_0^b, x_1, \lambda) \in I_\mathcal{D})) = 1] > \frac{1}{2} + \epsilon(\lambda)$$

Assume by contradiction that σ_0^{honest} is not a $\mathscr{C}$-PBE w.r.t. p_0. Then, we must have that:

$$\mu_0(\sigma_0^{\text{abort}}, \sigma_1^{\neg\text{deviate}}) > \mu_0(\sigma_0^{\text{honest}}, \sigma_1^{\neg\text{deviate}})$$

Clearly we have that $\mathcal{A}$'s strategies are limited to $\sigma_1^{\neg\text{deviate}} = \{\sigma_1^{\text{honest}}, \sigma_1^{\text{abort}}\}$, as by assumption π has indistinguishable initial information sets w.r.t. p_0, so no strategy in the set $\sigma_1^{\text{deviate}}$ exists by Equation 2. Consider first the strategy pair $(\sigma_0^{\text{abort}}, \sigma_1^{\text{abort}})$:

$$\mu_0(\sigma_0^{\text{abort}}, \sigma_1^{\text{abort}}) = \mu^0 = \mu_0(\sigma_0^{\text{honest}}, \sigma_1^{\text{abort}})$$

Thus, σ_0^{honest} is a $\mathscr{C}$-PBE w.r.t. p_0, contradicting the assumption. Similarly, consider the strategy pair $(\sigma_0^{\text{abort}}, \sigma_1^{\text{honest}})$:

$$\mu_0(\sigma_0^{\text{abort}}, \sigma_1^{\text{honest}}) = \mu^0 < \mu_0(\sigma_0^{\text{honest}}, \sigma_1^{\text{honest}}) = \mu^+$$

Thus, σ_0^{honest} is a $\mathscr{C}$-PBE w.r.t. p_0, contradicting the assumption.

5.2 Correctness

Asharov et al.'s [3] notion of correctness is similar to their notion of privacy: party p_i prefers to learn the correct output of the function f to learning an incorrect output. We modify their definition with respect to the utility gained from aborting before the protocol starts. Rather than specify this utility as $\mu_i^c(\mathcal{H}_{\pi,i}^{\varnothing}) = \mu^+$, we say that a party that does not participate in the protocol receives $\mu_i^c(\mathcal{H}_{\pi,i}^{\varnothing}) = \mu^0$, so that parties prefer to participate in the protocol. As defined in the original work, players receive the same utility for not participating as they do for receiving the correct output of the function. As we assume computation is costly, it seems more natural to assign greater utility to receiving the correct output of the function.

As previously specified when considering privacy, we consider the set of infinitely many input tuples $(x_0, x_1^0, x_1^1, \lambda)$ where we have that $|x_0| = |x_1^0| = |x_1^1| = \lambda$, and party p_0's input is fixed at x_0 while p_1's input is in the set $\{x_1^0, x_1^1\}$.

Definition 11. *Let f be a deterministic two-party function, and let π be a two-party protocol that computes f correctly. Then, for every x_0, x_1 as above the **utility function for correctness** for party p_i, denoted μ_i^c, is defined as:*

$$- \ \mu_i^c(\mathcal{H}_{\pi,i}^{\varnothing}) = \mu^0$$

$$- \ \mu_i^c(\text{output}_{\pi,i}, x_0, x_1) \mapsto \begin{cases} \mu^+ : \text{output}_{\pi,i} = f(x_0, x_1) \\ \mu^- \quad\quad : otherwise \end{cases}$$

We consider σ^{honest} to represent the strategy that follows the protocol specification of π, which by definition computes f correctly. Similarly, any other strategy $\sigma \in \{\sigma^{\text{deviate}}, \sigma^{\text{abort}}\}$ is assumed to compute f *incorrectly*. That is, we limit σ^{deviate} to those strategies that yield an incorrect output. Other strategies certainly exist in σ^{deviate} that will not alter the result, but these are handled when privacy and fairness are required. To satisfy the correctness condition, we need only consider those strategies in σ^{deviate} that yield incorrect outputs of f.

Definition 12. *Let f and π be as above. Then, π is $\mathscr{C}$-**Game-Theoretic Correct** for party p_i if $\mu_i^c(\sigma_0^{\text{honest}}, \sigma_1^{\text{honest}})$ is a $\mathscr{C}$-PBE with respect to $\mu_{i,i\in\{0,1\}}^c$, $\beta_{i,i\in\{0,1\}}$ and all $\mathscr{C}$-bounded adversaries $\mathcal{A}$.*

We now prove a theorem defining how protocol π may satisfy Definition 12:

Theorem 2. *Let f be a deterministic two-party function, and let π be a two-party protocol that computes f correctly. Then, π is $\mathscr{C}$-**Game-Theoretic Correct** w.r.t. p_0 (resp. p_1) if*

$$\forall \beta_0, \sigma_1^{\text{deviate}} \in I_0(\mathcal{H}^k) \implies I_0(\mathcal{H}^k) = \{\sigma_1^{\text{deviate}}\}$$

That is, all information sets containing strategy $\sigma_1^{\text{deviate}}$ are singleton nodes, distinguishable by any distinguisher $\mathcal{D}$ of bounded complexity $\mathscr{C}$.

Proof (Theorem 2). We demonstrate that if π is $\mathscr{C}$-Game-Theoretic Correct w.r.t. p_0, then $\forall \beta_0, \sigma_1^{\text{deviate}} \in I_0(\mathcal{H}^k) \implies I_0(\mathcal{H}^k) = \{\sigma_1^{\text{deviate}}\}$. Intuitively, this means that if π satisfies Definition 12, then p_0 must be able to differentiate p_1 selecting σ^{deviate} rather than $\sigma^{\neg\text{deviate}}$.

If π is $\mathscr{C}$-Game-Theoretic Correct w.r.t. p_0, then by definition we have that:

$$\mu_0^c(\sigma_0^{\text{honest}}|\beta_0, \mathcal{H}_0) + \epsilon(\lambda) \geq \mu_0^c(\sigma_0', \sigma_0^{\neg\text{honest}}|\beta_0, \mathcal{H}_0)$$

That is, if π is $\mathscr{C}$-Game-Theoretic Correct, then players receive greater utility by playing strategy σ^{honest} than any other strategy $\sigma^{\neg\text{honest}} = \{\sigma^U/\sigma^{\text{honest}}\}$. Assume by contradiction that

$$\forall \beta_0, \sigma_1^{\text{deviate}} \in I_0(\mathcal{H}^k) \not\implies I_0(\mathcal{H}^k) = \{\sigma_1^{\text{deviate}}\} : \exists I_0(\mathcal{H}^k) = \{\sigma_1^{\text{abort}}, \sigma_1^{\text{honest}}, \sigma_1^{\text{deviate}}\}$$

That is, $\sigma_1^{\text{deviate}}$ exists in *non-singleton* information sets for p_0. Thus, for some previous history $\mathcal{H}_0^j, j < k$, we have that p_0 cannot distinguish between $\mathcal{H}^j = \{\sigma_1^{\text{deviate}}\}$ and $\mathcal{H}^j = \{\sigma_1^{\text{honest}}\}$, where we do not consider $\mathcal{H}^j = \{\sigma_1^{\text{abort}}\}$ as p_1 would output $\perp$, and p_0 would know with probability 1 that this strategy was used. Recall that risk-averse participants assume σ^{deviate} when information sets are non-singletons. We have that

$$\mu_0^c(\sigma_0^{\text{honest}}, \sigma_1^{\text{deviate}}) = \mu^- < \mu_0^c(\sigma_0^{\text{abort}}, \sigma_1^{\text{deviate}}) = \mu^0$$

which contradicts the assumption that σ_0^{honest} is a $\mathscr{C}$-PBE, as aborting yields more utility than engaging in the protocol, and that π is $\mathscr{C}$-Game-Theoretic Correct w.r.t. p_0 by Definition 12.

5.3 Fairness

In Asharov et al.'s [3] original definitions for fairness, players are implicitly assumed to abide by the *exclusivity* property: a player prefers to be the *only* party to learn the output over a fair distribution of the function result. We argue that this assumption does not always hold.

Any framework constructed under the assumption of exclusivity is limited to the set of *non-cooperatively computable* (NCC) [2] functions. Let $f(\cdot, \cdot)$ be a two-party function, with party p_i holding input x_i, $i \in \{0, 1\}$. If party p_i provides an alternate input $x_i' \neq x_i$ to f, a fair protocol outputs $f(x_i', x_{1-i})$ to all parties. However, if p_i can compute $g(f(x_i', x_{1-i}), x_i) = f(x_i, x_{1-i})$, then p_i has no *rational* incentive to provide their true input x_i as p_i alone can now deduce the correct output of the function $f(x_i, x_{1-i})$ from the output $f(x_i', x_{1-i})$. Thus, any framework requiring the *exclusivity* requirement is limited to functions for which the correct output cannot be produced given knowledge of the function and its output on a different input.

As an example, consider auction scenarios. Clearly, any adversary requires that all parties learn the output of the protocol *even if it is not the correct output*, as the result induces others to perform the actual goal of the protocol: distributing goods or services to the winner. If an adversary was the only party to receive the output, no distribution occurs and the effort was pointless.

We modify Asharov et al.'s [3] original utility function for fairness to reflect the fact that the exclusivity assumption does not always hold. Let E denote the set of players whose utility functions for fairness μ^f value *exclusivity*:

Definition 13. *Let π be a two-party protocol and f be a two-party function. Then, for every x_0, x_1 as above the **utility function for fairness** for party p_i, denoted μ_i^f, is defined as:*

$$\mu_0^f(\sigma_0, \sigma_1) \mapsto \begin{cases} \mu^+ : \text{output}_{\pi,0} = f(x_0, x_1) \wedge \text{output}_{\pi,1} \neq f(x_0, x_1) \wedge p_0 \in E \\ \mu^+ : \text{output}_{\pi,0} = f(x_0, x_1) \wedge \text{output}_{\pi,1} = f(x_0, x_1) \wedge p_0 \notin E \\ \mu^- : \text{output}_{\pi,0} = f(x_0, x_1) \wedge \text{output}_{\pi,1} \neq f(x_0, x_1) \wedge p_0 \notin E \\ \mu^- \qquad : \text{output}_{\pi,0} \neq f(x_0, x_1) \wedge \text{output}_{\pi,1} = f(x_0, x_1) \\ \mu^0 \qquad\qquad\qquad\qquad\qquad\qquad\qquad : otherwise \end{cases}$$

We consider σ^{honest} to represent the strategy that follows the protocol specification of π. Similarly, fairness is only compromised when a party selects σ^{abort}, which deprives other players of information necessary to compute the output.

Definition 14. *Let f and π be as above. Then, π is $\mathscr{C}$-**Game-Theoretic Fair** for party p_i if $\mu_i^f(\sigma_0^{\text{honest}}, \sigma_1^{\text{honest}})$ is a $\mathscr{C}$-PBE with respect to $\mu_{i,i\in\{0,1\}}^f, \beta_{i,i\in\{0,1\}}$ and all $\mathscr{C}$-bounded adversaries $\mathcal{A}$.*

We now prove a theorem defining how protocol π may satisfy Definition 14:

Theorem 3. *Let f be a deterministic two-party function, and let π be a two-party protocol that computes f correctly. Then, π is $\mathscr{C}$-**Game-Theoretic Fair** w.r.t. p_0 (resp. p_1) iff $\forall \mathcal{H}^k$*

$$|Pr[\text{output}_{\pi,0}(\mathcal{H}^k) = f(x_0, x_1)] - Pr[\text{output}_{\pi,1}(\mathcal{H}^k) = f(x_0, x_1)]| \leq \epsilon(\lambda)$$

That is, at any round k, the strategy σ^{abort} yields a player at most a negligible advantage over other players at determining the correct function output $f(x_0, x_1)$.

Proof (Theorem 3). We first demonstrate that if π is $\mathscr{C}$-Game-Theoretic Fair w.r.t. p_0, then p_1 has a negligible advantage over p_0 at determining the correct function output $f(x_0, x_1)$ when playing strategy σ_1^{abort}.

If π is $\mathscr{C}$-Game-Theoretic Fair w.r.t. p_0, then by definition we have that:

$$\mu_0(\sigma_0^{\text{honest}}|\beta_0, \mathcal{H}_0) + \epsilon(\lambda) \geq \mu_0(\sigma_0', \sigma_0^{\text{abort}}|\beta_0, \mathcal{H}_0)$$

That is, if π is $\mathscr{C}$-Game-Theoretic Fair, then players receive more utility by playing strategy σ^{honest} than aborting and attempting to recover $f(x_0, x_1)$ on their own. Assume by contradiction that p_1 has a non-negligible advantage over p_0 at determining the correct function output $f(x_0, x_1)$ when playing strategy σ_1^{abort}. Without loss of generality, we assume $\mathcal{A}$ corrupts p_1. Then we have that

$$Pr[\text{output}_{\pi,1}(\mathcal{H}^k) = f(x_0, x_1)] > \frac{1}{2} + \epsilon(\lambda)$$

which is a non-negligible advantage. Thus, we have that:

$$\begin{aligned}
\mu_0(\sigma_0^{\text{honest}}, \sigma_1^{\text{abort}}) &= Pr[\text{output}_{\pi,1}(\mathcal{H}^k) = f(x_0, x_1)] \cdot \mu^- \\
&\quad + Pr[\text{output}_{\pi,0}(\mathcal{H}^k) = f(x_0, x_1)] \cdot \mu^+ \\
&< \mu^0 < \mu_0(\sigma_0^{\text{abort}}, \sigma_1^{\text{abort}}) = \mu^0
\end{aligned}$$

thus contradicting the assumption that σ_0^{honest} is a $\mathscr{C}$-PBE w.r.t. p_0, and that π is $\mathscr{C}$-Game-Theoretic Fair by Definition 14, as σ^{abort} yields more utility for p_0 than σ^{honest}.

Next, we show that if p_1 has at most a negligible advantage over p_0 at determining the correct function output $f(x_0, x_1)$ when playing strategy σ_1^{abort}, then π is $\mathscr{C}$-Game-Theoretic Fair. By definition, we have that

$$|Pr[\text{output}_{\pi,0}(\mathcal{H}^k) = f(x_0, x_1)] - Pr[\text{output}_{\pi,1}(\mathcal{H}^k) = f(x_0, x_1)]| \leq \epsilon(\lambda)$$

Assume by contradiction that σ_0^{honest} is not a $\mathscr{C}$-PBE w.r.t. p_0. Then, we must have that:

$$\mu_0(\sigma_0^{\text{abort}}, \sigma_1^{\text{abort}}) > \mu_0(\sigma_0^{\text{honest}}, \sigma_1^{\text{abort}})$$

Consider first the strategy pair $(\sigma_0^{\text{abort}}, \sigma_1^{\text{abort}})$:

$$\mu_0(\sigma_0^{\text{abort}}, \sigma_1^{\text{abort}}) = \mu^0 = \mu_0(\sigma_0^{\text{honest}}, \sigma_1^{\text{abort}})$$

Thus, σ_0^{honest} is a $\mathscr{C}$-PBE w.r.t. p_0, contradicting the assumption. Similarly, consider the strategy pair $(\sigma_0^{\text{abort}}, \sigma_1^{\text{honest}})$:

$$\mu_0(\sigma_0^{\text{abort}}, \sigma_1^{\text{honest}}) = \mu^0 < \mu_0(\sigma_0^{\text{honest}}, \sigma_1^{\text{honest}}) = \mu^+$$

Thus, σ_0^{honest} is a $\mathscr{C}$-PBE w.r.t. p_0, contradicting the assumption.

6   Conclusion

We have presented an expressive framework for reasoning about the security of cryptographic protocols in game theoretic terms, where all players are only assumed to be rational. We have demonstrated the ability of the perfect Bayesian equilibrium concept to model the inherent uncertainty and auxiliary information in cryptographic protocols, and translated this into the computational domain. Finally, we have provided novel definitions of privacy, correctness and fairness in game theoretic terms, and demonstrated the conditions under which they hold.

References

1. Katz, J.: Bridging game theory and cryptography: Recent results and future directions. In: Canetti, R. (ed.) TCC 2008. LNCS, vol. 4948, pp. 251–272. Springer, Heidelberg (2008)
2. Shoham, Y., Tennenholtz, M.: Non-cooperative computation: boolean functions with correctness and exclusivity. Theor. Comput. Sci. 343, 97–113 (2005)
3. Asharov, G., Canetti, R., Hazay, C.: Towards a game theoretic view of secure computation. In: Paterson, K.G. (ed.) EUROCRYPT 2011. LNCS, vol. 6632, pp. 426–445. Springer, Heidelberg (2011)
4. Dodis, Y., Halevi, S., Rabin, T.: A cryptographic solution to a game theoretic problem. In: Bellare, M. (ed.) CRYPTO 2000. LNCS, vol. 1880, pp. 112–130. Springer, Heidelberg (2000)
5. Halpern, J., Teague, V.: Rational secret sharing and multiparty computation: extended abstract. In: Proceedings of the Thirty-Sixth Annual ACM Symposium on Theory of Computing, STOC 2004, pp. 623–632. ACM, New York (2004)
6. Lysyanskaya, A., Triandopoulos, N.: Rationality and adversarial behavior in multi-party computation. In: Dwork, C. (ed.) CRYPTO 2006. LNCS, vol. 4117, pp. 180–197. Springer, Heidelberg (2006)
7. Gradwohl, R., Livne, N., Rosen, A.: Sequential rationality in cryptographic protocols. In: Proceedings of the 2010 IEEE 51st Annual Symposium on Foundations of Computer Science, FOCS 2010, pp. 623–632. IEEE Computer Society, Washington, DC (2010)
8. Halpern, J.Y.: Beyond nash equilibrium: solution concepts for the 21st century. In: Proceedings of the Twenty-Seventh ACM Symposium on Principles of Distributed Computing, PODC 2008, pp. 1–10. ACM, New York (2008)
9. Kol, G., Naor, M.: Games for exchanging information. In: Proceedings of the 40th Annual ACM Symposium on Theory of Computing, STOC 2008, pp. 423–432. ACM, New York (2008)
10. Nojoumian, M., Stinson, D.: Socio-rational secret sharing as a new direction in rational cryptography. In: Grossklags, J., Walrand, J. (eds.) GameSec 2012. LNCS, vol. 7638, pp. 18–37. Springer, Heidelberg (2012)
11. Halpern, J.Y., Pass, R.: Game theory with costly computation. In: Proceedings of the Behavioral and Quantitative Game Theory on Conference on Future Directions BQGT, vol. 10, p. 1 (2008)
12. Groce, A., Katz, J.: Fair computation with rational players. In: Pointcheval, D., Johansson, T. (eds.) EUROCRYPT 2012. LNCS, vol. 7237, pp. 81–98. Springer, Heidelberg (2012)

13. Osborne, M.J., Rubinstein, A.: A Course in Game Theory. MIT Press Books, vol. 1. The MIT Press (1994)
14. Fudenberg, D., Tirole, J.: Game Theory. MIT Press (August 1991)
15. Goldwasser, S., Micali, S.: Probabilistic encryption. J. Comput. Syst. Sci. 28, 270–299 (1984)
16. Harsanyi, J.C.: Games with incomplete information played by bayesian players, i-iii. part ii. bayesian equilibrium points. Management Science 14(5), 320–334 (1968)
17. Goldwasser, S., Micali, S., Rackoff, C.: The knowledge complexity of interactive proof systems. SIAM J. Comput. 18, 186–208 (1989)
18. Gordon, S.D., Katz, J.: Rational secret sharing, revisited. Cryptology ePrint Archive, Report 2006/142 (2006), http://eprint.iacr.org/
19. Micali, S., Shelat, A.: Purely rational secret sharing (extended abstract). In: Reingold, O. (ed.) TCC 2009. LNCS, vol. 5444, pp. 54–71. Springer, Heidelberg (2009)
20. Fuchsbauer, G., Katz, J., Naccache, D.: Efficient rational secret sharing in standard communication networks. In: Micciancio, D. (ed.) TCC 2010. LNCS, vol. 5978, pp. 419–436. Springer, Heidelberg (2010)
21. Groce, A., Katz, J., Thiruvengadam, A., Zikas, V.: Byzantine agreement with a rational adversary. In: Czumaj, A., Mehlhorn, K., Pitts, A., Wattenhofer, R. (eds.) ICALP 2012, Part II. LNCS, vol. 7392, pp. 561–572. Springer, Heidelberg (2012)
22. Zhang, Z., Liu, M.: Rational secret sharing as extensive games. Science China Information Sciences 56, 1–13 (2013)
23. Shamir, A.: How to share a secret. Commun. ACM 22(11), 612–613 (1979)
24. Kol, G., Naor, M.: Cryptography and game theory: designing protocols for exchanging information. In: Canetti, R. (ed.) TCC 2008. LNCS, vol. 4948, pp. 320–339. Springer, Heidelberg (2008)
25. Miltersen, P.B., Nielsen, J.B., Triandopoulos, N.: Privacy-enhancing auctions using rational cryptography. In: Halevi, S. (ed.) CRYPTO 2009. LNCS, vol. 5677, pp. 541–558. Springer, Heidelberg (2009)
26. Zhang, Z., Liu, M.: Unconditionally secure rational secret sharing in standard communication networks. In: Rhee, K.-H., Nyang, D. (eds.) ICISC 2010. LNCS, vol. 6829, pp. 355–369. Springer, Heidelberg (2011)
27. Gilboa, I., Zemel, E.: Nash and correlated equilibria: Some complexity considerations. Discussion Papers 777, Northwestern University, Center for Mathematical Studies in Economics and Management Science (June 1988)
28. Urbano, A., Vila, J.: Computationally restricted unmediated talk under incomplete information. Economic Theory 23, 283–320 (2004)
29. von Stengel, B., Forges, F.: Extensive-form correlated equilibrium: Definition and computational complexity. Math. Oper. Res. 33(4), 1002–1022 (2008)
30. Bonanno, G.: Agm-consistency and perfect bayesian equilibrium. part i: definition and properties. International Journal of Game Theory, 1–26 (2011)
31. González-Díaz, J., Meléndez-Jiménez, M.: On the notion of perfect bayesian equilibrium. TOP, 1–16 (2011), 10.1007/s11750-011-0239-z

Game-Theoretic Approach to Feedback-Driven Multi-stage Moving Target Defense

Quanyan Zhu[1] and Tamer Başar[2,*]

[1] Department of Electrical Engineering,
Princeton University, NJ, USA, 08544
quanyanz@princeton.edu
[2] Coordinated Science Laboratory and
Department of Electrical and Computer Engineering,
University of Illinois at Urbana Champaign,
1308 W. Main St., Urbana, IL, USA, 61801
basar1@illinois.edu

Abstract. The static nature of computer networks allows malicious attackers to easily gather useful information about the network using network scanning and packet sniffing. The employment of secure perimeter firewalls and intrusion detection systems cannot fully protect the network from sophisticated attacks. As an alternative to the expensive and imperfect detection of attacks, it is possible to improve network security by manipulating the *attack surface* of the network in order to create a *moving target defense*. In this paper, we introduce a proactive defense scheme that dynamically alters the attack surface of the network to make it difficult for attackers to gather system information by increasing complexity and reducing its signatures. We use concepts from systems and control literature to design an optimal and efficient multi-stage defense mechanism based on a feedback information structure. The change of attack surface involves a reconfiguration cost and a utility gain resulting from risk reduction. We use information- and control-theoretic tools to provide closed-form optimal randomization strategies. The results are corroborated by a case study and several numerical examples.

1 Introduction

The static nature of computing systems facilitates an attacker's capability of gathering information and executing attacks. Given sufficient amount of time, an attacker can map out the system, gain access to a node and spread to other hosts and services within the system [1]. Although heavily secured perimeter firewalls and intrusion detection systems are deployed to protect the network from outside attackers, in practice they are not effective for zero-day vulnerabilities or virus, and can be avoided by skilled attackers. In addition, while the attackers have only

* This research was partially supported by an NSERC Postdoctoral Fellowship (PDF) and partially by AFOSR MURI Grant FA9550-10-1-0573, and an NSA grant through the Information Trust Institute of the University of Illinois.

S.K. Das, C. Nita-Rotaru, and M. Kantarcioglu (Eds.): GameSec 2013, LNCS 8252, pp. 246–263, 2013.
© Springer International Publishing Switzerland 2013

to exploit one vulnerability to be successful, a firewall has to process millions of packets every minute and perform a sophisticated and timely analysis in order to detect software that exploits a previously unknown attack vector. Clearly, an attacker has an advantage over the defender, and the sole reliance on these technologies is not sufficient for assuring security.

As an alternative to the insufficient and expensive detection of attackers, the network security can be improved by changing the appearance of the system and creating a *moving target*. The availability of services will be time-varying under different system configurations, and the system can block dangerous network behaviors if an attacker does not follow the network dynamics. In addition, in order for an attack to succeed, an attacker has to spend a significant amount of resources to carefully guide his attacks.

In this paper, we introduce stochastic dynamics into multiple layers of computing systems for securing the system by shifting its attack surface. One challenge of achieving moving target defense is to understand the tradeoff between security and usability. A complete security could be achieved by frequently changing the network and thus making it completely unusable. Hence it is essential to take into account the reconfiguration cost of shifting the surface, and the attacker's cost for learning and changing its attack vector. Moreover, the interactions between an attacker and a defender can be seen as a game [14, 19] in which the system creates a moving target for minimizing its risk and maintaining its usability, while an attacker dynamically explores and exploits a vulnerability for causing maximum damage on the system.

With this motivation, we formulate a two-person zero-sum game to model the conflict of goals, and develop a feedback learning framework to implement a moving target defense based on real-time data and observations made by the system. To achieve this goal, we first decompose the system into multiple layers as it is often composed of multiple network zones, such as those in enterprise IT network [2] and cyber-physical networks [3]. An attacker has to launch a multi-stage attack starting from network scanning and packet sniffing to illegitimate authentication and service interruption. The infamous Stuxnet virus, for example, follows a sequence of attacks depicted in Fig. 1 before compromising the supervisory control and data acquisition (SCADA) system that controls and monitors uranium enrichment infrastructure [4, 5]. We build a multi-layer game model to capture the fact that the attack is carried out through multiple stages, and the defense mechanism is developed at each layer of the system. Built into the game model, the notion of *attack surface* is defined as the set of vulnerabilities exhibited by the system that can potentially be exploited by the attacker. The essential goal of moving target strategies is to find an optimal configuration policy for the defender to shift the attack surface that minimizes its risk and the damage inflicted by an attacker.

A natural solution for the zero-sum game is mixed strategy saddle-point equilibrium (SPE), i.e., the system defender randomizes its configuration and the associated attack surface, while an attacker randomizes over the set of vulnerabilities that he can exploit. The SPE mixed-strategy pair naturally leads to a

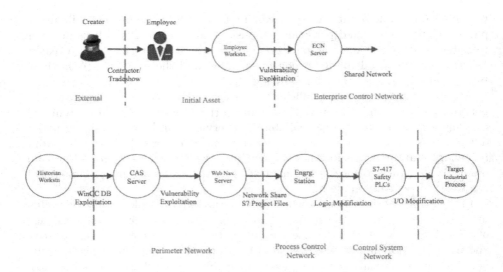

Fig. 1. Illustration of a sequence of attacks of Stuxnet: It is spread first using infected USB drive passed onto the employee. The virus exploits vulnerabilities of printer servers, WinCC database, service servers, and Siemens S7 project files at multiple stages, and propagates from employee workstation to the control system network through intermediate networks.

way of implementing moving target defense. However, it is an equilibrium solution concept which describes the steady-state outcome of a game of complete information after a sufficiently large number of repeated plays. Due to the uncertainties in the environment and limited knowledge of the players' own risk functions, it is pivotal for the system to implement a dynamic real-time defense mechanism driven by the data observed by the players. Hence we develop a feedback multi-stage defense strategy that enables the system to update its mixed strategy based on the risk it estimates online. In this case, the moving target defense is adaptive to the exogenous environment and less vulnerable than a static equilibrium strategy, which could be known to a resourceful attacker.

The contribution of this work can be summarized as follows.

(i) We formally develop a metric for quantifying attack surface, and establish a game-theoretic model for providing formal analysis of security strategies and guiding the design of moving target defense.
(ii) We develop a feedback system framework for strategically shifting attack surface based on observation.
(iii) We introduce reconfiguration and learning cost for the defender and attacker, respectively, and analyze the joint dynamics of strategy update and risk estimation.

The paper is organized as follows. Section 2 presents related works. Section 3 describes the two-person zero-sum game model and Section 4 provides a system framework for moving target defense based on learning mechanism of the game. In Section 5, we analyze the learning dynamics presented in Section 5. Section 6 illustrates the defense mechanism using numerical examples, and we conclude in Section 7.

2   Related Work

Moving target defense (MTD) is a broad class of proactive defense mechanisms, in which the defending system creates security strategies that change over time to limit the exposure of vulnerabilities and increase complexity and costs for attacks [7] and [8]. Many techniques have recently been developed for achieving MTD, including instruction set, and address space layout randomization [12], deceptive routing [23], and software diversity [13]. However, very few work has studied quantitative tradeoffs of MTD for guiding the design and analysis of the defense mechanism. In this work, we develop a feedback learning mechanism for designing MTD based on a game-theoretic framework.

Our work is related to the following existing literature. In [6], formal methods have been used to provide an attack surface metric as an indicator of the system's security. It develops an approach to reduce attack surface, which complements the software industry's traditional code quality improvement approach for security risk mitigation. Motivated by the concept, we define the notion of attack surface based on the set of existing vulnerabilities, which can be conveniently incorporated into the MTD game-theoretic model.

As depicted in Fig. 1, attackers often launch a sequence of attacks which exploits vulnerabilities at multiple stages of the system. The goal of dividing the system into multiple layers is to capture this fact. In addition, this approach is well-aligned with the research on attack graphs for assessing the cause-consequence relationships between various network states [9,10]. Based on attack graphs and trees, the system can be divided into logical layers, which correspond to a set of nodes of the same depth on a tree.

Game theory provides an appropriate theoretical framework for modeling the win-lose situation between a defender and an attacker [14]. The notion of mixed-strategy equilibrium is a natural solution concept for many applications of MTD. Our work is related to some recent works that apply game theory to MTD. In [21], a game-theoretic framework has been used to study a defense mechanism that strategically randomizes over a set of cryptographic keys that authenticate the commands from a system operator to PLC of a power system network. In [22], deceptive routing game is used to design defense strategies that mislead a jammer from attacking legitimate routes by deploying fake routes in wireless communication networks.

The feedback MTD defense mechanism is related to learning algorithms for games. In particular, the joint learning dynamics are related to the distributed learning algorithm described in [15–17], which have studied a class of distributed payoff and strategy learning for games of incomplete information. Different from best response dynamics and fictitious play algorithms [11], the strategy updates do not require players' knowledge of their own payoff functions and the observations of actions played by others. Strategy update in the MTD defense is related to a class of imitative learning dynamics in which a player imitates the strategy of other players [24,25]. Here, we develop imitative strategies for MTD through a cost on shifting attack surface for the defender and a cost for learning system vulnerabilities and changing attack vectors for the attacker.

3   System Model

In this section, we introduce a game-theoretic framework for moving target defense between an attacker and a system defender. Many networked computing systems nowadays can be decomposed into hierarchical layers, and there are defense mechanisms residing on each of these layers. An attacker has to launch a multi-stage attack, which exploits vulnerabilities of the system at different layers, in order to compromise its final target. Hence we partition the system to be defended into a finite number of layers, and let $l = 1, 2, \cdots, N$ be the index of system layers, and $\mathcal{V}_l := \{v_{l,1}, v_{l,2}, \cdots, v_{l,n_l}\}$ be the set of n_l existing system vulnerabilities at layer l. We assume that $\mathcal{V}_l$ is common knowledge to the attacker and the system. A vulnerability $v_{l,i}$ is a weakness of the system that an adversary can exploit and launch an attack to compromise the system. The system at layer l can be configured in different ways, and each configuration exhibits its own set of vulnerabilities. Let $\mathcal{C}_l : \{c_{l,1}, c_{l,2}, \cdots, c_{l,m_l}\}$ be the set of feasible configurations of the system at layer l, and $\pi_l : \mathcal{C}_i \to 2^{\mathcal{V}_l}$ be the vulnerability map which associates with each configuration a subset of vulnerabilities. We call $\pi_l(c_{l,j})$ the *attack surface* at stage l when the system is configured to $c_{l,j}$.

At each stage l, an attacker chooses a vulnerability in the set $\mathcal{V}_l$ to exploit and launch an attack $a_{l,j}$. Let $\mathcal{A}_l := \{a_{l,1}, a_{l,2}, \cdots, a_{l,n_l}\}$ be the set of n_l attacks at stage l, and $\gamma_l : \mathcal{V}_l \to \mathcal{A}_l$ be the attack map which associates vulnerability $v_{l,j} \in \mathcal{V}_l$ with attack $a_{l,j} \in \mathcal{A}_l$. The corresponding inverse map is denoted by $\gamma_l^{-1} : \mathcal{A}_l \to \mathcal{V}_l$. Here, without the loss of generality, we assume that there is a one-to-one correspondence between $\mathcal{A}_l$ and $\mathcal{V}_l$. An attack $a_{l,j}$ can successfully cause damage on the system at layer l if the exploited vulnerability $v_{l,j}$ resides in the configuration $c_{l,i}$, i.e., $v_{l,j} \in \pi_l(c_{l,i})$.

Denote by $r_l : \mathcal{A}_l \times \mathcal{C}_l \to \mathbb{R}_+$ the damage or cost caused by the attacker at stage l, given by

$$r_l(a_{l,j}, c_{l,i}) = \begin{cases} D_{ij}, & \gamma_l^{-1}(a_{l,j}) \in \pi_l(c_{l,j}) \\ 0, & \text{otherwise} \end{cases}, \tag{1}$$

where $D_{ij} \in \mathbb{R}_+$ is the (bounded) damage or risk quantified in terms of monetary values.

The goal of the attacker is to penetrate and compromise the system while the system aims to choose configurations that minimize the damage or risk. Hence we use a two-person zero-sum game to model this conflict between an attacker A and a defender S. Let Ξ_l be the game at stage l described by the triplet

$\langle\{A, S\}, \{\mathcal{A}_l, \mathcal{C}_l\}, \{r_l\}\rangle$. Since vulnerabilities are inevitable in modern computing systems, one approach for the system is to adopt moving target defense which randomizes between different system configurations, making it difficult for the attacker to learn and locate the system vulnerabilities to exploit. This naturally leads to the mixed strategy equilibrium solution concept of the game, where the defender chooses a randomized strategy $\mathbf{f}_l := (f_{l,1}, f_{l,2}, \cdots, f_{l,m_l})$ over the set $\mathcal{C}_l$, and the attacker at layer l chooses a randomized strategy $\mathbf{g}_l := (g_{l,1}, g_{l,2}, \cdots, g_{l,n_l})$ over the set $\mathcal{A}_l$, i.e.,

$$\mathbf{f}_l \in \mathcal{F}_l := \left\{ \mathbf{f}_l \in \mathbb{R}_+^{m_l} : \sum_{h=1}^{m_l} f_{l,h} = 1 \right\}, \quad \mathbf{g}_l \in \mathcal{G}_l := \left\{ \mathbf{g}_l \in \mathbb{R}_+^{n_l} : \sum_{h=1}^{n_l} g_{l,h} = 1 \right\}.$$

The game Ξ_l is a finite zero-sum matrix game with a bounded cost function. Hence there exists a mixed strategy saddle-point equilibrium (SPE) $(\mathbf{f}_l^*, \mathbf{g}_l^*)$, which satisfies the following inequality for all $\mathbf{f}_l \in \mathcal{F}_l$ and $\mathbf{g}_l \in \mathcal{G}_l$,

$$r_l(\mathbf{f}_l^*, \mathbf{g}_l) \le r_l(\mathbf{f}_l^*, \mathbf{g}_l^*) \le r_l(\mathbf{f}_l, \mathbf{g}_l^*), \tag{2}$$

where r_l is the expected cost given by

$$r_l(\mathbf{f}_l, \mathbf{g}_l) := \mathbb{E}_{\mathbf{f}_l, \mathbf{g}_l} r_l = \sum_{k=1}^{n_l} \sum_{h=1}^{m_l} f_{l,h} g_{l,k} r_l(a_{l,k}, c_{l,h}).$$

The expect cost r_l at SPE $(\mathbf{f}_l^*, \mathbf{g}_l^*)$ is the value of the game Ξ_l, denoted as $\mathrm{val}(\Xi_l) = \hat{r}_l(\mathbf{f}_l^*, \mathbf{g}_l^*)$ and it is unique for zero-sum games under mixed strategies. The static analysis of the game provides an insight into the system performance (value of the game) and its strategy against an attacker if the zero-sum game is played repeatedly.

Solving the game using (2) requires that each player have complete information of the game, including the knowledge of the cost function (1) and the strategy spaces of the players. In practice, the information available to the system can be limited and sometimes uncertain. Hence the players need to learn information online and adapt their defense strategies.

Example 1. *In Fig. 2, we illustrate the multi-stage moving target defense game. At each layer l, the system has a set of vulnerabilities: $\mathcal{V}_l = \{v_{l,1}, v_{l,2}, v_{l,3}\}$, $l = 1, 2, 3, 4$. At layer 1, two configurations are feasible, i.e., $\mathcal{C}_1 = \{c_{1,1}, c_{1,2}\}$. A configuration $c_{1,1}$ is chosen in Fig. 2 and it has an attack surface $\pi_1(c_{1,1}) = \{v_{1,1}, v_{1,2}\}$. In Fig. 3, a configuration $c_{1,2}$ is chosen and it has a corresponding attack surface $\pi_1(c_{1,2}) = \{v_{1,2}, v_{1,3}\}$. Likewise, in both Fig. 2 and Fig. 3, $\pi_2(c_{2,1}) = \{v_{2,1}, v_{2,2}\}$, $\pi_3(c_{3,1}) = \{v_{3,1}, v_{3,2}\}$, and $\pi_4(c_{4,2}) = \{v_{4,2}, v_{4,3}\}$. The static system configuration $\{c_{1,1}, c_{2,1}, c_{3,1}, c_{4,2}\}$ depicted in Fig. 2 allows an attacker to launch a sequence of attacks which exploit $v_{1,1} \to v_{2,2} \to v_{2,3} \to v_{4,3}$ if sufficient amount of time and resources are given to the attacker. Mixed strategies provide a mechanism for the system defender to randomize between different configurations so that the attack surface shifts at each layer of the system. Fig. 3 depicts a scenario where the system changes its configuration to $\{c_{1,2}, c_{2,1}, c_{3,1}, c_{4,2}\}$. Then, the original attack sequence will not succeed.*

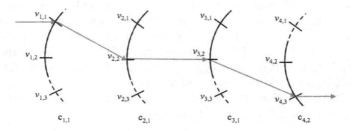

Fig. 2. A static configuration of attack surface that leads to a sequence of attacks on the physical system: An attacker can succeed in targeting the resources at the last layer by exploiting vulnerabilities $v_{1,1} \to v_{2,2}, \to v_{3,2} \to v_{4,2}$. Solid curves describe an attack surface containing existing vulnerabilities; dotted curves describe vulnerabilities circumvented by the current configuration; solid arrows refer to successful attacks.

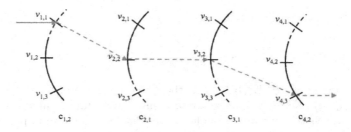

Fig. 3. Randomized configuration of attack surface that protects the system from attack exploitations at the first step. The attack is thwarted at layer 1. Solid curves describe an attack surface containing existing vulnerabilities; dotted curves describe vulnerabilities circumvented by the current configuration; dotted arrows refer to unsuccessful attacks.

Remark 1. *The attack surface at stage l for a given configuration $c_{l,h}$ is measured by $\pi_l(c_{l,h})$. We can further measure the level of vulnerability given $\mathbf{f}_l, \mathbf{g}_l$. Denote by $\eta_l = (\eta_{l,1}, \eta_{l,2}, \cdots, \eta_{l,n_l}) \in \mathcal{H}_l$ the mixed strategies for defending the set of vulnerabilities $\mathcal{V}_l$, where*

$$\mathcal{H}_l := \left\{ \eta_l \in \mathbb{R}^{n_l}_+ : \sum_{h=1}^{n_l} \eta_{l,h} = 1 \right\}.$$

Suppose that vulnerabilities on the attack surface are defended with equal probabilities. Then the mixed strategy $\mathbf{f}_l$ on $\mathcal{C}_l$ leads to the following mixed strategy η_l on $\mathcal{V}_l$:

$$\eta_{l,h} = \sum_{k \in \mathcal{N}_h} \frac{f_{l,k}}{|\pi_l(c_{l,k})|}, \quad h = 1, 2, \cdots, n_l, \tag{3}$$

where $\mathcal{N}_h = \{h' \in \mathbb{Z}_+ : v_{l,h} \in \pi_l(c_{l,h'}), c_{l,h'} \in \mathcal{C}_l\}$. *The level of vulnerability ψ at layer l can be quantified by the Kullback-Leibler (K-L) divergence between two distributions η_l, $\mathbf{g}_l$, i.e.,*

$$\psi_l(\mathbf{f}_l, \mathbf{g}_l) := d_{KL}(\eta_l \| \mathbf{g}_l) = \sum_{h=1}^{n_l} \eta_{l,h} \ln\left(\frac{\eta_{l,h}}{g_{l,h}}\right). \tag{4}$$

where d_{KL} is the K-L divergence, and $\eta_{l,h}$ can be obtained from $\mathbf{f}$ using (3). Following Example 1, we let $\mathbf{g}_1 = (\frac{1}{3}, \frac{1}{3}, \frac{1}{3})$ and $\mathbf{f}_1 = (\frac{1}{2}, \frac{1}{2})$, which randomizes between two configurations $c_{1,1}$ and $c_{1,2}$. We obtain $\eta_1 = (\frac{1}{4}, \frac{1}{2}, \frac{1}{4})$, which leads to $\psi = \ln\left(\frac{3\sqrt{2}}{4}\right)$. It is clear that the system is less vulnerable under moving target defense when ψ is large.

4   Moving Target Defense

Section 3 describes an ideal game model of complete information. It is common to see that the payoff function r_l can be subject to noise and disturbance, and the system/attacker cannot know each other's action spaces. This information needs to be learned over time and often there will be a cost associated with learning and adaptation. In this section, we introduce an adaptive moving target defense framework based on the system model in Section 3 in which the system dynamically updates its defense strategy through learning in an uncertain environment and without complete information. We use subscript t here to denote the strategy or cost at time t. At time t, each player independently chooses actions $c_{l,t} \in \mathcal{C}_l$ and $a_{l,t} \in \mathcal{A}_l$ according to strategies $\mathbf{f}_{l,t}$ and $\mathbf{g}_{l,t}$, respectively. The players cannot observe the action played by the other player but can observe the cost $r_{l,t}$ as an outcome of action pair $(c_{l,t}, a_{l,t})$ at time t. Based on the observed cost, the system and the attacker can estimate the average risk of the system $\hat{r}_{l,t}^S : \mathcal{C}_l \rightarrow \mathbb{R}_+$ and $\hat{r}_{l,t}^A : \mathcal{V}_l \rightarrow \mathbb{R}_+$, respectively, as follows:

$$\hat{r}_{l,t+1}^S(c_{l,h}) = \hat{r}_{l,t}^S(c_{l,h}) + \mu_t^S \mathbb{1}_{\{c_{l,t}=c_{l,h}\}}(r_{l,t} - \hat{r}_{l,t}^S(c_{l,h})), \tag{5}$$

$$\hat{r}_{l,t+1}^A(a_{l,h}) = \hat{r}_{l,t}^A(a_{l,h}) + \mu_t^A \mathbb{1}_{\{a_{l,t}=a_{l,h}\}}(r_{l,t} - \hat{r}_{l,t}^A(a_{l,h})). \tag{6}$$

In (5) and (6), μ_t^S and μ_t^A are payoff learning rates for the system and the attacker. Note that in the learning schemes above, the players do not know the actions played by the other players and each one estimates the average cost based on choices made in the past. The two updates are made in a distributed fashion; however, they are coupled through the fact that the observed cost depends on actions played by both players at time t.

The players can make use of the cost function learned online for updating the moving target defense strategies. The change of defense strategies from $\mathbf{f}_{l,t}$ to $\mathbf{f}_{l,t+1}$ involves a cost for the system to reconfigure by maneuvering its defense resources and altering its attack surface from $\pi_l(c_{l,t})$ to $\pi_l(c_{l,t+1})$, where $c_{l,t}$ and $c_{l,t+1}$ are selected according to distributions $\mathbf{f}_{l,t}$ and $\mathbf{f}_{l,t+1}$, respectively.

Hence we introduce the following switching cost for the system as the relative entropy between two strategies:

$$R_{l,t}^S = \epsilon_{l,t}^S \sum_{h=1}^{m_l} f_{l,h,t+1} \ln \left(\frac{f_{l,h,t+1}}{f_{l,h,t}} \right), \tag{7}$$

where $\epsilon_{l,t}^S > 0$. This cost is added onto the expected cost given by $\langle \mathbf{f}_{l,t+1}, \hat{\mathbf{r}}_{l,t}^S \rangle$, where $\langle \cdot, \cdot \rangle$ denotes the inner product between two vectors of appropriate dimensions, and $\hat{\mathbf{r}}_{l,t}^S = [\hat{r}_{l,t}^S(c_{l,1}), \hat{r}_{l,t}^S(c_{l,2}), \cdots, \hat{r}_{l,t}^S(c_{l,m_l})]'$. Hence at time t, with the learned cost vector $\hat{\mathbf{r}}_{l,t}^S$, the system solves the following system problem

$$\text{(SP)} \quad \sup_{\mathbf{f}_{l,t+1} \in \mathcal{F}_l} \langle \mathbf{f}_{l,t+1}, -\hat{\mathbf{r}}_{l,t}^S \rangle - \epsilon_{l,t}^S \sum_{h=1}^{m_l} f_{l,h,t+1} \ln \left(\frac{f_{l,h,t+1}}{f_{l,h,t}} \right). \tag{8}$$

Likewise, it takes an attacker resources (in terms of time and energy) to explore new vulnerabilities and exploit them. Hence we introduce a similar cost that capture the learning cost for the attacker

$$R_{l,t}^A = \epsilon_{l,t}^A \sum_{h=1}^{n_l} g_{l,h,t+1} \ln \left(\frac{g_{l,h,t+1}}{g_{l,h,t}} \right), \tag{9}$$

where $\epsilon_{l,t}^A > 0$. A similar problem for the attacker is

$$\text{(AP)} \quad \sup_{\mathbf{g}_{l,t+1} \in \mathcal{G}_l} \langle \mathbf{g}_{l,t+1}, \hat{\mathbf{r}}_{l,t}^A \rangle - \epsilon_{l,t}^A \sum_{h=1}^{n_l} g_{l,h,t+1} \ln \left(\frac{g_{l,h,t+1}}{g_{l,h,t}} \right). \tag{10}$$

Theorem 1. *The following statements hold for (SP) and (AP):*

(i) *The following strategies $f_{l,h,t+1}$ and $g_{l,h,t+1}$ are optimal for (SP) and (AP), respectively.*

$$f_{l,h,t+1} = \frac{f_{l,h,t} e^{-\frac{\hat{r}_{l,t}(c_{l,h})}{\epsilon_{l,t}^S}}}{\sum_{h'=1}^{m_l} f_{l,h',t} e^{-\frac{\hat{r}_{l,t}(c_{l,h'})}{\epsilon_{l,t}^S}}}, \tag{11}$$

$$g_{l,h,t+1} = \frac{g_{l,h,t} e^{\frac{\hat{r}_{l,t}(a_{l,h})}{\epsilon_{l,t}^A}}}{\sum_{h'=1}^{n_l} g_{l,h',t} e^{\frac{\hat{r}_{l,t}(a_{l,h'})}{\epsilon_{l,t}^A}}}. \tag{12}$$

(ii) *The optimal values achieved at (11) and (12) for (SP) and (AP) are given by*

$$W_{l,t}^S = \epsilon_{l,t}^S \ln \left(\sum_{h=1}^{m_l} f_{l,h,t} e^{\frac{-\hat{r}_{l,h,t}}{\epsilon_{l,t}^S}} \right), \tag{13}$$

$$W_{l,t}^A = \epsilon_{l,t}^A \ln \left(\sum_{h=1}^{n_l} g_{l,h,t} e^{\frac{\hat{r}_{l,h,t}}{\epsilon_{l,t}^A}} \right). \tag{14}$$

Sketch of Proof. The results are obtained by directly solving two concave constrained optimization problems. A complete proof can be found in [26]. ☐

The parameters $\epsilon^S_{l,t}$ and $\epsilon^A_{l,t}$ model the switching and learning costs for the system and the attacker respectively. The parameter values are also related to the degree of rationality of the players associated with the zero-sum game Ξ_l. When $\epsilon^S_{l,t}, \epsilon^A_{l,t}$ are close to zero, the players tend to be highly rational, whereas they are irrational when the parameters go to infinity. This observation is summarized in the following theorem.

Theorem 2. *The following statements hold for (SP) and (AP):*

(i) The players are of high rationality when $\epsilon^S_{l,t}$ and $\epsilon^A_{l,t}$ approach 0, i.e.,

$$\lim_{\epsilon^S_{l,t} \to 0} W^S_{l,t} = \min_{c_{l,h} \in \mathcal{N}_l} \hat{r}^S_{l,h,t} \tag{15}$$

$$\lim_{\epsilon^A_{l,t} \to 0} W^A_{l,t} = \max_{a_{l,h} \in \mathcal{A}_l} \hat{r}^A_{l,h,t} \tag{16}$$

(ii) The players are of low rationality when when $\epsilon^S_{l,t}$ and $\epsilon^A_{l,t}$ approach $+\infty$, i.e.,

$$\lim_{\epsilon^S_{l,t} \to \infty} W^S_{l,t} = \langle \mathbf{f}_{l,t}, -\hat{\mathbf{r}}^S_{l,t} \rangle \tag{17}$$

$$\lim_{\epsilon^A_{l,t} \to \infty} W^A_{l,t} = \langle \mathbf{g}_{l,t}, \hat{\mathbf{r}}^A_{l,t} \rangle \tag{18}$$

Sketch of Proof. The results follow directly from taking limits of the closed form solution of $W^S_{l,t}$ and $W^A_{l,t}$ in Theorem 1(ii). A complete proof can be found in [26]. ☐

The optimization problems (SP) and (AP) provide a mechanism for players to update their mixed strategies at time $t + 1$ based on the information learned at time t. The defender will choose a new configuration $l,t+1$ according to $\mathbf{f}_{l,t+1}$ to alter the attack surface from $\pi_l(\mathbf{c}_{l,t})$ to $\pi_l(\mathbf{c}_{l,t+1})$.

Fig. 4 summarizes the feedback-driven moving target defense from the defender's perspective. The system updates the mixed strategies for moving target defense by solving (SP) based on the risked learned online using (5). The defender reconfigures the system according to (11) and shifts the attack surface to minimize the damage and risk of the system. An intelligent attacker on the other hand can also follow the same procedure to explore and exploit existing vulnerabilities of the system.

Remark 2. *Note that there is a clear relation between the defense mechanism depicted in Fig. 4 and feedback control systems. The risk learning can be seen as a sensor which measures outputs of the system and estimates the system state. The "Shift Attack Surface" block in the diagram can be regarded as an actuator which sends input to command and control the system. The design of moving target defense is analogous to a feedback controller design. Acknowledging this connection allows to apply control-theoretic tools to analyze the system.*

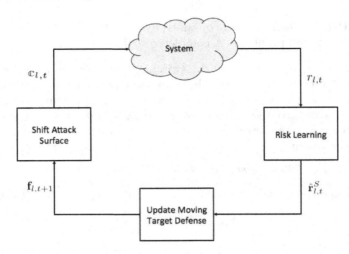

Fig. 4. System framework of the moving target defense: The defender learns online the risk of the system and updates the mixed strategy for moving target defense. The system shifts its attack surface according to the updated defense strategy.

Remark 3. *Compared to existing moving target defense which adopts a static distribution to randomize the attack surface, the defense mechanism described in Fig. 4 provides another layer of defense by changing the mixed strategy (or distribution) over time according to the environment. In this way, it creates more complexity and higher cost for an attacker to learn and gain system information.*

5   Learning Dynamics

In this section, we analyze the feedback moving target defense mechanism described in Section 4. The dynamics for mixed strategy update using (11) and (12) can be generalized by taking a convex combination of (11), (12) and the previous mixed strategy. They are described by

$$f_{l,h,t+1} = (1 - \lambda_{l,t}^S)f_{l,h,t} + \lambda_{l,t}^S \left(\frac{f_{l,h,t}e^{-\frac{\hat{r}_{l,t}(c_{l,h})}{\epsilon_{l,t}^S}}}{\sum_{h'=1}^{m_l} f_{l,h',t}e^{-\frac{\hat{r}_{l,t}(c_{l,h'})}{\epsilon_{l,t}^S}}} \right), \tag{19}$$

$$g_{l,h,t+1} = (1 - \lambda_{l,t}^A)f_{l,h,t} + \lambda_{l,t}^A \left(\frac{g_{l,h,t}e^{\frac{\hat{r}_{l,t}(a_{l,h})}{\epsilon_{l,t}^A}}}{\sum_{h'=1}^{n_l} g_{l,h',t}e^{\frac{\hat{r}_{l,t}(a_{l,h'})}{\epsilon_{l,t}^A}}} \right), \tag{20}$$

where $\lambda_{l,t}^S, \lambda_{l,t}^A \in [0,1]$ are learning rates. If $\lambda_{l,t}^S = \lambda_{l,t}^A = 1$, the dynamics correspond to the update procedure in Section 4. Note that the dynamics (19) and (20) are coupled with cost learning in (5) and (6).

The convergence of the coupled dynamics can be studied by its corresponding continuous-time dynamics. Let $e_{c_{l,h}} \in \mathcal{F}_l, e_{a_{l,h}} \in \mathcal{G}_l$ be a vector of proper dimension with $h-$th entry being 1 and others being 0, and all the variables as a continuous function of time. We let the learning rates satisfy the following conditions:

$$\sum_{t \geq 1} \lambda_{l,t}^S = +\infty, \quad \sum_{t \geq 1} (\lambda_{l,t}^S)^2 < +\infty, \quad \sum_{t \geq 1} \lambda_{l,t}^A = +\infty, \quad \sum_{t \geq 1} (\lambda_{l,t}^A)^2 < +\infty;$$

$$\sum_{t \geq 1} \mu_{l,t}^S = +\infty, \quad \sum_{t \geq 1} (\mu_{l,t}^S)^2 < +\infty, \quad \sum_{t \geq 1} \mu_{l,t}^A = +\infty, \quad \sum_{t \geq 1} (\mu_{l,t}^A)^2 < +\infty.$$

Theorem 3. *The joint cost and strategy learning algorithms (5) and (19), (6) and (20) converge to the following set of ordinary differential equations (ODEs).*

(i) System defender's dynamics

$$
\begin{cases}
\dfrac{d}{dt} f_{l,h,t} = f_{l,h,t} \left(\dfrac{e^{\frac{\hat{r}_{l,t}(c_{l,h})}{\epsilon_{l,t}^S}}}{\displaystyle\sum_{h'=1}^{m_l} f_{l,h',t} e^{\frac{\hat{r}_{l,t}(c_{l,h'})}{\epsilon_{l,t}^S}}} - 1 \right), \quad h = 1, 2, \cdots, m_l, \\[4ex]
\dfrac{d}{dt} \hat{r}_{l,t}^S(c_{l,h}) = -\mathbf{r}_{l,t}(e_{c_{l,h}}, \mathbf{g}_{l,t}) - \hat{r}_{l,t+1}^S(c_{l,h}), \quad c_{l,h} \in \mathcal{C}_l
\end{cases}
\tag{21}
$$

(ii) Attacker's dynamics

$$
\begin{cases}
\dfrac{d}{dt} g_{l,h,t} = g_{l,h,t} \left(\dfrac{e^{\frac{\hat{r}_{l,t}(a_{l,h})}{\epsilon_{l,t}^A}}}{\displaystyle\sum_{h'=1}^{n_l} g_{l,h',t} e^{\frac{\hat{r}_{l,t}(a_{l,h'})}{\epsilon_{l,t}^A}}} - 1 \right), \quad h = 1, 2, \cdots, n_l, \\[4ex]
\dfrac{d}{dt} \hat{r}_{l,t}^A(a_{l,h}) = \mathbf{r}_{l,t}(\mathbf{f}_{l,t}, e_{a_{l,h}}) - \hat{r}_{l,t+1}^A(a_{l,h}), \quad a_{l,h} \in \mathcal{A}_l
\end{cases}
\tag{22}
$$

Sketch of Proof. Under the assumptions of the learning rates, the results can be shown using stochastic approximation techniques, [27]. □

Theorem 4. *Let $\epsilon_{l,t}^S = \epsilon_{l,t}^A = \epsilon_l$ for all t. The following statements hold.*

(i) SPE of the game Ξ_l are steady states of the dynamics (21) and (22).
(ii) The interior stationary points of the dynamics are SPE of the game Ξ_l.

Sketch of Proof. The results follow from the properties of imitative Boltzmann-Gibbs dynamics ([25] and the references therein). A complete proof can be found in [26]. □

Remark 4. *The adaptive moving target defense illustrated in Fig. 4 can be employed at each layer of the system independently. This is reasonable if attackers can explore and exploit vulnerability at each layer of the system simultaneously. For the case where the attacker has to launch a stage-by-stage attack, the game-theoretic model in Section 3 can be extended to a stochastic dynamic game framework to capture the fact that the outcome of the game at layer l leads to the game at layer l + 1. In the stochastic game model, the transition probabilities can be taken to be attacker's success probabilities at each stage, and the systematic risk can be taken to be the aggregate potential damage on the system across all the layers.*

6   Numerical Example

In this section, we illustrate the feedback-driven moving target defense within the context of Example 1. We let $\mathcal{V}_1 = \{v_{1,1}, v_{1,2}, v_{1,3}\}$ be the set of three vulnerabilities at layer 1. They lead to a low (L), medium (M), and high (H) level of damage on the system, respectively, if it is compromised. The attacker can choose to exploit each of these vulnerabilities and launch an attack. The set of attacks is given by $\mathcal{A}_1 = \{a_{1,1}, a_{1,2}, a_{1,3}\}$ with attack $a_{l,h}$ corresponding to vulnerability $v_{l,h}$, $h = 1, 2, 3$. The success of an attack $a_{1,h}$ will result costs of 1p.u., 2p.u., 3p.u. of damage, respectively. Hence the game matrix can be written as follows:

$$\Xi_1 : \quad \begin{array}{c|c|c|c} & a_{1,1} & a_{1,2} & a_{1,3} \\ \hline c_{1,1} & 1 & 2 & 0 \\ \hline c_{1,2} & 0 & 2 & 3 \end{array}$$

The column player is the attacker (minimizer) and the row player is the defender (maximizer). The game has a pure Nash equilibrium where the attacker chooses $a_{1,2}$ while the defender chooses $c_{1,2}$, which results in the value 2 for the game. We set $\epsilon_{l,t}^S = \epsilon_{l,t}^A = \frac{1}{30}$. Fig. 5 and Fig. 6 illustrate the strategy and cost update dynamics (21) for the system defender. Fig. 7 and Fig. 8 illustrate the strategy and cost update dynamics (22) for an intelligent attacker. We see that the dynamics converge to a Nash equilibrium of the matrix game Ξ_1.

The numerical examples above have illustrated the convergence properties of the learning algorithm when the system and the attacker both adopt the same learning mechanism. In practice, two players can follow different dynamics, and the observations made by the system are noisy. Hence, the equilibrium of the game may not be attained as in Figs. 5 and 7. The distributed nature of the feedback system in Section 4 provides nevertheless a convenient framework for a defender to respond to its own observations. We let the mixed strategy of the attacker $\mathbf{g}_t$ be an i.i.d. random process, where $\mathbf{g}_t$ is uniformly chosen from $\mathcal{G}_1$. Assume that the payoff matrix Ξ_1 is subject to an additive noise v_1, which is uniformly distributed on $[0, 1/2]$. In Fig. 9, we show for different values of ϵ, the evolution of mixed strategy generated by (19) of the system when the attacker behaves randomly, while the system optimally switches between two configurations. When ϵ is the large, it is more costly for the system to change its

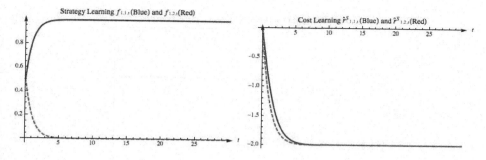

Fig. 5. System's strategy learning: Continuous-time evolution of defender's mixed-strategies $f_{1,1,t}$ and $f_{1,2,t}$

Fig. 6. System's payoff learning: Continuous-time evolution of defender's estimated cost $\hat{r}^S_{1,1,t}$ and $\hat{r}^S_{1,2,t}$

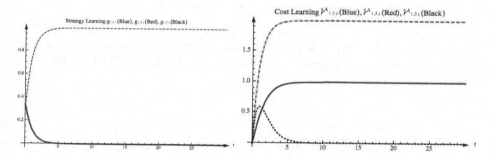

Fig. 7. Attacker's payoff learning: Continuous-time evolution of attacker's mixed strategies $g_{1,1,t}$, $g_{1,2,t}$, and $g_{1,3,t}$

Fig. 8. Attacker's payoff learning: Continuous-time evolution of attacker's estimated payoff $\hat{r}^A_{1,1,t}$, $\hat{r}^A_{1,2,t}$, and $\hat{r}^S_{1,3,t}$

attack surface regularly. Hence, the evolution of mixed strategy $f_{1,1,t}$ is smoother than the one with smaller ϵ.

In Fig. 10, we show the risk measured by the defender, which depends on attacker's action and defender's attack surface. From $t = 99$ to $t = 115$, the system sees an unusual peak of risk under $c_{1,2}$. This exogenous input is used to model unexpected malicious events or alerts that have been detected by or alerted to the system due to the potential risk imposed by $v_{1,1}$. In Fig. 9, we can see that the randomized strategy $f_{1,1}$ reacts to the surge of risk at $t = 99$ and the probability of choosing $c_{1,1}$ starts to increase until the alert is over. The mixed strategy at steady state is found to be $\mathbf{f}_1 = (0.61, 0.39)$. The feedback mechanism allows the defense system to adapt to unanticipated events and enable emergency response that enhances the resiliency of the system.

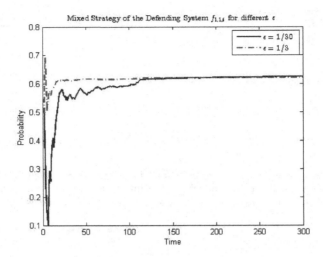

Fig. 9. Mixed strategy of the defending system for different values of ϵ: $f_{1,1,t}$ is the probability of choosing configuration $c_{1,1}$, and $f_{1,2,t} = 1 - f_{1,1,t}$ is the probability of choosing $c_{1,2}$

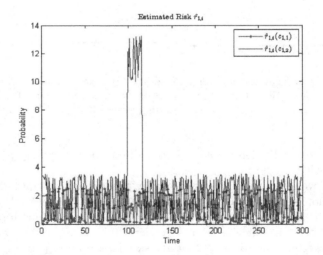

Fig. 10. Risk estimated by the defender for employing configurations $c_{1,1}$ and $c_{1,2}$, respectively. An unexpected event is detected during the period from $t = 99$ to $t = 115$.

In Fig. 11, the vulnerability metric ψ is computed at each time instant. We observe that the moving target defense outperforms a static randomized strategy $(1/3, 1/3, 1/3)$ as its ψ value is constantly higher than the one under the static strategy. The feedback mechanism provides a way to misalign the vulnerability the attack intend to exploit and the vulnerabilities on system's attack surface.

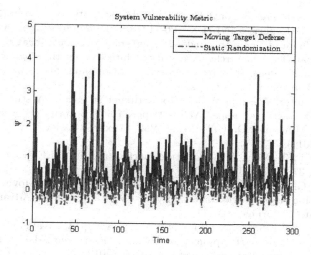

Fig. 11. System vulnerability metric ψ measures the level of vulnerability when the system uses f_1 to defend against an attacker who adopts g_1. A higher value of ψ indicates a lower level of vulnerability. In comparison to a static randomized strategy $f_1 = (1/3, 1/3, 1/3)$, the moving target defense yields a better performance.

7   Conclusions

Moving target defense is an alternative solution to the current expensive and imperfect detection of an intelligent attacker. It is a defense mechanism that dynamically varies the attack surface of the system being defended, and provides probabilistic protections despite exposed vulnerabilities. In this paper, we have developed a game-theoretic framework for guiding the quantitative design of moving target defense as a tradeoff between security and usability. Based on the model, we have proposed a feedback mechanism that allows the system to monitor its current system state and update its randomized strategy based on its observation. We have analyzed the equilibrium stochastic joint strategy and payoff dynamics by studying its associated continuous-time dynamics. As discussed in Remark 4, this work could be further extended to a stochastic game framework where transition probabilities between games capture strategy interdependencies across the layers. Instead of finding equilibrium for games at each layer, we would be interested in a security policy that minimizes the overall risk of the multi-layer system.

References

1. Bowers, K.D., van Dijk, M., Griffin, R., Juels, A., Oprea, A., Rivest, R.L., Trian-dopoulos, N.: Defending against the unknown enemy: Applying FlipIt to system security. In: Grossklags, J., Walrand, J. (eds.) GameSec 2012. LNCS, vol. 7638, pp. 248–263. Springer, Heidelberg (2012)

2. Department of Energy, "Control systems cyber security: defense in depth strategies," External Report # INL/EXT-06-11478, http://energy.gov/sites/prod/files/oeprod/DocumentsandMedia/Defense_in_Depth_Strategies.pdf

3. Zhu, Q., Başar, T.: A hierarchical security architecture for the smart grid. In: Hossain, E., Han, Z., Poor, H.V. (eds.) Smart Grid Communications and Networking, Cambridge University Press (2012)

4. Byres, E., Ginter, A., Langill, J.: "How Stuxnet spreads – A study of infection paths in best practice systems," White Paper, Tofino Security (February 22, 2011)

5. Falliere, N., Murchu, L.O., Chien, E.: "W32. Stuxnet Dossier," Symantec Reports (February 2011)

6. Manadhata, P.K., Wing, J.M.: An attack surface metric. IEEE Trans. on Software Engineering 37(3), 371–386 (2011)

7. Jajodia, S., Ghosh, A.K., Swarup, V., Wang, C., Wang, X.S.: Moving Target Defense: Creating Asymmetric Uncertainty for Cyber Threats. In: Advances in Information Security. Springer (2011)

8. Jajodia, S., Ghosh, S.K., Subrahmanian, V.S., Swarup, V., Wang, C., Wang, X.S.: Moving Target Defense II: Application of Game Theory and Adversarial Modeling. In: Advances in Information Security. Springer (2012)

9. Poolsappasit, N., Dewri, R., Ray, I.: Dynamic Security Risk Management Using Bayesian Attack Graphs. IEEE Transactions on Dependable and Secure Computing 9(1), 61–74 (2012)

10. Ten, C.-W., Liu, C.-C., Manimaran, G.: Vulnerability assessment of cybersecurity for SCADA systems using attack trees. In: Proc. IEEE Power Eng. Soc. Gen. Meeting, Tampa, FL, June 24-28, pp. 1–8 (2007)

11. Fudenberg, D., Levine, D.K.: The Theory of Learning in Games. The MIT Press (1998)

12. Kc, G.S., Keromytis, A.D., Prevelakis, V.: Countering code-injection attacks with instruction-set randomization. In: Proceedings of the 10th ACM Conference on Computer and Communications Security (CCS 2003), New York, NY, USA, pp. 272–280 (2003)

13. Neti, S., Somayaji, A., Locasto, M.E.: Software diversity: security, entropy and game theory. In: Proceedings of the 7th USENIX Conference on Hot Topics in Security, HotSec 2012 (2012)

14. Manshaei, M.H., Zhu, Q., Alpcan, T., Başar, T., Hubaux, J.-P.: Game theory meets network security and privacy. ACM Computing Survey 45(3), 25:1–25:39 (2013)

15. Zhu, Q., Tembine, H., Başar, T.: Hybrid learning in stochastic games and its applications in network security. In: Lewis, F., Liu, D. (eds.) Reinforcement Learning and Approximate Dynamic Programming for Feedback Control, ch. 14. Computational Intelligence Series, pp. 305–329. IEEE Press, Wiley (2013)

16. Zhu, Q., Tembine, H., Başar, T.: Distributed strategic learning with application to network security. In: Proc. 2011 American Control Conference (ACC 2011), San Francisco, CA, June 29-July 1, pp. 4057–4062 (2011)

17. Zhu, Q., Tembine, H., Başar, T.: Heterogeneous learning in zero-sum stochastic games with incomplete information. In: Proc. 49th IEEE Conference on Decision and Control (CDC 2010), Atlanta, Georgia, December 15-17, pp. 219–224 (2010)

18. Zhu, Q., Clark, A., Poovendran, R., Başar, T.: Deceptive routing games. In: Proc. 51st IEEE Conference on Decision and Control (CDC 2012), Maui, Hawaii, December 10-13 (2012)

19. Başar, T., Olsder, G.J.: Dynamic Noncooperative Game Theory. SIAM Series in Classics in Applied Mathematics (January 1999)

20. Zhu, Q., Başar, T.: Dynamic policy-based IDS configuration. In: Proc. 48th IEEE Conference on Decision and Control (CDC 2009), Shanghai, China, December 16-18 (2009)
21. Clark, A., Zhu, Q., Poovendran, R., Başar, T.: An impact-aware defense against Stuxnet. In: Proc. 2013 American Control Conference (ACC 2013), Washington, DC, June 17-19, pp. 4146–4153 (2013)
22. Zhu, Q., Clark, A., Poovendran, R., Başar, T.: Deceptive routing games. In: Proc. 51st IEEE Conference on Decision and Control (CDC 2012), Maui, Hawaii, December 10-13, pp. 2704–2711 (2012)
23. Clark, A., Zhu, Q., Poovendran, R., Başar, T.: Deceptive routing in relay networks. In: Grossklags, J., Walrand, J. (eds.) GameSec 2012. LNCS, vol. 7638, pp. 171–185. Springer, Heidelberg (2012)
24. Sandholm, W.H.: Excess payoff dynamics and other well-behaved evolutionary dynamics. Journal of Economic Theory 124(2), 149–170 (2005)
25. Weibull, J.W.: Evolutionary game theory. MIT Press (1997)
26. Zhu, Q., Başar, T.: "Feedback-Driven Multi-Stage Moving Target Defense", CSL Technical Report
27. Borkar, V.S.: Stochastic approximation: A dynamical systems viewpoint. Cambridge University Press (2008)
28. Franklin, G.F., Powell, D.J., Emami-Naeini, A.: Feedback Control of Dynamic Systems, 5th edn. Prentice Hall PTR, Upper Saddle River (2001)

99. Z[...], June 7, 19[..], "A[...]s based D[...]Compression T[...]ing Data Tools: Applications of Compression after R[...]," B[...]a M[...], 5(5):[...]phes C[...]amics [...]er II,
19[...].

4. Ma[...]f [...]te Carbon Dioxide [...]Body p[...] B[...]ent [...]e M[...]ilc [...]on [...]e,
Colloic [...]C[...] M[...]cro [...]port [...]ong D[...]ig S[...]r[...] 1997, G[...] Supplement in
the thin [...]fil[...] 1970 for [...]pgr chologie

5. Al[...]f Chi[...]r [...]pacifi[...]nal [...] Bene[...]s Co[...]mi[...]ogr the [...]amer [...]ory B[...]v
Mar 1[...]4 R[...]m, [...]e [...]0 O and an A B[...] n[...] [...]ch[...]e [...]t, A M[...]le[...]hall
a[...]iversity 19[...] Sc[...]UL mag[...]

6. [...]t[...]r [...]ey M[...]ogography [...]a Pr[...]s [...]5 p[...]here [...]a[...]a[...] [...] s[...]s [...]schw[...]y
[...]ologi[...]ed Mi[...]od[...] c[...]sf[...] 6, D[...]h[...] Tr[...]c[...] [...]er[...]y [...]er O[...]ie [...]F Ba-
berg, C[...] edition, 19[...]0

7. Ma[...] 19[...]a W[...]omog[...]ne al[...]cr[...]ifter a[...]c[...]em [...]yssion[...] S[...]t[...] [...]mpo[...]c[...] p-
am[...]ta, [...]or th[...] B[...]ietrie [...]Datac[...] l[...]c[...]s[...]l Le[...]al L[...]ang
Sect[...]ral 1[...]7 [...]d. dan[...] min[...], [...]ess [...]h[...]sa S[...] Ca[...]orth [...]ms[...] the [...]
c[...] dio[...]e [...] M[...]str ce[...] S[...]ffe [...]e en[...]aft[...]on o[...] D[...] der [...]wa[...]sop[...]erze[...]le [...]a
wh[...] [...] s[...]o[...] k[...]

8. L[...]y [...]. [...] and Mar[...]bart [...] [...]hlf[...]om[...], [...]cp[...]otio[...] in [...] mit[...]ng [...]s [...] [...]fer a[...]r
b[...]tio[...] ofb[...]pia[...] 2[...] M[...]0

9. Ber[...]h, S[...]. Erik[...]al. J. [...]lau[...] eu[...]ft, [...] R[...]mcd K[...] L[...]b[...] a[...] [...]. F. [...] N[...]mu[...]e,
[...]91 [...]5 c, [...]omp[...]ti[...]n [...]f [...]e [...]7 M[...]r[...] M[...] [...]th [...] B[...]ul[...] B[...]ly

Author Index